Advances in Wind Energy and Wind Turbines

Advances in Wind Energy and Wind Turbines

Editor

Paweł Ligęza

MDPI • Basel • Beijing • Wuhan • Barcelona • Belgrade • Manchester • Tokyo • Cluj • Tianjin

Editor
Paweł Ligęza
Laboratory of Flow
Metrology, Strata Mechanics
Research Institute, Polish
Academy of Sciences,
Krakow, Poland

Editorial Office
MDPI
St. Alban-Anlage 66
4052 Basel, Switzerland

This is a reprint of articles from the Special Issue published online in the open access journal *Energies* (ISSN 1996-1073) (available at: https://www.mdpi.com/journal/energies/special_issues/AWE_WT).

For citation purposes, cite each article independently as indicated on the article page online and as indicated below:

LastName, A.A.; LastName, B.B.; LastName, C.C. Article Title. *Journal Name* **Year**, *Volume Number*, Page Range.

ISBN 978-3-0365-7306-9 (Hbk)
ISBN 978-3-0365-7307-6 (PDF)

Contents

About the Editor

Paweł Ligęza

Paweł Ligęza, Ph.D., D.Sci., born in 1959 in Krakow, Poland, is the Head of the Laboratory of Flow Metrology in the Strata Mechanics Research Institute of Polish Academy of Sciences in Krakow. He graduated from the Faculty of Electrical Engineering, Automatics, IT and Electronics of the AGH University of Science and Technology (1983) with M. Eng. degree. Ph.D. was conferred to him in the SMRI of PAS (1993) and D.Sc. in the AGH University of Science and Technology (2002). The title of the Professor of technical sciences he obtained in 2011. Paweł Ligęza's research interests include metrology, electronics, mechanics, mechatronics, computer sciences, mining sciences, modeling and control, especially measurement systems design with application to flow metrology. He is an expert in anemometry. He published over 200 scientific papers of a domestic and international range, 20 patents and 5 monographs. He was awarded with the Golden Cross of Merit by the President of the Republic of Poland. He got the Research Award of the VII Department PAS Scientific Secretary twice. He is interesting in cognitive science, and his main hobby is amateur radio (SP9IIG).

energies

MDPI

Review

Basic, Advanced, and Sophisticated Approaches to the Current and Forecast Challenges of Wind Energy

Paweł Ligęza

Strata Mechanics Research Institute, Polish Academy of Sciences, Reymonta 27, 30-059 Krakow, Poland; ligeza@imgpan.pl

Abstract: The article is a synthetic review of contemporary wind energy issues. It was created on the basis of a survey of literature from the last two years, with mainly review articles. This work is intended to be a source of information for a wide group of scientists and students from various fields. The aim is to interest them in a wide range of topics related to wind energy and wind turbines. This may allow for the selection of an area and the undertaking of research in this interesting and future-oriented field.

Keywords: wind energy; wind turbines; development; operation; maintenance; monitoring; challenges

Citation: Ligęza, P. Basic, Advanced, and Sophisticated Approaches to the Current and Forecast Challenges of Wind Energy. *Energies* **2021**, *14*, 8147. https://doi.org/10.3390/en14238147

Academic Editor: Antonio Rosato

Received: 28 October 2021
Accepted: 30 November 2021
Published: 5 December 2021

1. Introduction

"Saddle the Wind" is a classic, if somewhat forgotten, western directed by Robert Parrish from the year 1958. Who could have thought then that after half a century, "saddling the wind" would become one of the key challenges for people? Certainly nowadays, no one dares to deny that the development of renewable energy is another giant leap for mankind.

This publication is intended to provide an overview of selected wind energy issues and problems. In this introduction I consciously omit the analysis of the causes and the great importance of the development of renewable energy, including wind energy, in the contemporary world. Virtually every publication in this field begins with such introductions. The number of publications on wind energy published so far is so great that it certainly exceeds the number of wind turbines currently installed all over the world. Therefore, in my review, I only included publications from the last two years. However, such an assumption is not sufficient as the number of important articles related to wind energy is still huge. Therefore, as my paper is limited mainly to the survey of the review articles, it is a kind of meta review. It also applies only to selected issues that I have arbitrarily considered representative. The aim of this review is to familiarize and interest a wide group of scientists and students from various backgrounds and thematic areas in the issues of wind energy. I discuss individual topics in terms of phrases, at a relatively general level, so they are understandable to specialists in various fields. Hence, for this reason, I resign from introducing acronyms denoting various methods, technologies, and indicators to the article. I use full names so that they are understandable to representatives of various fields. Increasingly, the same acronyms, even in similar fields, mean something else. The number of acronyms introduced in scientific publications is growing at an alarming pace. In many cases, creating new acronyms is simply unreasonable and introduces more chaos than order.

With this review article, I would like to encourage the global scientific community to find an area where they can conduct research and contribute to the development of renewable energy. The approach I propose in this review publication may be controversial for many experienced scientists, so I invite you to polemics but, above all, to share your achievements in the field of wind energy. Such an opportunity was created when I guest edited the special issue of, "Advances in Wind Energy and Wind Turbines" in the journal *Energies*.

2. Overview of Selected Basic Issues

The subjects of the article were divided into three substantive parts. The choice of topics and divisions were made on the basis of the research interests, experiences, and knowledge of the authors. Thus, these are subjective divisions based on heuristic premises aimed at introducing a certain systematic orientation of the review of research issues. This review does not pretend to be a complete review of wind power problems. However, the divisions used should be treated as conventional ones, despite other researchers being of the opinion that the hierarchy may be completely different. This section discusses the conventionally basic topics that apply to most typical wind turbines.

2.1. Wind Turbine Locations

Designing a wind farm requires determining its location and optimizing the placement of turbines in the selected area. A review of works on these issues is included in [1]. The article presents the historical background regarding the locations of wind farms. Then, a review of wind farm modeling and optimization methods are discussed, taking into account heuristic methods, artificial intelligence, and mathematical programming algorithms. The main factor of optimization in the design of wind farms is the achievement of maximum efficiency while minimizing the expenditure. The locations of the turbines should take into account the decreases in the efficiencies of the turbines placed in the aerodynamic shadows and the increases in the levels of turbulence generated by adjacent turbines. Apart from the technical and economic aspects, when locating and designing the placements of turbines, a number of other extremely important factors, such as environmental impact and sociological aspects, must of course be taken into account. It is postulated to develop wind farm optimization methods using computational fluid dynamics, deep machine learning, and image recognition and processing techniques. It is possible to further improve the efficiencies of the wind farms while reducing operating costs as well as reducing the environmental impact.

2.2. Wind Parameters Forecasting

In wind energy, forecasting wind parameters is an extremely important and key issue. This process is necessary for the management of individual components and the entire energy network, current operations, energy distribution, planning for switching systems on and off, maintenance, and repairs. Wind forecasting also allows for effective exploitation and technical–economic optimization. A review of the methods of forecasting wind parameters can be found in the article [2]. Physical and statistical issues, the use of artificial intelligence, and hybrid methods are considered here. Various prediction time scales and methods of accuracy were considered. Most often included in wind parameters are velocity, direction, fluctuations, temperature, humidity, and air pressure. The development of forecasting methods uses the huge available databases and complex algorithms such as neural networks or deep machine learning. Due to the development of offshore wind farms, wind forecasting methods apply not only to land areas but to an increasing extent of the open sea.

2.3. Low Speed Wind Turbines

One of the general factors influencing the location of wind farms is the characteristics of the wind in a given area. The optimal location is an area where winds have stable high velocity and slight fluctuations in direction. However, it is also necessary to use locations with lower wind velocities. The problem of turbines for generating electricity at low wind velocities is discussed in [3]. The design of such turbines requires optimization of the geometry, twist, and blade positioning, turbine size, number of blades, reduction of rotor inertia, selection of start characteristics, and improvement and optimization of energy conversion systems. The article presents problems related to the design of standard turbines dedicated to low wind velocities and selected unconventional solutions. These are vertical axis wind turbines, turbines with wind boosters, turbines with nozzles and diffusers, multi-wing turbines, wind concentration systems, and other solutions. An important issue is also the development of algorithms controlling the operations of the turbines in order to obtain maximum efficiency for variable wind velocity. Installing turbines in low wind velocity areas also requires economic consideration. The authors state that these types of turbines are an important complement to other energy systems, and new solutions should be constantly researched.

2.4. *Bearings for Wind Turbines*

Wind turbines are based on rotation movement and the associated torque transmission. Therefore, one of the important issues regarding the reliability of turbines is the issue of bearings. The review article [4] is devoted to this very issue. The operating conditions of bearings in wind turbines are extremely difficult, mainly due to the magnitude of the loads, their variability over time, and environmental conditions. At the same time, the reliability of these elements is of decisive importance for long term, uninterrupted operation and economic effects. The costs associated with the possible need to replace the bearings in the event of premature wear can have a decisive impact on the profitability of the investment. In the article, the authors present an overview of tribological solutions related to the problem of the bearing arrangement of the various elements of a wind turbine. Bearings used in the main shaft, gearbox, or generator are subject to different, but always the highest, requirements. The issues of bearing design, materials used, friction surface technology, lubrication, and condition monitoring during operations are considered. The authors emphasize the necessity of further verification of the basic causes of bearing damage, with particular emphasis on the parts of the gearboxes. It is also important to pay attention to the various environmental conditions related to the geographical locations of the turbines. It is necessary to conduct research in the field of innovative bearing structures, for example, fluid film bearings.

2.5. *Vibrations Problems*

The article [5] provides an overview of problems related to vibrations of wind turbines as well as control and reduction techniques. The features of six main categories of systems used are discussed and compared: advanced blade pitch control, variable rotor diameter, flow control, tuned dampers, active tendons, and piezoelectric actuators. The authors point out that many of these systems are used in other fields, but their transfer to wind energy requires solving many problems, for example, scaling. The development of wind turbines is heading towards the creation of facilities with gigantic dimensions, high slenderness and flexibility, and weight reduction. This requires both adaptation works in the field of conventional solutions as well as innovative conceptions and research.

Wind turbines are mechanical constructions of considerable height and high slenderness. They are found on various bases in various geological conditions, both on land and offshore. They are subject to the wind impact of varying intensities and forces related to the rotations of the rotors with the huge blades. They are exposed to constant vibrations, which are a significant destructive factor. This issue is exacerbated by the possibility of various mechanical resonances. Counteracting this influence is difficult due to the complexity of the process of generating vibrations. The article [6] provides an overview of antivibration techniques and strategies as applied to wind turbines. Theoretical and experimental research in this field is constantly carried out. Passive, active, and hybrid vibration reduction methods are used. Passive methods focus on the optimization of the turbine design and the use of appropriate damping and ballast elements. Active methods require sensors, actuators, and complex control systems as well as other elements that adapt dynamic properties to the current conditions. In addition, an effective and comprehensive facility vibration monitoring system is essential.

2.6. Turbines Control Systems

Due to the variable nature of the air flow, wind energy requires dedicated control systems. The issues related to this problem are discussed in the article [7]. An overview of the currently used control systems for fixed speed, variable slip, doubly fed induction generator, and full converter turbines is presented here. Currently, the last two types are mainly used for large wind farms. Block diagrams and the structure of control systems for all types of turbines were presented; their characteristics and the quality of the generated power were compared. The development of control techniques and the challenges in this area mainly relate to the problems of nonlinearity of the objects and the use of adaptive techniques and state prediction to ensure reliable and effective operations.

Optimizing the efficiency of wind power requires the constant development of methods for controlling, managing, monitoring, and maintaining turbines. The use of artificial intelligence methods becomes effective here. An interesting scientometric review in this area is presented in the article [8]. It provides a critical overview of techniques utilized in the wind industry with the use of artificial intelligence. The literature on this subject in the last decade is huge. Authors in the area of network visualization present recent publications on data-driven, decision-making techniques. They focus on the verification and assessment of various artificial intelligence methods used in wind energy. Statistical methods were used here to analyze the available literature knowledge. This allows for the definition of thematic groups in this field and for setting trends and development directions. The authors indicate that deep machine learning methods in particular constitute the area of potentially the highest scope of application. According to the authors, the current level of application of artificial intelligence methods in wind energy is relatively low compared to other fields. This is a challenge for research teams and designers of renewable energy systems.

2.7. Wind Power Storage and Smoothing

An important issue in the process of wind energy generation as well as with the use of other renewable energy sources is temporary storage. A review of issues related to energy storage systems is presented in the article [9]. It provides a general overview of such systems using a variety of technologies while the mechanical systems are thoroughly discussed. The structures and comparisons of flywheel energy storage systems, pumped hydro energy storage, and compressed air energy storage are presented. The advantages and disadvantages of the applications of these systems in renewable energy storage have been considered. The authors also present their recommendations in this regard.

The generation of wind energy is characterized by a strong variation with time, depending on the current velocity of movement of air masses. The possibilities for anticipation and planning in this regard are limited. This is a severely disadvantageous feature of this energy source. Therefore, one of the important issues in this area is smoothing the energy stream with the use of its storage. A review of the developed strategies in this regard can be found in the work [10]. The process of hydrogen production, air compression, rotating massive objects, supercapacitors, and superconductor coils are used to store energy, but the basic is electrochemical batteries. Optimizing a power generation system requires an appropriate control strategy. The paper [10] reviews such strategies related to the filtration, charging, and discharging processes and wind velocity prediction. Solutions using classic proportional-integral-derivative controllers, fuzzy logic, and the predictive control model are discussed. The authors note that thus far the systems based on deep learning technology have been used to a small extent.

A comprehensive review of publications related to the problem of wind power smoothing in high power systems is included in the work [11]. The development of energy storage methods for short and long term applications is discussed here. Attention was paid to hybrid solutions and cooperation with the network. Reducing wind power fluctuations is a constant challenge that must be constantly developed. Thanks to the conducted research, the cost of energy storage systems is reduced. On the other hand, the parameters related to capacity, lifetime, and reaction speed are improving. This enables the development of complex control systems for wind turbines and the improvement of the quality of wind energy smoothing systems.

3. Advanced Problems of Wind Turbines

This part of the article is intended by the authors to discuss more advanced and complex problems related to wind power. It is focused on certain specialized issues related to the locations of wind turbines in difficult environmental conditions as well as the issues of operations, maintenance, and repairs.

3.1. Urban Wind Turbine Locations

The issue of installing wind turbines in urbanized areas is discussed in the review [12]. The analysis of the current state of wind energy in urban areas in technical, economic, and environmental aspects was made here. The foundation of wind turbines in urban areas is a separate issue with specific features. Roofs and skyscraper, industrial, recreation, and communication areas can be used here. The undoubted advantage of urban locations is the use of energy on site without the need for long distance transmission and thus energy losses. However, winds in built-up areas are variable in direction and force and are difficult to predict. The nature of the air flows is different from that of open areas with many shadow areas and considerable levels of turbulence. The limitation is also the cost of installation and environmental nuisance. Due to the specificity of urban areas, various turbine solutions are used in terms of scale and technology. It is important to choose the optimal turbine for the available location conditions and wind characteristics. However, the authors conclude that urban distributed wind energy can be an important supplement to the constantly growing energy demand.

3.2. Offshore Wind Turbines

Offshore turbines and wind farms are both the current and future-oriented direction of renewable energy development. This area of wind energy placement has many advantages over onshore farms but poses many technological challenges. An overview of this subject is provided in the article [13]. The stage of construction and placement of both anchoring and floating turbines is decisive for the success of the project. The article provides a comprehensive overview of the methods and equipment used to install offshore turbines. It focuses on the technological, economic, legal, and environmental aspects. The types of foundations used for wind turbines and the assembly techniques of their components are presented. The authors state that while the problem of the permanent foundations is relatively well developed, the issues related to floating foundations require further work and studies. Installing turbines in the open sea is also a great challenge. Constant development of numerical computation tools at the level of project development and optimization is necessary. On the other hand, in the field of turbine installation, the subject of improving techniques and equipment is also subject to constant improvement.

The operations and maintenance of offshore wind turbines are important technological and economic problems in the field of modern wind energy. Regarding the operations of turbines, a continuous process of condition monitoring, event prediction, and making current and preventive decisions is carried out. The article [14] contains a summary of currently used conventional strategies in this field and indicates the directions of development. The use of modeling and computer simulation techniques becomes essential. This allows for the prediction of events and the use of preventive actions. An important direction of development is the use of IT methods, with particular emphasis on artificial intelligence algorithms. This will allow for the automation of decision-making processes and the introduction of autonomous systems. For this purpose, it is necessary to create operational databases, which constitute the basis for the development and improvement of operational and decision-making algorithms.

Floating offshore wind turbines are potentially the most developing direction of wind energy but also the greatest technological and research challenge. Their development allows the use of huge wind resources; hence, research in this area is very important. In the article [15], the authors reviewed the methods of testing floating turbines. This applies to computer simulations, experimental tests in laboratory pools, and in situ measurements. Thanks to the development of informatics hardware and software, numerical research is now an important tool for solving complex, nonlinear aerodynamic–hydrodynamic problems. However, numerical simulations require the determination of the boundary—initial conditions and experimental verification. In the field of laboratory tests, attention was paid to the important problem of scaling. The results of actual measurements on selected offshore facilities, available in the literature, were discussed and analyzed. The development of hybrid methods combining computer model research and experiments in a laboratory pool is also presented. This process uses a virtual–experimental feedback loop. According to the authors, this is the most promising direction of research.

3.3. Testing and Damage Detection

The rotor blades play a key role in the performance of wind turbines; however, as active elements they are susceptible to damage. The damages may be related to construction errors and defects arising in the production, transport, and assembly processes. They can occur as a result of factors during the operations process such as high loads, transients, vibrations, collisions, emergency states, and others. Therefore, the process of blade condition monitoring and damage detection is important. The article [16] presents an overview of characteristic damage to blades and the detection methods. The techniques used include strain measurement, acoustic and ultrasound emissions acquisition, vibration monitoring, thermography, and optical monitoring methods. The review presents the theoretical basis, the used physical quantities, and the methodology of damage detection for the discussed technologies. A comparison of selected features of the methods as well as prospects and development directions of damage detection techniques are also included. This problem is extremely important for the proper, safe, and reliable operations of wind turbines.

The article [17] provides an extension to the review of nondestructive diagnostic methods for wind turbine blades. In addition to those already mentioned, radiographic and electromagnetic testing methods and shearography are also discussed. The strengths and weaknesses of diagnostic methods are presented, with references to the relevant literature items. Particular attention was paid to the technical, utilitarian, economic, and logistical aspects of applying the discussed methods and the criteria for their selection.

3.4. Advanced Techniques of Air Velocity Measurement

Wind energy requires anemometric measurements, [18] both at the turbine design stage and during the operations. The variety of methods of measuring the air flow velocity, depending on the metrological requirements, is significant. Laboratory studies often require the measurement of three-dimensional velocity fields around objects. It is also important to measure the level of turbulence and the frequency energy spectrum of the flow [19]. Optical methods are used in such measurements—laser doppler anemometry and particle image velocimetry. Hot-wire anemometry is also used in this area. However, tachometric, pressure, and ultrasonic anemometers are used in situ during turbine exploitation. Anemometric measurements always require the selection of an optimal measurement method and instrument that ensure the required measurement parameters [20].

3.5. Increasing the Efficiency of Turbines

One of the interesting directions of wind energy development are diffuser augmented wind turbines. This solution, which uses a specially constructed aerodynamic turbine casing, allows for the optimization of the energy production process. The theoretical analysis and a review of the mathematical models of this solution are discussed in [21]. The developed models are used to design and optimize the performance of such turbines.

The analysis presented in [21] is of a critical nature, paying attention to the simplifying assumptions adopted and the shortcomings of the models used thus far. This allows for the opening of new research directions in this field. An important problem indicated is estimating the maximum power that can be obtained from these types of turbines for specific dimensions. This is one of the key issues associated with optimizing the performance of these turbines.

4. Some Sophisticated Challenges

The last substantive part discusses selected issues related to wind power that do not directly relate to energy production. They were referred to as sophisticated by the authors. Although theoretically without considering these issues, the operations of wind farms are possible, the development of civilization requires the consideration of these issues on par with technological issues. The examples of issues selected in this part of the article represent this matter.

4.1. Effects on Human Health

The impact of wind turbines on the health of the inhabitants of nearby areas is subject to monitoring and constant research. A review of the literature on this subject has been compiled in [22]. The main factor to be analyzed is the low frequency sounds and infrasound generated by the turbines' rotors. Additionally, the visual impact is considered as a disturbing landscape element and other potential threats. The influence on the emotional state of people, irritation and irritability, sleep disturbance, and metabolism and blood circulation as well as on mental disorders and behavioral disorders is investigated. The literature review mainly points to the irritation and irritability of people as a result of the acoustic emissions of turbines. Other health effects are not convincingly and unequivocally confirmed. Rather, they are a derivative of the emotional, mental, and personal attitudes of residents exposed to the acoustic impact of wind farms. In the process of planning wind farms, attention should be paid to relations with local communities, their participation in the project, reliable information and compliance with ethical requirements, and respect for local customs and traditions.

An extensive literature review on the impact of wind turbine noise on sleep disturbance and insomnia [23] does not confirm a direct relationship between these elements. The conducted comprehensive meta-analysis of the available sources did not show any confirmation in the form of objective indicators of the impact of turbine noise on the sleep patterns of the local inhabitants. However, due to the variety of research methodologies, the use of nonobjective indicators and the lack of a uniform methodology of conducting research to date, the authors leave the final solution as an open problem. Well-prepared and objectified research on the influence of turbines on the organisms of people living in the immediate vicinity is a scientific challenge for the future.

4.2. Social Perception

In addition to technological, economic, and legal issues, social perception is an important aspect of development, evolution, and expansion of wind energy. A review of the literature in this area is included in [24]. The authors point out that despite the widespread support for the idea of renewable energy, local acceptance is at a very low level. This is a significant barrier to the development of this branch of energy. The article analyzes specific case studies in various regions of Europe described in the literature. Comparative and both qualitative and quantitative analyses were carried out. It was indicated that an important factor in the development of renewable energy is gaining social trust. It is necessary to implement information strategies based on high quality technical, economic, legal, social, environmental and health data.

4.3. Impact on the Natural Environment

The impact of wind energy on the ecosystem is subject to constant research and monitoring. One problem is the noise generated by wind turbines. This issue concerns not only the acoustic waves generated in the air but also the underwater noise produced by offshore wind farms. The article [25] discusses the most important problems related to this issue. Based on the collected literature, the results of measurements made in various locations were reviewed. This allowed for the analysis and comparison of noise parameters generated by offshore turbines. The noise level was analyzed depending on the type and number of turbines, size, wind velocity, and distance from the noise source. Based on the survey of measurements, it has been found that the level of low frequency underwater noise produced by turbines is approximately 10 to 20 dB lower than that generated by other facilities such as large cargo ships. While the noise level is relatively high near the turbines themselves, it decreases quickly with distance. The size of the turbines or the velocity of the wind have less influence here. Currently, turbine noise has a significant environmental impact in areas where the levels of other sources are low. However, the constant development of offshore wind energy will result in the intensification of underwater noise and the necessity to take action to reduce the noise impact on the environment.

One of the very important issues related to wind energy is the impact on the natural environment. One of the aspects of this topic is the mortality of flying fauna, i.e., birds and bats. This issue is studied and monitored in many areas of the world. The article [26] reviews this subject in Latin America. This is particularly interesting because most of the research is focused on Europe and North America. The authors of the article draw attention to the lack of reviews in Latin America so far and are trying to fill this gap. It is important because Latin America is potentially a region of rapid and intensive development of wind energy in the coming years. The research conducted so far in this area indicates that sparrows constitute the largest number of victims among the birds, while there are no endangered species. However, one endangered species was reported among bats. The available amount of information in this area is limited; therefore, it is difficult to make an unambiguous and complete assessment of this phenomenon. However, it is necessary to develop and intensify activities related to reducing the possibility of collisions between animals and wind turbines. It would be helpful to use the experience from other areas of the world where research in this field is more advanced.

4.4. Economic Issues and Indicators

The departure from traditional methods of electricity production based on the combustion of fossil resources has become a civilization necessity. Therefore, various methods of generating energy are being developed. It becomes imperative to compare and economically evaluate various energy sources. Levelized costs of electricity are a widely used indicator for comparing the costs of energy production. Along with the development of wind power and other modern energy technologies, it was necessary to adapt the methods of determining this indicator to the properties of these technologies. The article [27] reviews these methods as applied to renewable energy. Economic considerations related to investments, operations, performance, and risks are included here. The uncertainty in determining economic indicators was analyzed. On the basis of the literature review, the authors of the work [27] indicate the problems that should be taken into account when determining economic indicators for renewable energy sources. These include variable efficiency, location and seasonal conditions, cooperation with the grid, and renewable energy policies. The authors point out the need to take into account the uncertainty of data and parameters in the modeling of economic indicators.

An important exploitation indicator is also life cycle assessments. A systematic review of the methods for determining this indicator, based on the literature, is included in [28]. The review covers all major methods of electricity generation, including those based on the combustion of fossil fuels, nuclear energy, and renewable energy. The authors indicate that despite the development of methods for determining life cycle assessments, there are still many unresolved problems that require further development. This applies to the improvement of data processing methods, standardization, and the wide use of results by decision makers.

5. Conclusions

As a conclusion of this survey, I would like to once again encourage readers to take up challenges and undertake research in the field of broadly understood wind energy. We can see that this theme applies not only to technical and engineering sciences, but also to natural sciences, humanities, medical, social, agricultural, economic, and possibly even art. What will be the future of wind energy development and its share in global energy production? One possible answer to this question can be found in an article published in the journal *Science* [29]. However, experience shows that simple trend extrapolation does not always produce correct results. Apart from the enthusiastic development of diverse renewable energy technologies, there are also questions. I recently received a challenge to include a bold estimation of the ratio of all energy invested in wind turbines to all energy gained from wind turbines so far worldwide. I do not know a simple answer to this question and possibly someone from the younger generation of scientists will create an appropriate model and be able to estimate this parameter with acceptable uncertainty.

Wind energy is one of the areas of renewable energy; however, it is important to realize that there are no perfect, unlimited sources of usable energy. Contemporary energy based on solar nuclear fusion, i.e., wind energy, hydropower, and photovoltaic, are not a panacea for the risks associated with traditional energy based on raw materials [30,31]. In addition to the development and improvement of renewable energy sources, an equally important issue is to prevent unnecessary dissipation of utility energy. Let us remember this and consciously use the available energy by treating it not as a commodity but as a common good.

Funding: This research received no external funding.

Conflicts of Interest: The author declare no conflict of interest.

References

1. Azlan, F.; Kurnia, J.C.; Tan, B.T.; Ismadi, M.-Z. Review on optimization methods of wind farm array under three classical wind condition problems. *Renew. Sustain. Energy Rev.* **2021**, *135*, 110047. [CrossRef]
2. Hanifi, S.; Liu, X.; Lin, Z.; Lotfian, S. A Critical Review of Wind Power Forecasting Methods—Past, Present and Future. *Energies* **2020**, *13*, 3764. [CrossRef]
3. Venkatramakrishnan, S.R.; Pandey, J.K.; Mondal, A.K.; Karn, A. Low Speed Wind Turbines for Power Generation: A Review. *J. Adv. Res. Fluid Mech. Therm. Sci.* **2020**, *67*, 146–169.
4. Dhanola, A.; Garg, H.C. Tribological challenges and advancements in wind turbine bearings: A review. *Eng. Fail. Anal.* **2020**, *118*, 104885. [CrossRef]
5. Awada, A.; Younes, R.; Ilinca, A. Review of Vibration Control Methods for Wind Turbines. *Energies* **2021**, *14*, 3058. [CrossRef]
6. Zuo, H.; Bi, K.; Hao, H. A state-of-the-art review on the vibration mitigation of wind turbines. *Renew. Sustain. Energy Rev.* **2020**, *121*, 109710. [CrossRef]
7. Ghaffarzadeh, H.; Mehrizi-Sani, A. Review of Control Techniques for Wind Energy Systems. *Energies* **2020**, *13*, 6666. [CrossRef]
8. Chatterjee, J.; Dethlefs, N. Scientometric review of artificial intelligence for operations & maintenance of wind turbines: The past, present and future. *Renew. Sustain. Energy Rev.* **2021**, *144*, 111051.
9. Mahmoud, M.; Ramadan, M.; Olabi, A.G.; Pullen, K.; Naher, S. A review of mechanical energy storage systems combined with wind and solar applications. *Energy Convers. Manag.* **2020**, *210*, 112670. [CrossRef]
10. Silva de Siqueira, L.M.; Peng, W. Control strategy to smooth wind power output using battery energy storage system: A review. *J. Energy Storage* **2021**, *35*, 102252. [CrossRef]
11. Barra, P.H.A.; Carvalho, W.C.; Menezes, T.S.; Fernandes, R.A.S.; Coury, D.V. A review on wind power smoothing using high-power energy storage systems. *Renew. Sustain. Energy Rev.* **2021**, *137*, 110455. [CrossRef]
12. Tasneem, Z.; Noman, A.A.; Das, S.K.; Saha, D.K.; Islam Md, R.; Ali Md, F.; Badal Md, F.R.; Ahamed Md, H.; Moyeen, S.I.; Alam, F. An analytical review on the evaluation of wind resource and wind turbine for urban application: Prospect and challenges. *Dev. Built Environ.* **2020**, *4*, 100033. [CrossRef]
13. Jiang, Z. Installation of offshore wind turbines: A technical review. *Renew. Sustain. Energy Rev.* **2021**, *139*, 110576. [CrossRef]

14. Rinaldi, G.; Thies, P.R.; Johanning, L. Current Status and Future Trends in the Operation and Maintenance of Offshore Wind Turbines: A Review. *Energies* **2021**, *14*, 2484. [CrossRef]
15. Chen, P.; Chen, J.; Hu, Z. Review of Experimental-Numerical Methodologies and Challenges for Floating Offshore Wind Turbines. *J. Mar. Sci. Appl.* **2020**, *19*, 339–361. [CrossRef]
16. Du, Y.; Zhou, S.; Jing, X.; Peng, Y.; Wu, H.; Kwok, N. Damage detection techniques for wind turbine blades: A review. *Mech. Syst. Signal Process.* **2020**, *141*, 106445. [CrossRef]
17. Marquez FP, G.; Chacon AM, P. A review of non-destructive testing on wind turbines blades. *Renew. Energy* **2020**, *161*, 998–1010. [CrossRef]
18. Ligęza, P.; Jamróz, P.; Ostrogórski, P. Methods for dynamic behavior improvement of tachometric and thermal anemometers by active control. *Measurement* **2020**, *166*, 108147. [CrossRef]
19. Ligęza, P. Constant-Temperature Anemometer Bandwidth Shape Determination for Energy Spectrum Study of Turbulent Flows. *Energies* **2021**, *14*, 4495. [CrossRef]
20. Ligęza, P. Static and dynamic parameters of hot-wire sensors in a wide range of filament diameters as a criterion for optimal sensor selection in measurement process. *Measurement* **2020**, *151*, 107177. [CrossRef]
21. Bontempo, R.; Manna, M. Diffuser augmented wind turbines: Review and assessment of theoretical models. *Appl. Energy* **2020**, *280*, 115867. [CrossRef]
22. Van Kamp, I.; Van den Berg, F. Health Effects Related to Wind Turbine Sound: An Update. *Int. J. Environ. Res. Public Health* **2021**, *18*, 9133. [CrossRef]
23. Liebich, T.; Lack, L.; Hansen, K.; Zajamšek, B.; Lovato, N.; Catcheside, P.; Micic, G. A systematic review and meta-analysis of wind turbine noise effects on sleep using validated objective and subjective sleep assessments. *J. Sleep Res.* **2020**, *30*, e13228. [CrossRef]
24. Segreto, M.; Principe, L.; Desormeaux, A.; Torre, M.; Tomassetti, L.; Tratzi, P.; Paolini, V.; Petracchini, F. Trends in Social Acceptance of Renewable Energy across Europe—A Literature Review. *Int. J. Environ. Res. Public Health* **2020**, *17*, 9161. [CrossRef]
25. Tougaarda, J.; Hermannsen, L.; Madsenb, P.T. How loud is the underwater noise from operating offshore wind turbines? *J. Acoust. Soc. Am.* **2020**, *148*, 2885. [CrossRef]
26. Agudelo, M.S.; Mabee, T.J.; Palmer, R.; Anderson, R. Post-construction bird and bat fatality monitoring studies at wind energy projects in Latin America: A summary and review. *Heliyon* **2021**, *7*, e07251. [CrossRef] [PubMed]
27. Shen, W.; Chen, X.; Qiu, J.; Hayward, J.A.; Sayeef, S.; Osman, P.; Meng, K.; Dong, Z.Y. A comprehensive review of variable renewable energy levelized cost of electricity. *Renew. Sustain. Energy Rev.* **2020**, *133*, 110301. [CrossRef]
28. Jordaan, S.M.; Combs, C.; Guenther, E. Life cycle assessment of electricity generation: A systematic review of spatiotemporal methods. *Adv. Appl. Energy* **2021**, *3*, 100058. [CrossRef]
29. Veers, P.; Dykes, K.; Lantz, E.; Barth, S.; Bottasso, C.L.; Carlson, O.; Clifton, A.; Green, J.; Green, P.; Holttinen, H.; et al. Grand challenges in the science of wind energy. *Science* **2019**, *366*, 6464. [CrossRef] [PubMed]
30. Skotniczny, P. Aerological factors favouring the occurrence of endogenous fires on coal waste stockpiles. *Arch. Min. Sci.* **2020**, *64*, 901–916.
31. Dutka, B.; Godyń, K. Coalification as a Process Determining the Methane Adsorption Ability of Coal Seams. *Arch. Min. Sci.* **2021**, *66*, 181–195.

energies

Article

Optimized Extreme Learning Machine-Based Main Bearing Temperature Monitoring Considering Ambient Conditions' Effects

Zhengnan Hou [1], Xiaoxiao Lv [2,*] and Shengxian Zhuang [1]

1 School of Electrical Engineering, Southwest Jiaotong University, Chengdu 610031, China; zhengnanhou@my.swjtu.edu.cn (Z.H.); zhuangshengxian@swjtu-leeds.com.cn (S.Z.)
2 School of Economics and Management, Southwest Jiaotong University, Chengdu 610031, China
* Correspondence: xiaoxiaolv@my.swjtu.edu.cn

Abstract: Wind Turbines (WTs) are exposed to harsh conditions and can experience extreme weather, such as blizzards and cold waves, which can directly affect temperature monitoring. This paper analyzes the effects of ambient conditions on WT monitoring. To reduce these effects, a novel WT monitoring method is also proposed in this paper. Compared with existing methods, the proposed method has two advantages: (1) the changes in ambient conditions are added to the input of the WT model; (2) an Extreme Learning Machine (ELM) optimized by Genetic Algorithm (GA) is applied to construct the WT model. Using Supervisory Control and Data Acquisition (SCADA), compared with the method that does not consider the changes in ambient conditions, the proposed method can reduce the number of false alarms and provide an earlier alarm when a failure does occur.

Keywords: Wind Turbine; temperature monitoring; ambient condition; Extreme Learning Machine; genetic algorithm; SCADA

Citation: Hou, Z.; Lv, X.; Zhuang, S. Optimized Extreme Learning Machine-Based Main Bearing Temperature Monitoring Considering Ambient Conditions' Effects. *Energies* **2021**, *14*, 7529. https://doi.org/10.3390/en14227529

Academic Editors: Davide Astolfi and Paweł Ligeza

Received: 26 September 2021
Accepted: 4 November 2021
Published: 11 November 2021

Publisher's Note: MDPI stays neutral with regard to jurisdictional claims in published maps and institutional affiliations.

1. Introduction

Since wind energy is renewable and pollution-free, many governments cite wind energy as a primary future energy source [1]. However, the high maintenance costs of Wind Turbines (WTs) seriously restrict the development of the wind energy industry [2]. The most effective way to reduce maintenance costs is to monitor the working state of WTs to sound alarms when failures occur. Thus, model-based WT monitoring has attracted widespread attention, and different methods have been proposed. Existing model-based WT monitoring methods can be roughly divided into two categories [3]: theoretical and data-driven. The advantage of theoretical methods is that fewer data are required. Reference [4] focused on the main physical mechanisms responsible for temperature changes. Reference [5] proposed an advanced annual energy computation model.

However, WTs are complex electromechanical systems, and the relationships between the various parameters are primarily nonlinear; thus, the construction of a theoretical model is difficult and often inaccurate [6]. Reference [7] proposed an approach that identifies turbines with weakened power generation performance through assessing the wind power curve profiles. Reference [8] discussed the short-horizon prediction of wind speed and power. Reference [9] used sophisticated models to understand the complex WT component degradation processes and to facilitate maintenance decision-making. Reference [10] proposed a robust data-driven fault detection approach. For data-driven methods, model accuracy relies on the quality and quantity of the data. A Supervisory Control and Data Acquisition (SCADA) system, which can record electrical parameters (active power, phase current, etc.), temperature parameters (main bearing temperature, generator rotor temperature, etc.), and operating parameters (motor speed, etc.), has been widely applied in wind farms [11–13].With hardware improvements, the abilities of SCADA are increasing, making the data-driven methods more suitable for WT monitoring [14].

A typical data-driven WT monitoring method analyzes real-time data using a model with one parameter as the output and other parameters as inputs [15]. When building data-driven models, intelligent algorithms are applied. By using Nonlinear State Estimate Technology (NSET), Reference [16] and Reference [17] built a tower vibration model and gearbox oil temperature model, respectively. Based on logistic regression (LR), Reference [18] and Reference [19] analyzed the direct-drive wind power generation set and the bearing performance condition, respectively. With Support Vector Machine (SVM), Reference [19] classified and diagnosed the possible fault of the bearing of WT. Reference [20] optimized SVM for Wind Turbine fault diagnosis based on a diagonal spectrum and clustering binary tree. Reference [21] presented a multi-sensory system for fault diagnosis using SVM. Reference [22] presented a two-stage fault detection and classification scheme for electric motor drives in Wind Turbine pitch systems using SVM. In addition to the above methods, Neural Network (NN) is widely applied due to its high accuracy and short training time. Reference [23] achieved a robust simultaneous estimate of actuator faults and system states using NN. Reference [24] applied an artificial NN approach to the gearbox bearings. Reference [25] improved NN by mapping the original samples into feature vectors in an embedding space. With NN, Reference [26] identified the wavelet-transformed power components' open-circuit faults accurately. Reference [27] captured dynamic equations modeling wind power output, vibration of the drive train, and vibration of the tower. Reference [28] attempted to assess the prediction intervals of time-series predictions with NN and Extreme Learning Machine (ELM).

Since SCADA of wind farms can obtain very large datasets, NN would entail high computational costs [29]. In contrast to NN, the input weights and hidden layer biases in ELM are assigned randomly, and the output weights of the hidden layer are directly calculated by a Moore–Penrose (MP) generalized inverse operation [30–32]. Consequently, ELM is more computationally efficient than NN [33]. However, the ELM model is based on one set of initial weights and biases, which are mostly random. With the randomness of the initial coefficients and large quantity of datasets, the model may easily fall into local minima.GA is a global random search optimization algorithm based on the genetic mechanism and evolution. Through genetic operations, the individuals with good fitness are selected. With the strong global search capability, the initial coefficients can be optimized by GA and the solution can avoid local minima [34]. Reference [35] improved the accuracy and efficiency of a prediction algorithm by adopting the Genetic Algorithm.

Another issue is that a WT is typically exposed to harsh field conditions year-round, including extreme weather (blizzards, cold waves, etc.). Different ambient conditions have different influences on the WT's temperature parameters. On one hand, the extreme weather causes the internal temperature to fluctuate, which can easily trigger a false alarm. On the other hand, the internal temperature shows a difference in the same normal working state due to the change in ambient temperature. For example, the ambient temperature difference can reach 50 °C between summer noon and winter midnight in the wind farm studied in this paper. Due to this huge ambient temperature difference, the main bearing temperature of a WT can differ by 20 °C in a normal working state. If the WT temperature model is constructed without considering the ambient temperature, the model can be inaccurate, which would easily cause false alarms in a normal working state, or fail to sound alarms when the internal temperature is abnormal.

Furthermore, many studies have confirmed that the WT performance is directly related to wind speed conditions [36], and thus WTs' internal temperature parameters [37]. Therefore, it is necessary to consider the wind speed conditions in WT monitoring. In existing studies, the wind speed condition division is primarily based on the absolute value of the wind speed. This type of division is effective for WT power monitoring but not accurate for WT temperature monitoring. Through research on real data, the internal temperature is found to exhibit certain differences when wind speeds increase and decrease. For example, the temperature difference in the same main bearing between a wind speed increase and decrease may be more than 5 °C under the same wind speed of 10m/s and the same ambient temperature of 15 °C. If the wind speed condition only depends on the absolute value, the temperature monitoring result may be inaccurate.

To address these issues, a novel WT monitoring method is proposed in this paper. Considering the effects of ambient conditions on the WT's internal temperature, the ambient temperature and wind speed change are used as inputs in the optimized ELM model, along with other related parameters. The model is developed by training with real SCADA data. The WT monitoring is achieved by analyzing the real-time data with the model.

In general, the primary innovations of this paper are as follows:

1. The effects of ambient conditions (ambient temperature and wind speed change) on a WT's internal temperature are investigated.
2. Changes in the ambient conditions are used as the input of the WT temperature model. To our knowledge, this study is the first to use wind speed change as the input of a WT model.
3. GA is used to optimize ELM to avoid local minima due to the irregularity of initial input weights and hidden layer bias.

The rest of this paper is organized as follows. Section 2 investigates the influence of ambient conditions on WT monitoring. In Section 3, the framework of the proposed method is presented in detail. In Section 4, the WT model is developed and verified. Section 5 presents the case study and the monitoring results of various analyses. Conclusions are summarized in Section 6.

2. Effects of Ambient Conditions

A WT is a complex electromechanical system with many subsystems and components (gearbox, generator and converter, etc. [27]), and every key component has a temperature sensor. However, a WT is exposed to harsh weather conditions year-round and the ambient conditions of a WT can be very different. This huge difference in the ambient conditions can cause the internal temperatures to perform quite differently, even under the same working state. Therefore, the effects of the ambient conditions on the internal temperatures must be investigated. This paper investigates wind speed changes and ambient temperature. The data in this paper come from the Damianshan Wind Farm in Wanyuan City, Sichuan Province, China. The wind farm has a total of 33 1.5MW-WTs, with an annual power of 90.489 million kWh. The SCADA in this wind farm records data every 1 min.

2.1. Wind Speed Change

Figure 1a shows the active power–wind speed curve under a normal working state, and Figure 1b shows the main bearing temperature–wind speed curve. To reduce the impact of the ambient temperature, the ambient temperature of the two datasets shown in the two curves is between 14.5 °C and 15.5 °C.

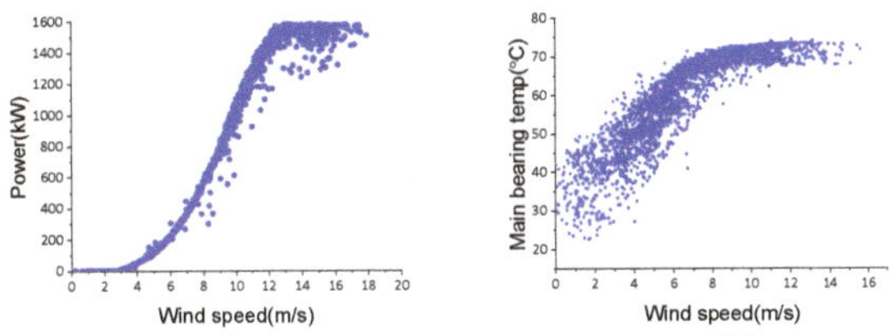

Figure 1. WT in normal state: (**a**) active power–wind speed; (**b**) main bearing temperature–wind speed.

Figure 1a shows that the active power is directly related to the wind speed. Therefore, WT temperature monitoring should consider wind speed. Most existing methods only divide the absolute value of wind speed into three regions, (0–3, 3–12, and ≥12) m/s, without considering wind speed change. However, the fact is that wind speed change also affects the internal temperatures.

Figure 1b shows that the main bearing temperature increases as the wind speed increases; this result occurs because wind speed has a direct positive correlation with rotor speed, which is directly related to the heat generated inside. Considering the progress of the heat conduction, there may be some delay between the wind speed change and internal temperature change. Due to this delay, at the same wind speed, the main bearing temperature during a wind speed increase is smaller than during a wind speed decrease. Figure 2 shows the main bearing temperature–wind speed curve of a WT during a wind speed increase and decrease. To accurately describe this phenomenon, despite the wind speed change, the working state and conditions are similar, which are an active power of 900–1000 kW, an ambient temperature of 14–16 °C, and a wind speed of 9–11 m/s.

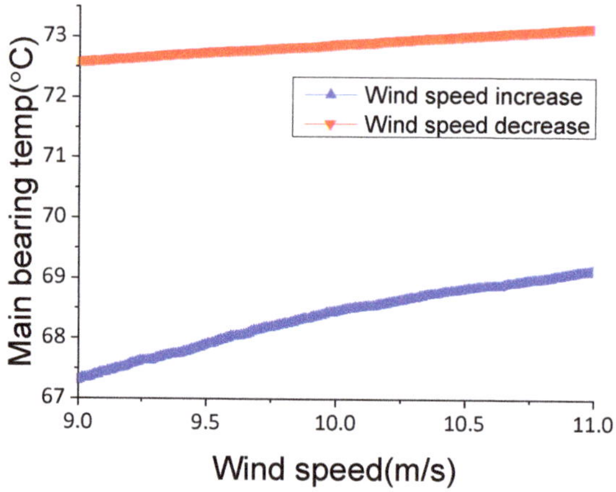

Figure 2. Main bearing temperature during wind speed increase and decrease.

Figure 2 shows that, under the same wind speed, the main bearing temperature experiences a significant difference during a wind speed increase and decrease. The average difference is 4.6 °C, and the maximum difference can reach 5.4 °C. These results indicate that wind speed change affects the internal temperatures. Therefore, in order to improve the accuracy of WT monitoring, it is necessary to consider not only the absolute value of the wind speed but also the wind speed change as one of the ambient conditions.

2.2. Ambient Temperature Change

Besides wind speed change, ambient temperature may also directly affect the internal temperature. Figure 3 shows the main bearing temperature of a WT at midnight (2:00–3:00 a.m.) in January (winter) and during the afternoon (14:00–15:00 p.m.) in August (summer). It is worth mentioning that, to enable effective comparison, the wind speed during the two periods was maintained at around 15m/s.

As shown in Figure 3, the difference in ambient temperature between winter and summer can reach 50.6 °C. Due to this ambient temperature difference, the average difference in the main bearing temperature is 19.2 °C. The internal temperature of the WT may change with the ambient temperature.

Additionally, extreme weather may cause large fluctuations in internal temperature in a short time. Figure 4 shows the main bearing temperature changes in one hour during a cold wave in November. During this cold wave, the WT maintains normal full-load power generation.

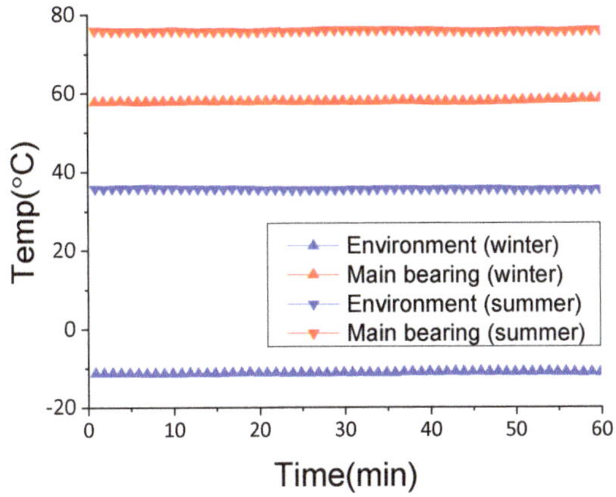

Figure 3. Main bearing temperature within different ambient temperatures.

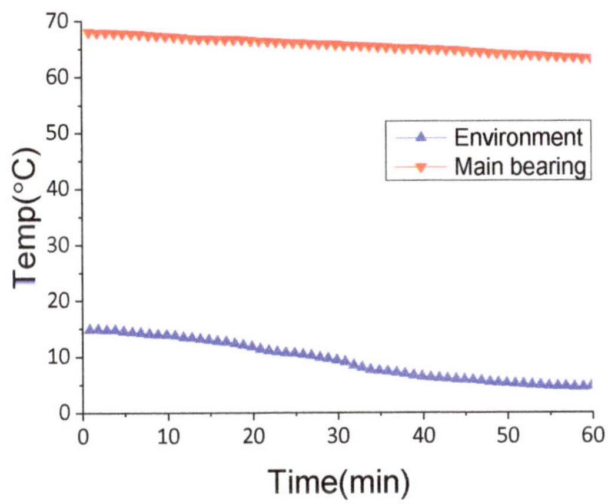

Figure 4. Main bearing temperature during cold wave.

Figure 4 shows that the ambient temperature undergoes a significant drop of 10.2 °C due to the cold wave and the main bearing temperature undergoes a drop of 4.4 °C. The internal temperature may fluctuate due to extreme weather. Therefore, temperature monitoring must consider the ambient temperature change.

3. WT Monitoring Framework

3.1. Overview of the Proposed Framework

To address the issues noted above, a novel WT monitoring method is proposed in this paper. The framework of the proposed method is shown in Figure 5. The monitoring process is composed of two parts: offline model construction and online data analysis. Offline model construction is to build a model that simulates normal WT behaviors, and online data analysis determines whether the WT is in a normal working state.

Compared with existing methods, the proposed method has two key advantages.

Ambient conditions are used as the input of the WT model. As mentioned earlier, both wind speed change and ambient temperature affect internal temperatures. Therefore, the proposed method uses wind speed change and ambient temperature as ambient conditions in the model input.

ELM is optimized for the randomness of initial weights and bias. Due to the randomness of initial coefficients, ELM may fall into local minima when constructing the WT model. To solve this problem, GA is applied to optimize the ELM to improve the model's accuracy.

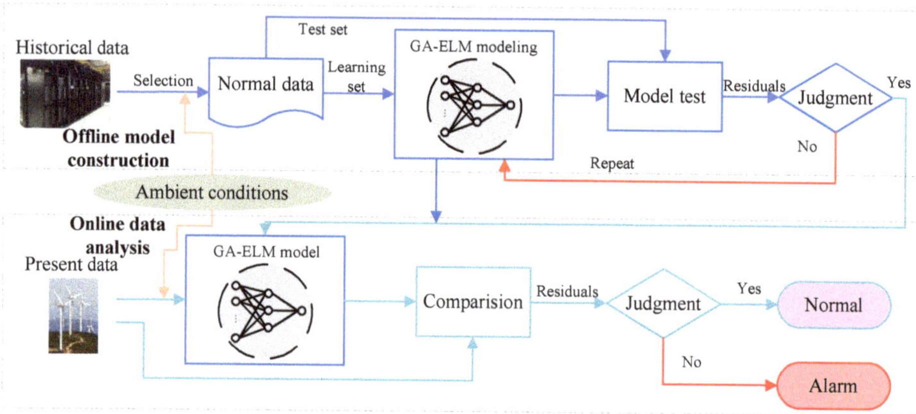

Figure 5. Framework of the proposed method.

3.2. Input Parameter Selection

The output of the model should directly describe the WT's working state and have a strong impact on maintenance. Among the various WT failures, the main bearing failure costs the most [18]. Because the main bearing temperature is closely related to the health of the main bearing, the main bearing temperature is chosen as the output of the model in this paper.

The input should be directly related to the main bearing and WT, which are: (a) the production parameters, such as active/reactive power; (b) the parameters that are near the main bearing temperature, such as gearbox front/rear bearing temperature; and (c) ambient conditions. In this study, the input of the model contains 10 parameters: active power, rotor speed, gearbox front bearing temperature, gearbox rear bearing temperature, nacelle ambient temperature, tower vibration, ambient temperature, ambient temperature change, wind speed, and wind speed change.

3.3. GA-ELM Modeling

The intelligent algorithm in model construction is the key to WT temperature monitoring. Compared with NN, Extreme Learning Machine (ELM) has the advantages of fast training speed and high accuracy. ELM is composed of a single hidden layer feed-forward Neural Network.

In ELM, $\boldsymbol{X} = [X_1, X_2, \ldots, X_n]^T \in R^n$ and $\boldsymbol{Y} = [Y_1, Y_2, \ldots, Y_m]^T \in R^m$ are the input and output of the model, respectively; ω_{ij} and ω_{ij} are the input and output weights, respectively. For n distinct samples $\boldsymbol{X}$, the ELM can approximate the target as

$$\hat{Y}_k = \sum_{i=1}^{\tilde{n}} \omega_{jk} \cdot g(\omega_{ij} \cdot X_i + b_i) \; j = 1, 2, \ldots, \tilde{n} \tag{1}$$

where $g(\cdot)$ represents the activation function, $\tilde{n}$ is the number of hidden nodes, and b_i is the hidden layer bias.

If ELM can fit n distinct samples with zero error, the matrix form of approximation can be expressed as

$$\boldsymbol{Y} = \hat{\boldsymbol{Y}} = \boldsymbol{H}\boldsymbol{\omega}_{\tilde{n}m} \tag{2}$$

where the output weights $\boldsymbol{\omega}_{\tilde{n}m} = \left[\boldsymbol{\omega}_1^T, \boldsymbol{\omega}_2^T, \ldots, \boldsymbol{\omega}_m^T\right]^T$, $\boldsymbol{\omega}_k = [\omega_{1k}, \omega_{2k}, \ldots, \omega_{\tilde{n}k}]^T$ and the hidden layer output matrix $\boldsymbol{H}$ can be expressed as

$$\boldsymbol{H} = \begin{bmatrix} g(\omega_{11} \cdot X_1 + b_1) & \ldots & g(\omega_{\tilde{n}1} \cdot X_1 + b_n) \\ \ldots & \ldots & \ldots \\ g(\omega_{1m} \cdot X_m + b_1) & \ldots & g(\omega_{\tilde{n}m} \cdot X_m + b_{\tilde{n}}) \end{bmatrix}_{m \times \tilde{n}} \tag{3}$$

With given input weights ω_{ij} and hidden layer bias b_j, the output weight can be analytically calculated by a least squares method as

$$\|\boldsymbol{H}\hat{\omega}_{\tilde{n}m} - \boldsymbol{Y}\| = \|\boldsymbol{H}\boldsymbol{H}^+\boldsymbol{Y} - \boldsymbol{Y}\| = \min_{\boldsymbol{\omega}_{\tilde{n}m}} \|\boldsymbol{H}\boldsymbol{\omega}_{\tilde{n}m} - \boldsymbol{Y}\| \tag{4}$$

where $\boldsymbol{H}^+$ is the generalized Moore–Penrose inverse of $\boldsymbol{H}$.

Then, the solution can be expressed as

$$\hat{\omega}_{\tilde{n}n} = \boldsymbol{H}^+\boldsymbol{Y} \tag{5}$$

However, the model is based on one set of initial input weights ω_{ij} and hidden layer bias b_j, which are mostly random. With the randomness of the initial coefficients, the ELM model could easily fall into local minima. To solve this problem, GA is applied in this paper to optimize ELM.

In GA, the coefficients to be optimized are coded as individual chromosomes. In this paper, the individual chromosome code contains the input weights and hidden layer bias of ELM. The fitness $\boldsymbol{F}$, which can judge whether the code is a good solution, is calculated as

$$F = 1 \Big/ \left(\sum_{i=1}^{m} |e_k|\right) \tag{6}$$

where e_k is the error of ELM as $e_k = Y_k - \hat{Y}_k$ and m is the number of output layer nodes.

The GA optimization process proceeds as follows:

Step 1, selection. GA selection is based on fitness; the probability of selection is calculated as

$$p_i = \frac{k/F_i}{\sum_{j=1}^{N} k/F_i} \tag{7}$$

where N is the number of individuals.

Step 2, crossover. GA crossover of two chromosomes at gene j is calculated as

$$\left.\begin{aligned} \alpha_{kj} &= \alpha_{kj}(1-\beta) + \alpha_{lj}\beta \\ \alpha_{lj} &= \alpha_{lj}(1-\beta) + \alpha_{kj}\beta \end{aligned}\right\} \tag{8}$$

where α_{kj} and α_{lj} are the gene j of chromosome k and chromosome l, respectively, and β is the cross-coefficient, which is a random number in the range (0–1).

Step 3, evolution. GA evolution of α_{ij} is as

$$\alpha_{ij} = \begin{cases} \alpha_{ij} + (\alpha_{ij} - \alpha_{max}) f(g), \gamma > 0.5 \\ a_{ij} + (\alpha_{min} - \alpha_{ij}) f(g), \gamma \leq 0.5 \end{cases} \tag{9}$$

$$f(g) = \gamma(1 - g/G_{max})^2 \tag{10}$$

where α_{max} and α_{min} are the upper and lower threshold of a_{ij}, respectively, g and G_{max} are the current and maximum number of GA evolutions, respectively, and γ is the evolution coefficient, which is in the range (0–1).

It is necessary to repeat the GA optimization until the maximum fitness is obtained. The individual chromosome code with the maximum fitness is the optimal solution. By decoding the optimal solution, optimal initial input weights and hidden layer bias can be obtained for the ELM. The flowchart of building the WT model based on GA-ELM is shown in Figure 6.

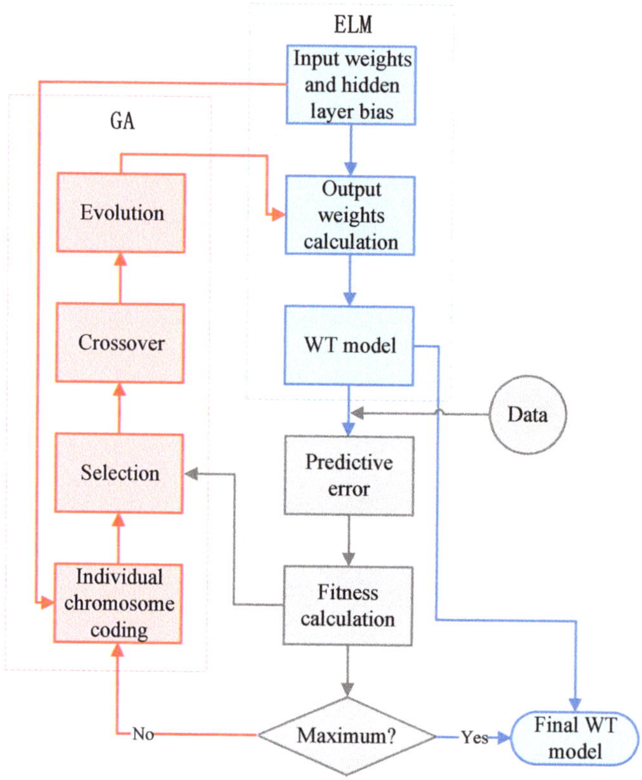

Figure 6. Topology of Extreme Learning Machine (ELM).

4. Model Development and Validation

4.1. SCADA Datasets

To build the WT model and verify its accuracy, the learning set and test set are shown in Table 1.

Table 1. Description of the learning and test sets.

Dataset	Start and End Time	Number of Data	Ambient Temperature	Wind Speed
Learning set	1 May 00:00 20 May 23:59	28,800	(8.41, 31.79) °C	(0.23, 23.62) m/s
Testing set	21 May 00:00 21 May 23:59	1440	(12.45, 20.02) °C	(4.63, 16.09) m/s

To ensure model accuracy, the learning set should cover the working conditions and state as much as possible only without failures. Similarly, the test set should also contain a variety of working conditions and states without failures.

4.2. Model Testing Result

To verify the GA optimization, the original ELM and Back-Propagation Neural Network (BPNN) are used for comparison with GA-ELM. The residuals of GA-ELM, ELM, and BPNN are shown in Figure 7a; the ambient temperature and wind speed of the testing set are shown in Figure 7b.

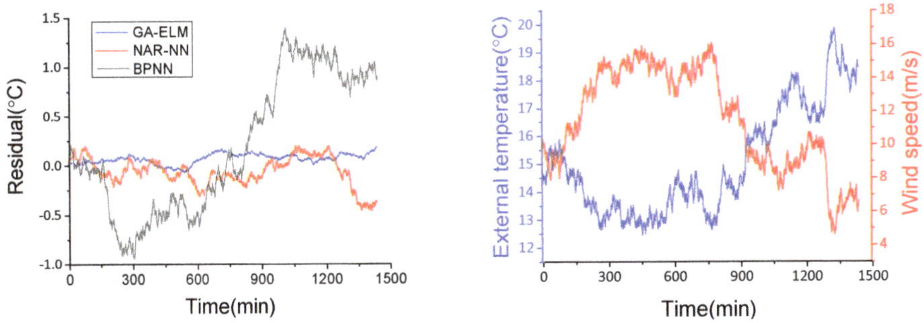

(a)

(b)

Figure 7. (a) Model testing results of different intelligent algorithms; (b)ambient conditions of testing set.

For the testing set, as shown in Figure 7b, the ambient temperature rises wavily and the wind speed drops rapidly after remaining stable for a period of time. This kind of irregularity may not fit the ELM and BPNN model at local minima. Thus, as shown in Figure 7a, the residual temperature of GA-ELM is smaller than that of ELM and BPNN.

To quantitatively compare the performance of the three algorithms, Mean Square Error (MSE), Mean Absolute Error (MAE), and Mean Absolute Percentage Error (MAPE) are used to analyze the results and calculated as

$$\text{MSE}= \frac{1}{s}\sum_{k=1}^{s}(e_k)^2 \tag{11}$$

$$\text{MAE} = \frac{1}{s}\sum_{k=1}^{s}|e_k| \tag{12}$$

$$\text{MAPE} = \frac{1}{s}\sum_{k=1}^{s}\left|\frac{e_k}{Y_k}\right| \tag{13}$$

Statistical indicators are shown in Table 2. In this study, all calculations are performed with the MATLAB R2017b environment on a 64-bit Windows operating system installed on a computer with an Intel Core i7-6700 CPU with 4GHz.

Table 2. Statistical indicators of different intelligent algorithms.

Criteria	GA-ELM	ELM	BPNN
MSE	0.10	0.21	0.58
MAE	0.19	0.59	0.91
MAPE (%)	0.26	0.73	1.84
Time (s)	3.46	3.22	3.15

Compared to ELM and BPNN, GA-ELM achieves a smaller MSE, MAE, and MAPE, demonstrating that the GA optimization is effective and the accuracy of the WT model is improved.

5. Case Study

To verify the proposed method, this paper sets up experiments for different ambient conditions and real failure. For an extreme ambient temperature, cases of winter midnight, summer noon, and a cold wave in normal state are presented. For wind speed change, cases of wind speed increase and decrease in normal state are presented. Meanwhile, for failure detection, a main bearing offset case is presented as well.

In these experiments, a comparison method is also applied to analyze the effects of the ambient conditions. The framework of the comparison method is the same as that of the proposed method, as shown in Figure 5. The comparison method uses the same GA-ELM model as the proposed method. The only difference between the proposed method and the comparison method is that the inputs of the comparison method, as with existing methods, do not contain the ambient conditions. To compensate for this, four other parameters (pitch motor 1, 2 and 3 temperature and hub ambient temperature) are added to the inputs of the comparison method. It should be noted that the residual in this paper is the difference in the actual value minus the predicted value. Moreover, for concise description, the proposed method in this paper is referred to as Method I, and the comparison method is referred to as Method II.

5.1. Ambient Temperature Change in Normal State

For ambient temperature, the data in winter, summer, and the cold wave are shown in Table 3. It should be noted that, during these three periods, the WT is in a normal working state.

Table 3. Description of datasets under different weather conditions.

Dataset	Start and End Time	Number of Data	Ambient Temperature	Wind Speed
Winter	28 January 00:00–23:59	1440	(−10.58,−0.02) °C	(9.75, 19.38) m/s
Summer	30 July 00:00–23:59	1440	(25.93, 36.59) °C	(4.24, 8.61) m/s
Cold wave	24 Feberary 18:30–20:30	120	(2.84, 16.32) °C	(15.63, 17.85) m/s

The residual results are shown in Figure 8, and the statistical indicators are shown in Table 4.

Table 4. Statistical indicators of Method I and II under a normal working state under different weather conditions.

Criteria	Winter		Summer		Cold Wave	
	Method I	Method II	Method I	Method II	Method I	Method II
MSE	0.18	0.91	0.16	0.62	0.18	1.39
MAE	0.14	0.98	0.11	0.73	0.19	1.07
MAPE (%)	0.25	1.79	0.16	1.04	0.30	1.96

It can be seen from Figure 8a that the Method I residual results are between −0.45 °C and 0.23 °C, but the Method II residual results are between −2.52 °C and 1.86 °C. Similarly, in Figure 8b, the Method II residual results (between −2.12 °C and 1.67 °C) are much greater than the Method I residual results (between−0.19 °C and 0.28 °C) in summer. Table 4 also shows that Method I achieves smaller statistical indicators than Method II. This proves that, with ambient temperature and ambient temperature change, Method I achieves better performance when facing cyclical and seasonal ambient temperature changes.

Furthermore, comparing Figure 8a,b the effect of ambient temperature in summer is smaller than in winter. A reasonable explanation is that the difference between the main bearing temperature and the ambient temperature is smaller in summer than in winter, since the range of the main bearing temperature in a normal working state is approximately 55–70 °C.

During the cold wave shown in Figure 8c, Method I has some reasonable fluctuations, but Method II shows a downward trend. If the monitoring is based on Method II, the main bearing temperature continues to decline for 2 h, which may lead to false alarms. However, the actual situation is that the WT is under a normal working state. This proves that, although Method II has more temperature parameters as compensation, Method I is more sensitive to the rapid ambient temperature change than Method II.

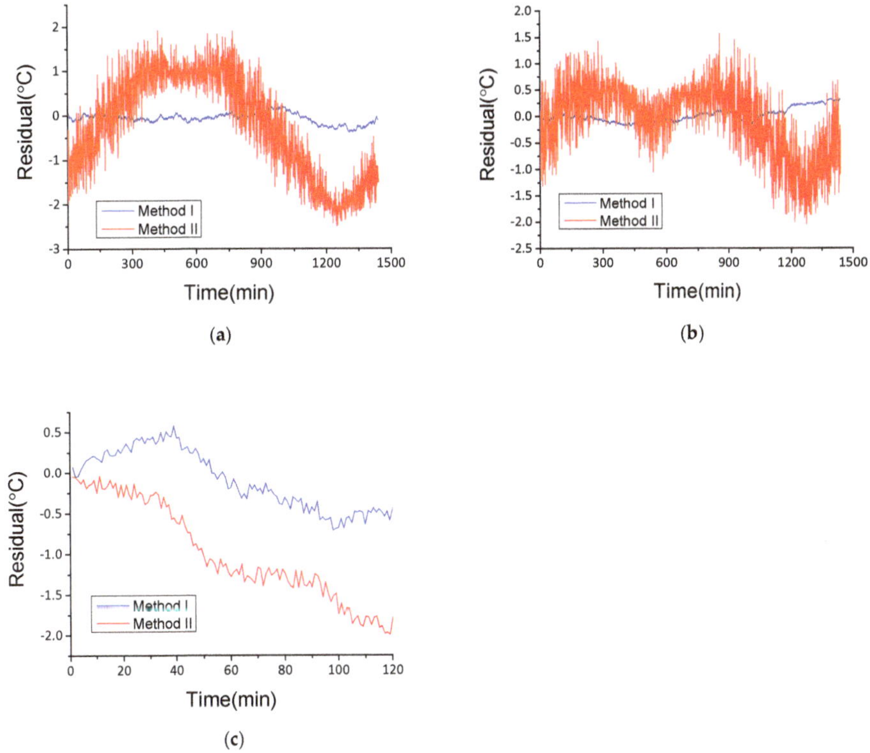

Figure 8. Residual results of Method I and II under a normal working state under different weather conditions: (**a**) winter; (**b**) summer; (**c**) cold wave.

5.2. Wind Speed Change in Normal State

For wind conditions, the datasets of wind speed increase and decrease are shown in Table 5. Similarly, in these two periods, the WT is in a normal working state. At the same time, in the event that the ambient temperature would have an influence, the ambient temperature is around 15 °C.

Table 5. Datasets of wind speed increase and decrease.

Dataset	Start and End Time	Number of Data	Ambient Temperature	Wind Speed
Wind speed increase	12 April 09:00–10:39	100	(13.92, 15.01) °C	(4.64, 15.12) m/s
Wind speed decrease	15 April 14:00–16:59	180	(14.45, 15.89) °C	(3.97, 14.83) m/s

The residual results of Method I and II are shown in Figure 9, and the statistical indicators are shown in Table 6.

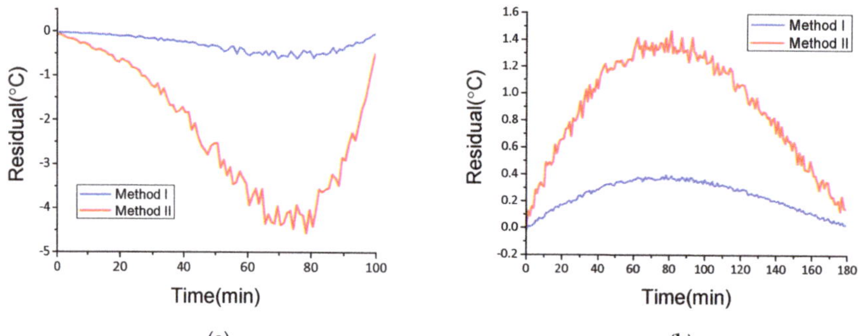

Figure 9. Residual results of Method I and II in normal working state within wind conditions: (**a**) wind speed increase; (**b**) wind speed decrease.

Table 6. Statistical indicators of Method I and II in normal working state during wind speed increase and decrease.

Criteria	Wind Speed Increase		Wind Speed Decrease	
	Method I	**Method II**	**Method I**	**Method II**
MSE	0.15	2.85	0.12	0.95
MAE	0.31	2.19	0.26	0.89
MAPE (%)	0.47	3.48	0.38	1.54

As shown in Figure 9a, during the wind speed increase, the residual results of Method I and Method II are negative. As mentioned in Section 2, these results are due to the delay between the wind speed change and internal temperature change. Since the wind speed change is used as a model input of Method I, the delay is reduced effectively, and the residuals' absolute value of Method I is much smaller than that of Method II. The amplitude of Method I is 0.48 °C, and that of Method II is 4.48 °C. The same reasoning can explain Figure 9b: during the wind speed decrease, the residuals' absolute value of Method I is much smaller than that of Method II (0.39 °C vs. 1.48 °C, respectively). The statistical indicators in Table 6 also prove that Method I achieves better performance than Method II during the wind speed change.

Additionally, both Method I and II achieve better performance during a wind speed decrease than during an increase. These results occur because the speed of wind speed change can directly determine the delay between the wind speed change and internal temperature change. In these two periods, the wind speed increase is faster than the decrease, which means that the delay during the increase is larger than during the decrease. Thus, the absolute values of the residual results during the wind speed increase are generally larger than those during the decrease.

5.3. Main Bearing Failure Detection

To verify the failure detection ability of the proposed method, a serious main bearing offset that occurred in the wind farm, which is shown in Figure 10, is used as a failure case. The dataset of 5 h before the failure happened is shown in Table 7, and the residual results are shown in Figure 11.

Table 7. Datasets of main bearing offset failure.

Dataset	Start and End Time	Number of Data	Ambient Temperature	Wind Speed
Failure	18 Mar 05:40–10:39	300	(−5.58, 0.02) °C	(3.64, 17.86) m/s

Figure 10. Main bearing offset failure.

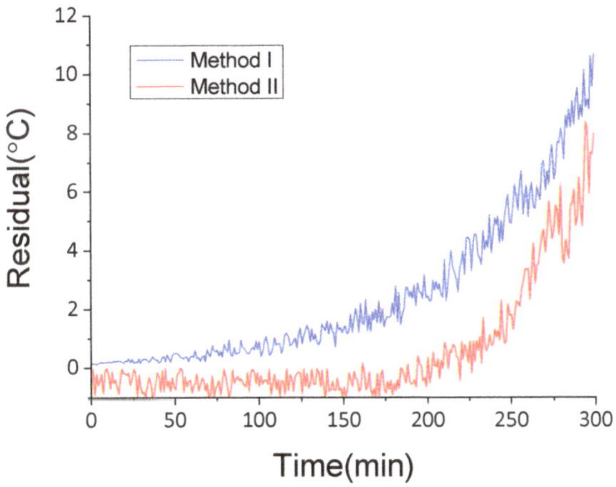

Figure 11. Residual results of Method I and II in main bearing offset failure.

The residual results of Method I and Method II both show upward trends, which means that both methods could predict the failure. However, during the 5h period, the ambient temperature continued to drop and the wind speed continued to increase, which caused the residual results of Method II at first to be negative, falling behind Method I. In particular, at 10:16 (time point 276), the wind speed increased rapidly, which caused the residual results of Method II to decrease over a short time. Comparing the two curves, it can be seen that Method II generally falls behind Method I by more than 50 min, reaching approximately90 min at a residual result of 2 °C.

Considering the conclusions of the previous experiments, Method I exhibits stable performance during extreme ambient temperatures and wind speed change, and the residual results of Method I are generally less than 0.5 °C. However, for Method II, due to the change in the ambient conditions, residual results could be more than 1 °C, sometimes reaching 4 °C, under a normal working state. Thus, the safe range of Method I is narrower than that of Method II. If the safe range of Method I is set to ±1 °C and that of Method II is set to ±2 °C, the alarm from Method I would be approximately120 min earlier than in Method II. If the safe range of Method I is set to ±0.5 °C and that of Method II is set to ±4 °C, the alarm of Method I would be more than 180 min earlier than that of Method II. These results demonstrate that Method I, with ambient conditions, can achieve higher monitoring accuracies and earlier failure alarms.

6. Conclusions

WTs are exposed to harsh conditions year-round, and the variability of ambient conditions can affect WT monitoring directly. The main bearing temperature can be different in the same working state but different under ambient temperature and wind speed changes. Conversely, a data-driven model may fall into local minima due to the randomness of the initial coefficients, which can also affect the monitoring accuracy.

To solve these problems, a novel WT monitoring method is proposed. The changes in ambient conditions are used as inputs of the WT model, and the GA-ELM is applied to construct the WT model. Model testing shows that, compared to ELM and BPNN, GA-ELM can avoid local minima and achieve smaller MSE, MAE, and MAPE. With cases in different ambient conditions and real failure, it proves that, compared to the method without considering the change in ambient conditions, the proposed method could reduce false alarms when WT is in a normal working state under extreme ambient conditions, and generate an earlier alarm when a failure is about to occur. This means that the proposed method can reduce operation and maintenance costs, thereby improving the economic efficiency of wind power generation.

It should be noted that this study focuses on ambient temperature and wind speed to describe ambient conditions. However, ambient conditions also include other factors, such as air humidity and air pressure. In future research, these factors should also be considered in order to improve the accuracy of WT monitoring.

Author Contributions: Methodology, Z.H. and X.L.; formal analysis, Z.H.; writing—original draft preparation, Z.H.; writing—review and editing, X.L.; project administration, S.Z.; funding acquisition, S.Z. All authors have read and agreed to the published version of the manuscript.

Funding: This research was funded by the National Key R&D Program of China, grant number 2017YFB1200703.

Data Availability Statement: Not applicable.

Conflicts of Interest: The authors declare no conflict of interest.

References

1. Yaramasu, V.; Wu, B.; Kouro, S. High-power wind energy conversion systems: State-of-the-art and emerging technologies. *Proc. IEEE* **2015**, *103*, 740–788. [CrossRef]
2. Garrigle, E.M.; Leahy, P.G. Cost Savings from Relaxation of Operational Constraints on a Power System with High Wind Penetration. *IEEE Trans. Sustain. Energy* **2017**, *6*, 881–888. [CrossRef]
3. Hou, Z.; Zhuang, S.; Lv, X. Monitoring and Analysis of Wind Turbine Condition based on Multivariate Immunity Perception. In Proceedings of the 2019 International Energy and Sustainability Conference (IESC), Farmingdale, NY, USA, 17–18 October 2019; pp. 1–5.
4. Xia, G.; Zhou, L.; Minder, J.; Fovell, R.; Jimenez, P. Simulating impacts of real-world wind farms on land surface temperature using the WRF model: Physical mechanisms. *Clim. Dyn.* **2019**, *53*, 3. [CrossRef]
5. Al-Masri, H.; Ab dullah, A.; AlmehiziaEhsani, M. Accurate Wind Turbine Annual Energy Computation by Advanced Modeling. *IEEE Trans. Ind. Appl.* **2017**, *3*, 1. [CrossRef]
6. Bakdi, A.; Kouadri, A.; Mekhilef, S. A data-driven algorithm for online detection of component and system faults in modern wind turbines at different operating zones. *Renew. Sustain. Energy Rev.* **2019**, *103*, 546–555. [CrossRef]
7. Long, H.; Wang, L.; Zhang, Z.; Song, Z.; Xu, J. Data-Driven Wind Turbine Power Generation Performance Monitoring. *IEEE Trans. Ind. Electron.* **2015**, *62*, 6627–6635. [CrossRef]
8. Kusiak, A.; Zhang, Z. Short-Horizon Prediction of Wind Power: A Data-Driven Approach. Energy Conversion. *IEEE Trans. Ind. Electron.* **2010**, *25*, 1112–1122.
9. Reder, M.; Yurusen, N.Y.; Melero, J.J. Data-driven learning framework for associating weather conditions and wind turbine failures. *Reliab. Eng. Syst. Saf.* **2018**, *169*, 554. [CrossRef]

10. Yin, S.; Wang, G.; Karimi, H.R. Data-driven design of robust fault detection system for wind turbines. *Mechatronics* **2014**, *24*, 298–306. [CrossRef]
11. Yang, W.; Court, R.; Jiang, J. Wind turbine condition monitoring by the approach of SCADA data analysis. *Renew. Energy* **2013**, *53*, 365–376. [CrossRef]
12. Mazur, D.C.; Entzminger, R.A.; Kay, J.A. Enhancing Traditional Process SCADA and Historians for Industrial & Commercial Power Systems with Energy (Via IEC 61850). *IEEE Trans. Ind.* **2016**, *52*, 76–82.
13. MDIlicXie, L.; Joo, J.Y. Efficient Coordination of Wind Power and Price-Responsive Demand—Part I: Theoretical Foundations. Power Systems. *IEEE Trans.* **2011**, *26*, 1875–1884.
14. Velandia-Cardenas, C.; Vidal, Y.; Pozo, F. Wind Turbine Fault Detection Using Highly Imbalanced Real SCADA Data. *Energies* **2021**, *14*, 1728. [CrossRef]
15. Kusiak, A.; Li, W. Virtual Models for Prediction of Wind Turbine Parameters. *IEEE Trans. Energy Convers.* **2010**, *25*, 245–252. [CrossRef]
16. Peng, G.; Infield, D. Wind Turbine Tower Vibration Modeling and Monitoring by the Nonlinear State Estimation Technique (NSET). *Energies* **2012**, *5*, 5279–5293.
17. Wang, Y.; Infield, D.G. Multi-machine Based Wind Turbine Gearbox Condition Monitoring Using Nonlinear State Estimation Technique. *EWEA* **2014**, *2014*. [CrossRef]
18. Zhao, W.S.; Qu, C.Y.; Zhang, H.B. Direct-Drive Wind Turbine Fault Diagnosis Based on Logistic Regression. In Proceedings of the 15th International Computer Conference on Wavelet Active Media Technology and Information Processing (ICCWAMTIP), Chengdu, China, 14–16 December 2018.
19. Wu, F.T.; Wang, C.C.; Liu, J.H.; Chang, C.-M.; Lee, Y.-P. Construction of Wind Turbine Bearing Vibration Monitoring and Performance Assessment System. *J. Signal Inf. Process.* **2013**, *4*, 430–438. [CrossRef]
20. Wenyi, L.; Zhenfeng, W.; Jiguang, H.; Guangfeng, W. Wind turbine fault diagnosis method based on diagonal spectrum and clustering binary tree SVM. *Renew. Energy* **2013**, *50*, 1–6. [CrossRef]
21. Santos, P.; Villa, L.; Renones, A.; Bustillo, A. An SVM-Based Solution for Fault Detection in Wind Turbines. *Sensors* **2015**, *15*, 5627–5648. [CrossRef]
22. Kandukuri, S.T.; Senanayaka, J.; Huynh, K.V.; Robbersmyr, K.G. A Two-Stage Fault Detection and Classification Scheme for Electrical Pitch Drives in Offshore Wind Farms Using Support Vector Machine. *IEEE Trans. Ind. Appl.* **2019**, *55*, 5109–5118. [CrossRef]
23. Rahimilarki, R.; Gao, Z.; Zhang, A.; Binns, R. Robust neural network fault estimation approach for nonlinear dynamic systems with applications to wind turbine systems. *IEEE Trans. Ind. Inform.* **2019**, *15*, 6302–6312. [CrossRef]
24. Bangalore, P.; Tjernberg, L.B. An Artificial Neural Network Approach for Early Fault Detection of Gearbox Bearings. *IEEE Trans. Smart Grid* **2017**, *6*, 980–987. [CrossRef]
25. Qu, F.; Liu, J.; Liu, X.; Jiang, L. A Multi-fault Detection Method with Improved Triplet Loss Based on Hard Sample Mining. *IEEE Trans. Sustain. Energy* **2020**, *12*, 127–137. [CrossRef]
26. Zhang, J.; Sun, H.; Sun, Z.; Dong, W.; Dong, Y. Fault Diagnosis of Wind Turbine Power Converter Considering Wavelet Transform, Feature Analysis, Judgment and BP Neural Network. *IEEE Access* **2019**, *7*, 179799–179809. [CrossRef]
27. Kusiak, A.; Zhang, Z.; Li, M. Optimization of Wind Turbine Performance with Data-Driven Models. *IEEE Trans. Sustain. Energy* **2010**, *1*, 66–76. [CrossRef]
28. Ak, R.; Fink, O.; Zio, E. Two Machine Learning Approaches for Short-Term Wind Speed Time-Series Prediction. *IEEE Trans. Neural Netw. Learn. Syst.* **2016**, *27*, 1734–1747. [CrossRef]
29. Li, S.; Wang, P.; Goel, L. Short-term load forecasting by wavelet transform and evolutionary extreme learning machine. *Electr. Power Syst. Res.* **2015**, *122*, 96–103. [CrossRef]
30. Wan, C.; Xu, Z.; Pinson, P. Probabilistic forecasting of wind power generation using extreme learning machines. *IEEE Trans. Power Syst.* **2014**, *29*, 1033–1044. [CrossRef]
31. Wan, C.; Xu, Z.; Wang, Y. A hybrid approach for probabilistic forecasting of electricity prices. *IEEE Trans. Smart Grid* **2014**, *5*, 463–470. [CrossRef]
32. Wan, C.; Zhao, X.; Pierre, P. Optimal prediction intervals of wind power generation. *IEEE Trans. Power Syst.* **2014**, *29*, 1166–1174. [CrossRef]
33. Liu, S.; Hou, Z.; Yin, C. Data-Driven Modeling for UGI Gasification Processes via an Enhanced Genetic BP Neural Network with Link Switches. *IEEE Trans. Neural Netw. Learn. Syst.* **2015**, *27*, 2718–2729. [CrossRef] [PubMed]
34. Shen, X.; Zheng, Y.; Zhang, R. A Hybrid Forecasting Model for the Velocity of Hybrid Robotic Fish Based on Back-Propagation Neural Network with Genetic Algorithm Optimization. *IEEE Access* **2020**, *8*, 111731–111741. [CrossRef]
35. Kaleeswaran, V.; Dhamodharavadhani, S.; Rathipriya, R. A Comparative Study of Activation Functions and Training Algorithm of NAR Neural Network for Crop Prediction. In Proceedings of the 2020 4th International Conference on Electronics, Communication and Aerospace Technology (ICECA), Coimbatore, India, 5–7 November 2020.
36. Villanueva, D.; AFeijó, A. Normal-Based Model for True Power Curves of Wind Turbines. *IEEE Trans. Sustain. Energy* **2016**, *7*, 1005–1011. [CrossRef]
37. Xie, K.; Jiang, Z.; Li, W. Effect of Wind Speed on Wind Turbine Power Converter Reliability. *IEEE Trans. Energy Convers. Ec.* **2012**, *27*, 96–104. [CrossRef]

energies MDPI

Article

Ultimate Limit State Scour Risk Assessment of a Pentapod Suction Bucket Support Structure for Offshore Wind Turbine

Young-Jin Kim [1], Duc-Vu Ngo [1], Jang-Ho Lee [2] and Dong-Hyawn Kim [3,*]

1 Department of Ocean Science and Engineering, Kunsan National University, Gunsan 54150, Korea; ioi1937@nate.com (Y.-J.K.); ngoducvubk@gmail.com (D.-V.N.)
2 School of Mechanical Convergence System Engineering, Kunsan National University, Gunsan 54150, Korea; jhlee@kunsan.ac.kr
3 School of Architecture and Coastal Construction Engineering, Kunsan National University, Gunsan 54150, Korea
* Correspondence: eastlite@kunsan.ac.kr

Abstract: Scour risk assessment considering reaction force at foundation was proposed and applied to newly developed pentapod suction bucket support structures for a 5.5 MW offshore wind turbine under ultimate limit state environmental load. Scour hazard was obtained according to scour depth by using an empirical formula, which is the function of marine environmental conditions such as significant wave height, significant period, and current velocity. Fragility of the pentapod support structure was evaluated using the bearing capacity limit state criterion under ultimate limit state load case. Scour risk was assessed by combining the scour hazard and the fragility. Finally, scour risk of the developed pentapod suction bucket support structure under ultimate limit state has been assessed.

Keywords: offshore wind turbine; suction bucket; pentapod suction bucket; scour; scouring fragility; scour risk

Citation: Kim, Y.-J.; Ngo, D.-V.; Lee, J.-H.; Kim, D.-H. Ultimate Limit State Scour Risk Assessment of a Pentapod Suction Bucket Support Structure for Offshore Wind Turbine. *Energies* **2022**, *15*, 2056. https://doi.org/10.3390/en15062056

Academic Editor: Paweł Lizęga

Received: 23 January 2022
Accepted: 7 March 2022
Published: 11 March 2022

1. Introduction

Suction Bucket (SB) is a highly competitive foundation solution among the several types of foundations currently implemented to support offshore wind turbines (OWT) due to their quick and noise-free installation. They are also a more economical solution compared to other foundations, especially for a large-scale wind farm [1]. The main design factors of such a SB support structure are the horizontal bearing capacity and the stiffness of the foundation. Latini and Zania, 2017 [2] investigated the dynamic behavior of SB and concluded that the skirt length of a SB is an important parameter that determines the dynamic behavior, and the horizontal bearing capacity is greatly affected by the ratio of the bucket diameter and length. Foundation stiffness is strongly dependent on the relative density of sand and the bucket's geometry and has been investigated in a series of studies [3–5]. When existing SB support structures are installed in area where the geological structure is a shallow soft-layer soil on top of a hard-layer soil, there is a problem that the SB diameter must be abnormally large to fulfill the required bearing capacity. To solve this problem, a pentapod suction bucket (PSB) support structure was developed by Ngo et al. [6], and a seismic fragility analysis was performed on the developed support structure.

Scouring is recognized as a risk factor that weakens the bearing capacity around the turbine support structures. Accordingly, local scour around OWT foundations has been studied by many researchers [7–9], and most studies have used scale model tests and numerical simulations. A series of laboratory experiments were conducted by Hu et al. [10] to investigate the scour development around tripod foundation in combined waves and the current. The influence of installation angles, KC number, and the ratio of velocities U_{cw} on the scour depth were also examined. In another study, Hu and his colleagues proposed a method to predict the equilibrium scour depth around the umbrella suction anchor foundation [11]. In order to reduce the risk of scouring, a prevention method is installed around the support structures [12]. Since the scouring protection method incurs costs, a reasonable risk assessment by scouring is necessary for the development of an economical power generation complex. Studies on the behavior of supporting structures by scouring have not been actively performed. A few studies have been recently reported. Yi et al., 2013 studied the effect of scour on dynamic instability of monopile offshore wind turbines [13]. In the study, the natural frequency change according to seabed scour depth (SD) was analyzed and possible resonance due to scour was discussed. Ma et al., 2018 [14] studied the effect of local scour around foundation on the dynamic behavior of the support structure. Both studies revealed that scour had little effect on frequencies but had some effect on dynamic behavior. Scour risk assessment around marine structures has also been performed in many studies. Yanmaz and Salamak [15] evaluated the risk of the scouring process at bridge piers. They established a probabilistic model for the scour depth predictions through empirical equations in the dimensional state. Khalid et al. [16] applied a non-linear regression-based technique to calculate the reliability of scour depth values at bridge piers installed in cohesive bed sediments. A study that evaluated a reliability-based probabilistic of the wave-induced scour depth around marine structure piles was performed by Homaei et al. [17]; they developed a probabilistic model by using an artificial intelligence method so as to predict the scour depth at pile groups under regular waves. However, these studies were conducted on monopiles, and the risk of scouring could not be evaluated quantitatively. Kim et al., 2020, proposed a scouring risk assessment method by combining the scouring hazard and fragility of the support structure [18]. They showed, for the first time, that scour risk can be evaluated by modifying the seismic risk assessment approach. Scour fragility of a tripod suction bucket (TSB) was found by defining a critical displacement at foundation. Ngo et al. recently reported a seismic fragility analysis of PSB support structure and showed that the seismic performance of the pentapod is superior to that of the TSB [6].

In this study, a scour risk analysis was performed to evaluate the stability of the newly developed PSB support structure at an ultimate limit state. For this, the hazard according to the scour depth was calculated. For the scour depth, an empirical formula defined as a function of marine environmental variables was used. As for the fragility, the bearing capacity limit state according to the ultimate load was used instead of the fragility function based on displacement [18]. The safety factor (SF) was applied to the reaction force of the PSB to determine how the scour risk was affected.

In the second chapter, theoretical background of scour risk assessment is explained. How to calculate scour hazard, fragility and risk are shown in detail. In the third chapter, why a PSB is developed and features of PSB are explained. After that, numerical analysis and conclusions are described in turn.

2. Scour Risk

2.1. Probability of SD

During the design process, the capacity of OWT substructure (SB foundation in this study) is designed against external loads such as wind, wave, and so on. The magnitude of these external loads is governed by wind speed, wave height or current velocity, etc. and their effects have also been verified through various design load cases (DLCs). Nevertheless, the initial OWT foundation's design capacity may be decreased due to several reasons, with local scouring around its foundation being the most common. In previous studies, maximum equilibrium SD was suggested as a keyword to evaluate the influence of scouring on the structure. This leads to unnecessary costs in the design and operation of an OWT, especially when the scale of the wind farm project is large. Therefore, it is necessary to consider scouring as a probabilistic parameter in risk analysis.

In this study, the empirical formula of Equation (1) proposed by Sumer and Fredsoe, 2001 [19] is used to obtain the probability distribution of SD. In Equation (1), the SD is a function of the Keulegan–Carpenter (KC) number and the parameter U_{cw}, in which KC was defined by the peak spectrum period (T_p), pile diameter (D), and maximum value of the undisturbed orbital velocity at the sea bottom just above the wave boundary layer (U_m).

$$\frac{S}{D} = \frac{S_C}{D}[1 - \exp\{-A(KC - B)\}],\ KC \geq 4 \tag{1}$$

where: S: SD; S_C: SD in the case of steady current alone; D: pile diameter; $KC = U_m T_p/D = 2\pi a/D$; $A = 0.03 + 0.75U_{cw}^{2.6}$; $B = 6\exp(-4.7U_{cw})$; $U_{cw} = U_c/(U_c + U_m)$; U_C: the undisturbed current velocity at the distance, y = D/2 from the bed; U_m: maximum value of the undisturbed orbital velocity at the bed; T_p: peak spectrum period; a: the amplitude of the motion of water particles at the bed.

In Equation (1), the SD probability distribution can be obtained by considering the variability of KC parameter. The variability of KC is governed by the variability of height and period of the significant wave. The probability density function (PDF) of SD obtained here is a scour hazard for risk analysis and denoted by $f_{SD}(x)$.

2.2. Scour Fragility

The fragility curve of structures under earthquakes has been proposed by Shinozuka et al., 2000 [20]. These fragility curves are represented by logarithmic normal distribution functions, and the two coefficients of the lognormal distribution function, median and logarithmic standard deviation, are obtained by the maximum likelihood estimation method. For the k-th damage among various damage stages, the fragility curve can be expressed as:

$$F_k(x) = \Phi\left|\frac{\ln\left(\frac{x}{c_k}\right)}{\zeta_k}\right| \tag{2}$$

where Φ is the standard normal cumulative distribution function, c_k is the median, and ζ_k is the logarithmic standard deviation.

When assessing structural fragility, a limit state should be determined first. There are various limit states such as serviceability limit state (SLS), fatigue limit state (FLS) and ultimate limit states (ULS). Kim et al. found fragility curves of TSB based on displacement [10]. Because the displacement of the bucket causes a serviceability problem rather than the destruction of the support structure, it is a fragility based on SLS. However, the purpose of this study is to evaluate bearing capacity of a newly designed PSB under ultimate load case. Therefore, an event in which the reaction force of the bucket exceeds the allowable bearing capacity was defined as the failure. Therefore, the fragility of this study can be called ULS based.

2.3. Scour Risk

By combining the probability of SD (hazard) and the scour fragility curve of the structure, the scour risk can be calculated as in Equation (3).

$$P_f = \int_{x_0}^{x_{max}} f_{SD}(x)F_k(x)dx \tag{3}$$

where x_{max} and x_0 is the maximum possible SD and the lowest SD, respectively.

3. Development of PSB Support Structure

3.1. Background

A ground survey was carried out on the coast of Gunsan, Southwest of Korea, prior to the design and installation of a 5.5 MW OWT. Figure 1a is the expected installation location, while Figure 1b and Table 1 show the ground investigation results. It is observed in Figure 1b that a bedrock layer appears at a depth of 7.5 m, and when an offshore wind turbine supported by suction bucket foundation is installed here, the length of the suction bucket skirt must be less than 7.5 m. Since the skirt length is limited, it is necessary to increase the bucket diameter to reinforce the overturning resistance and improve bearing capacity. However, various problems include the lack of manufacturing facilities for large-diameter suction buckets, transportation and installation equipment, etc. Moreover, increasing the diameter can make the spacing between the buckets too narrow, leading to overlap stress. Hence, developing a multi-pod suction bucket support structure that exhibits equivalent support is necessary.

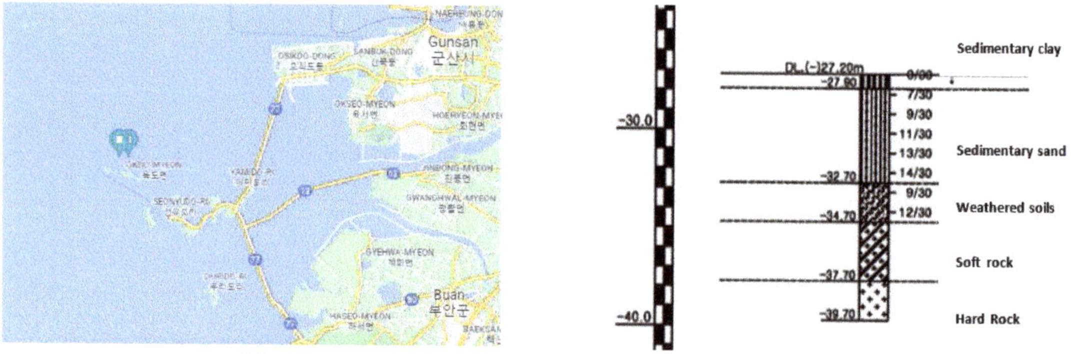

(**a**) (**b**)

Figure 1. Installation area: (**a**) map; (**b**) soil layer.

Table 1. Soil profile of survey site.

Soil Layer	Depth (m)	Unit Weight (kN/m^3)	Elasticity (MPa)	Internal Friction Angle (deg)	Cohesion Yield Stress (kPa)	Poisson's Ratio
Upper clay	0.0~0.7	17	25.00	32.3	-	0.491
Upper sand	0.7~4.8	17.5	35.56	32.3	5	0.400
Lower sand	4.8~7.5	17.5	67.48	37.0	5	0.400
Weathered rock	7.5~	20.0	76.00	32.0	-	0.450

3.2. Development of PSB Support Structure

Figure 2 shows the top view and thrust direction of the PSB model. The spacing (R) between each bucket and the center of the tower is assumed to be the same. According to the thrust direction, as shown in Figure 2, the pull-out resistance (V_t) and compression resistance (V_c) are indicated by red and blue dots, respectively. Then, the resistance moment (M_R) can be calculated as in Equations (4)–(7).

$$M_R = V_t b_1 + 2V_t b_2 \tag{4}$$

$$b_1 = R + R\cos(36^\circ) = 1.809R \tag{5}$$

$$b_2 = R\cos(36^\circ) + R\cos(72^\circ) = 1.118R \tag{6}$$

$$M_R = 1.809V_t R + 2.236V_t R = 4.045V_t R \tag{7}$$

When the number of buckets is 3 to 6, the moment of resistance (M_R) can be calculated the same way, and the corresponding normalized resistance moments are shown in Figure 3 [6].

It can be seen from Figure 3 that the PSB has better pull-out and compression resistance than other multi-pod support structures. Therefore, a PSB support structure using five buckets was finally developed. Figure 4 shows the developed PSB support structure.

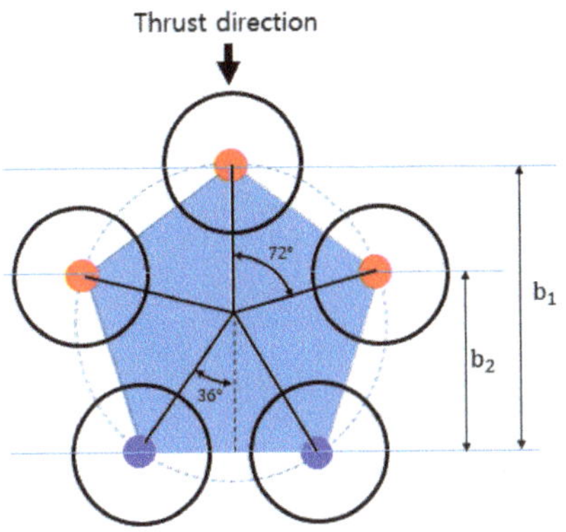

Figure 2. Top view of PSB [6].

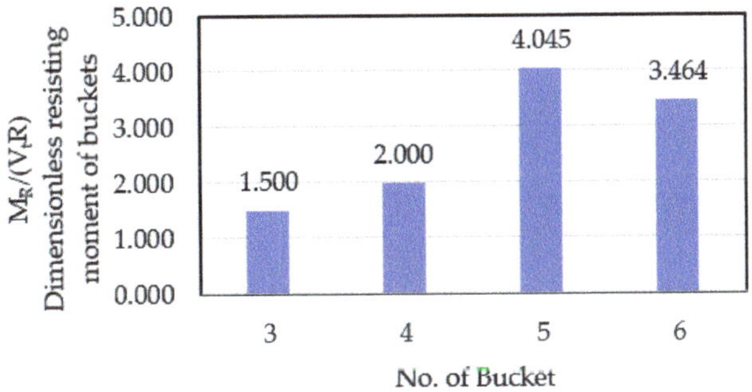

Figure 3. Dimensionless resisting moment of buckets.

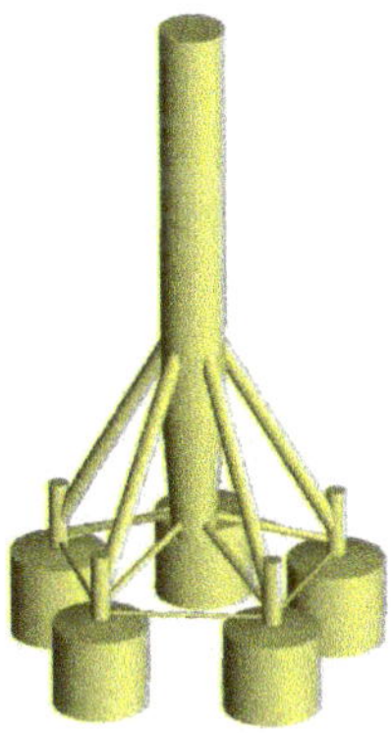

Figure 4. Pentapod support structure.

4. Numerical Example

4.1. Wind Turbine and Support Structure Model

Figure 5 is a 5.5 MW OWT with PSB support structure modeled by Bladed [21]. The hub is located at the height of 110 from sea level and the water depth is 27.723 m. The total height of the offshore wind turbine is 137.723 m. Substructure is a pentapod with five single suction buckets. The geometric and material properties of the supporting tower structure are referred from a previous study [6]. The mechanical characteristics of the seabed connected to a substructure is represented by a soil stiffness matrix.

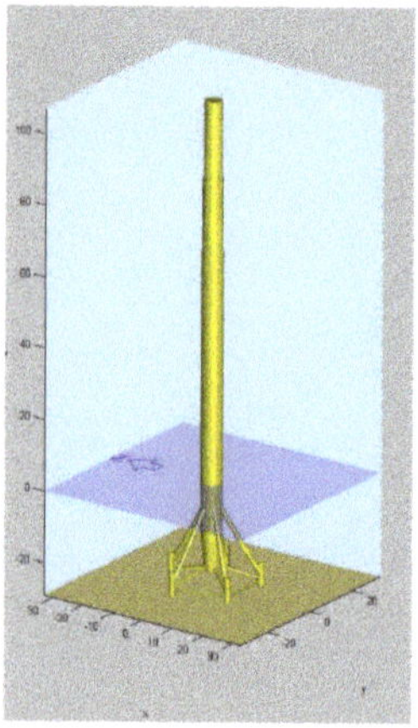

Figure 5. Analysis target.

4.2. Soil-Structure-Interaction Simulation

In order to express the interaction behavior of the suction bucket and the contact ground, a three-dimensional displacement analysis of the bucket-ground model caused by an external force was performed. Then, a stiffness matrix from the analysis was applied to Bladed. The commercially available Finite Element (FE) software ABAQUS [22] was used. The finite element model was shown in Figure 6, the soil was modeled with C3D8R elements, and the bucket foundation modeled by shell element with a diameter (D) of 9 m and a skirt length (L) of 7 m. The center of the bucket top was set as the reference point to apply an external load. Displacements were obtained after sequentially applying external loads in the direction of 6 degrees of freedom (*Fx, Fy, Fz, Mx, My, Mz*) to the reference point. In order to consider the effect of ground nonlinearity, the load in each direction was divided into 10 steps and applied gradually. From the load-response relationship the following equation can be written

$$[\mathrm{d}] = [F][p] \tag{8}$$

where $[\mathrm{d}] = \begin{bmatrix} \Delta_x & \Delta_y & \Delta_z & \theta_x & \theta_y & \theta_z \end{bmatrix}^{\mathrm{T}}$ is the displacement vector; $[p] = [F_x \;\; F_y \;\; F_z \;\; M_x \;\; M_y \;\; M_z]^{\mathrm{T}}$ the load vector; $[F]$ the flexibility matrix from numerical analysis [23]. Then, the stiffness matrix for ground can be derived as

$$[p] = [K][\mathrm{d}] \tag{9}$$

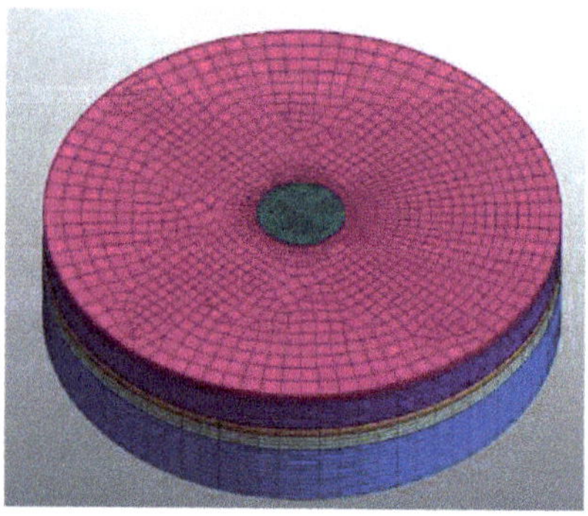

Figure 6. Stiffness matrix analysis model.

The stiffness matrix $[K]$ is the inverse of $[F]$ and obtained in this study for Bladed input as in Figure 7.

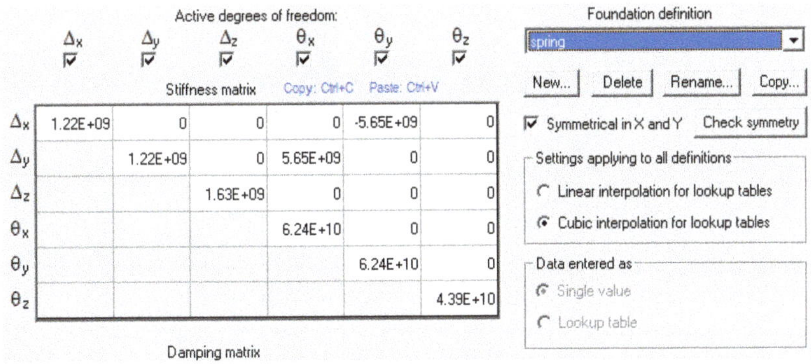

Figure 7. Bladed input for stiffness of foundation spring.

4.3. Wind Thrust

Thrust force by wind is automatically calculated in Bladed. At first, a wind field was generated using the Kaimal model, and a total of 30 cases of thrust force set was calculated by changing the phase of the wind field. Figure 8 shows a time history of thrust force for 10 min calculated by Bladed.

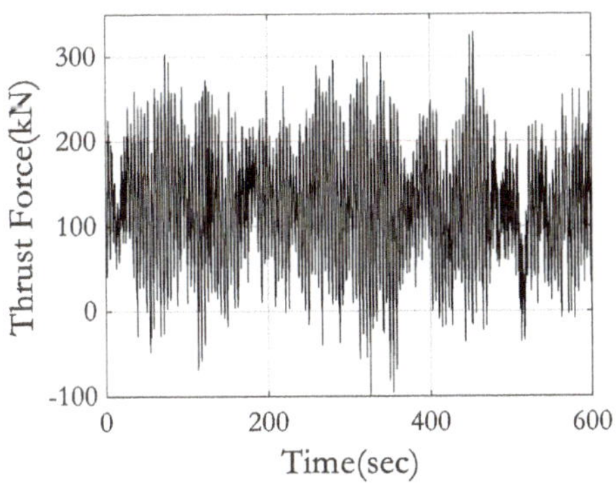

Figure 8. Time history of thrust force.

4.4. Calculation of Wave Load

The Morrison Equation [24] was used to calculate the wave loading acting on the structure in Bladed. To evaluate the risk due to scouring, the most hazardous event caused by the environment during the life cycle was considered. Significant wave height (H_s) and wave period (T_s) are adopted from the HYPA model of the Korea Oceanic Research and Development Institute [25] from 1979 to 2003. Figure 9 shows the probability density function estimated using the Weibull distribution and it can be expressed as Equation (10) where a and b are 5.56 and 9.66, which mean scale and shape parameters, respectively. To obtain the ultimate limit state (ULS) wave load, the significant wave height corresponding to a 50-year return period, H_{s50}, was estimated to be 6.64 m from the PDF. It is the wave height at which an excess probability is 1/50. The significant period was estimated to be 12.9 s using Equation (11) [26].

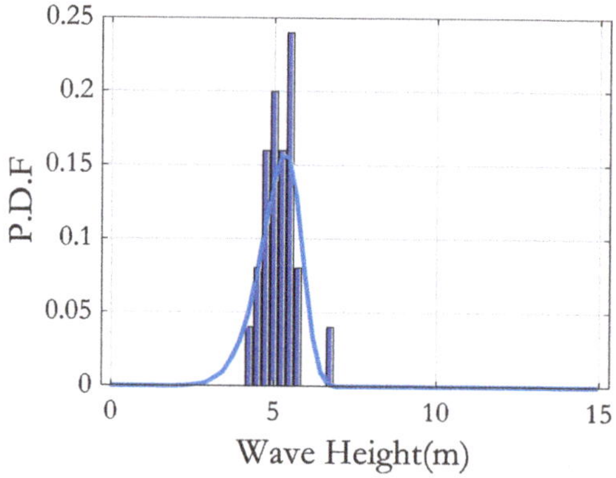

Figure 9. PDF of annual maximum H_s.

$$f_X(x) = \frac{b}{a}\left(\frac{x}{a}\right)^{b-1} e^{-\left(\frac{x}{a}\right)^b} \tag{10}$$

$$T_s = 3.3H_s^{0.63} \tag{11}$$

To simulate the wave acting on the OWT substructure, Bretschneider's wave spectrum with significant wave height (H_s), period (T_s) and frequency (f) is adopted to generate the sea surface elevation time history, it was defined as

$$S(f) = 0.257H_s^2 T_s^{-4} f^{-5} exp\left[-1.03(T_s f)^{-4}\right] \tag{12}$$

Using the Bretschneider spectrum mentioned above, the surface wave elevation profile can be obtained and was shown as Figure 10.

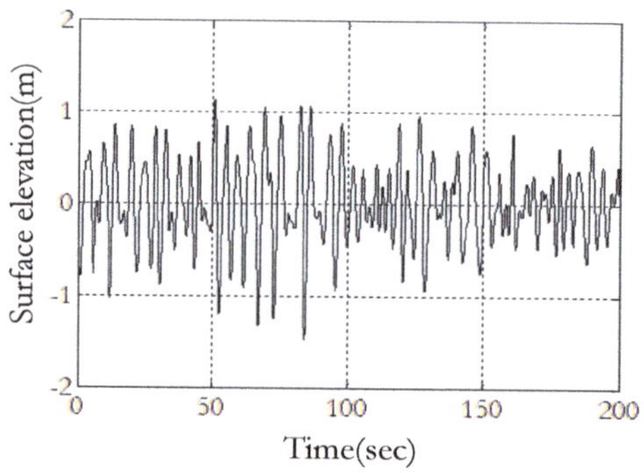

Figure 10. Surface wave elevation.

4.5. Probability Distribution of SD

The probability distribution of SD can be obtained by giving random variability to the variable in Equation (1). The most important among them is the distribution of KC. From the Weibull distribution of Equation (10), the annual maximum significant wave, H_s, is generated and T_s is estimated from Equation (11). Then, peak period T_p corresponding to the H_s and T_s is calculated from the spectrum of Equation (12). Once T_p is calculated scour depth S can be obtained using Equation (1). With 50,000 times random sampling for H_s, KC distribution was developed as Figure 11.

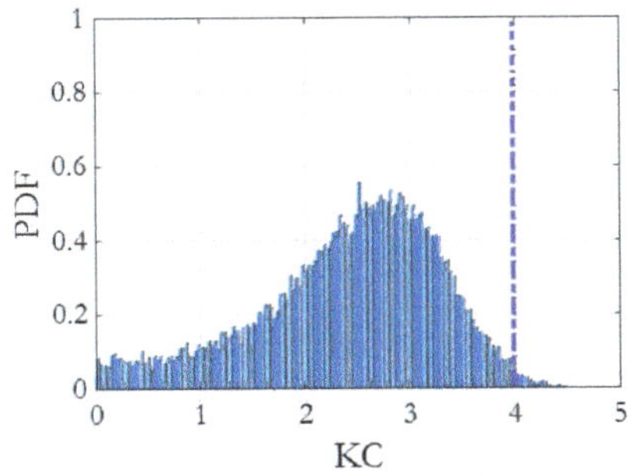

Figure 11. Probability distribution of KC.

The distribution of current speed was collected from the National Oceanic and Atmospheric Research Institute [27]. The estimated distribution of current speed fits the normal distribution, and the mean and standard deviation are 1.34 and 0.19, respectively.

Since only values of KC greater than 4.0 are effective for scour generation, the depth of scour was calculated using the distribution of KC greater than 4.0 and the tidal flow distribution. As a result, the distribution of scour depth was obtained as shown in Figure 12. It was found that a log normal distribution fit well, as shown in Figure 12. Two parameters λ and ζ were 0.75 and 0.55, respectively.

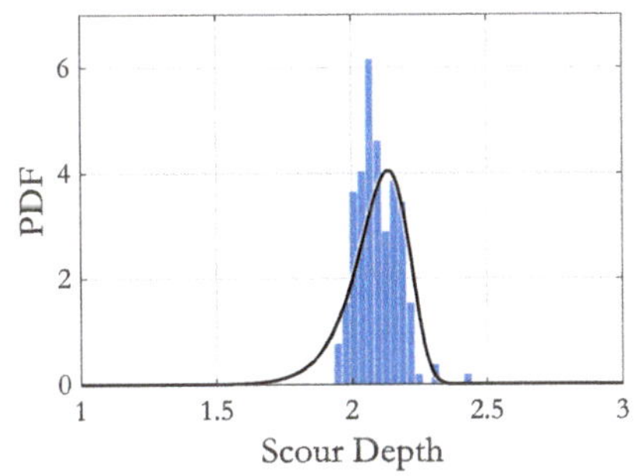

Figure 12. Probability distribution of SD.

4.6. Scouring Fragility Curve

To obtain the fragility curve, the limit state function of PSB is defined as the following equation.

$$g(X) = R_a(SD) - R_{max} \tag{13}$$

where R_a is the allowable bearing capacity of a bucket; R_{max} the maximum reaction force at each bucket. The R_a is a function of scour depth SD since the contact area of a bucket with ground is dependent on SD.

The bearing capacity of a bucket can be calculated numerically in horizontal and vertical directions [28]. For the analysis, the suite of 30 cases were analyzed by changing the seed of the wind field. Based on the structural responses at the mudline location for each case (i.e., reaction force to tension, compression, and horizontal force), the scour fragility was obtained. Accordingly, the maximum reaction members at the mudline obtained by performing a dynamic analysis in the non-sour state were compared with the allowable bearing capacity by SD to determine to what extent SD is safe.

Figure 13a shows the allowable pull-out force of a bucket for each SD. From the figure, the allowable pull-out force rapidly decreases according to SD increase. The tension-bearing capacity of the SB is mainly provided by vertical friction between bucket wall and soil. Therefore, scour reduces bearing capacity in tension direction. Figure 13b compares the allowable tension force at SD of 6.5 m, with the maximum pull out load corresponding to seed variation in wind field. The maximum pull out loads are smaller than the allowable tension capacity. Therefore, no failure is expected in tensional bearing capacity. Figure 14 shows the same results for compressive mode. While Figure 14a shows the allowable compressive force for each SD, Figure 14b compares the maximum compressive force calculated from the structural analysis results and the allowable compressive force at SD of 6.5 m. Both Figures 13 and 14 show that the tension and the compressive forces do not exceed the allowance one because the structural responses are far smaller than the allowable tension and compressive forces in all cases. Therefore, the fragility assessment of support structure in terms of tension and compression force here is not necessary.

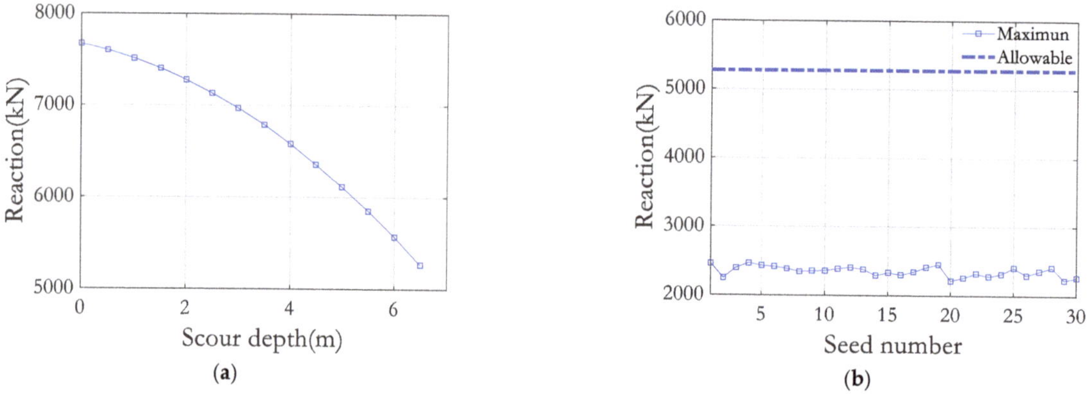

Figure 13. Variation of pull-out reaction: (**a**) allowable reaction; (**b**) reaction at SD = 6.5 cm.

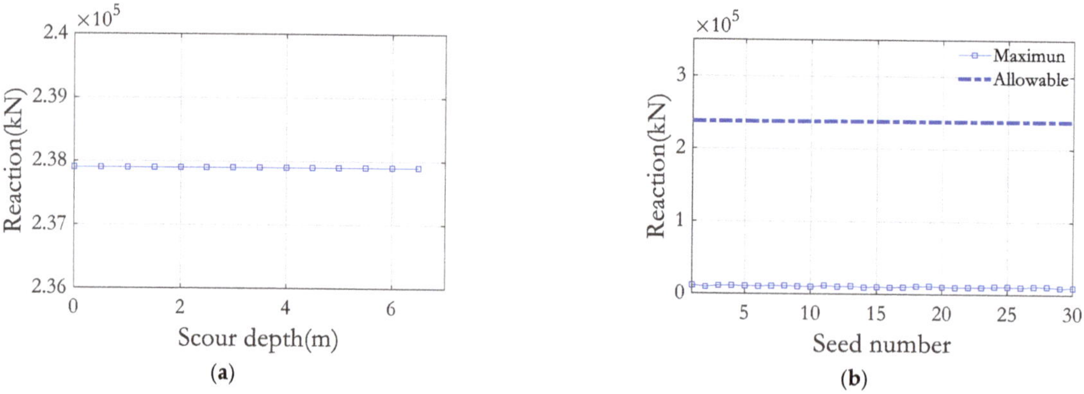

Figure 14. Variation of compressive reaction: (**a**) allowable reaction; (**b**) reaction at SD = 6.5 cm.

The results of horizontal force analysis are given in Figure 15. As shown in Figure 15a, the horizontal bearing capacity decreased as the SD increased. Figure 15b shows the result of comparing the maximum horizontal reaction force and the allowable horizontal force at SD of 4 m. It can be seen from Figure 15b that most of the horizontal reaction force exceeded the allowable horizontal force at the SD of 4.0 m. If more than 4.0 m scour occurs, the probability of failure will increase.

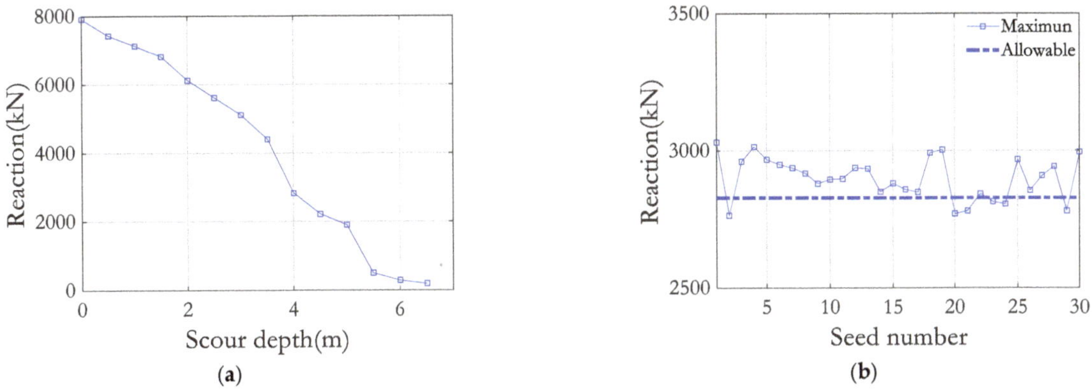

Figure 15. Variation of horizontal reaction: (**a**) allowable reaction; (**b**) reaction at at SD = 4 m.

To see how the safety margin affects the fragility, five cases of safety factors (SFs) were applied to the maximum horizontal reaction force. The fragility curves of the five safety factor cases are shown in Figure 16. The corresponding median and standard deviation values of the fragility curves are listed in Table 2. As can be seen from Figure 16 and Table 2, when the safety factor was not considered, the fragility was more than 50% at SD of around 3.93 m, and when the safety factor 2.0 was considered, the fragility was more than 50% at SD of approximately 2.43 m.

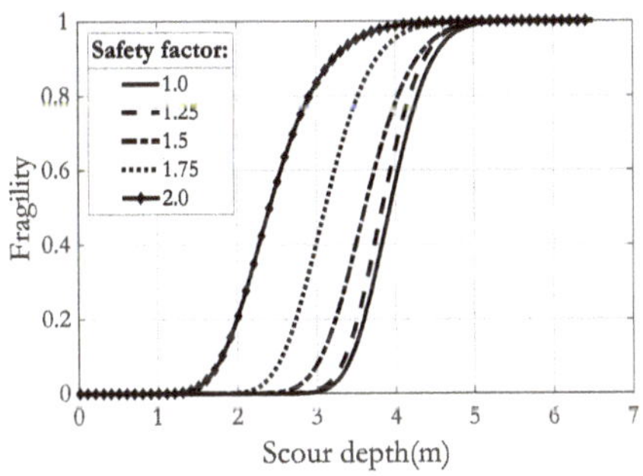

Figure 16. Scour fragility curve.

Table 2. Median and log-standard deviation.

Safety Factor (SF)	Median (m)	Log-Std. (m)
1.00	3.93	0.05
1.25	3.84	0.05
1.50	3.62	0.05
1.75	3.12	0.05
2.00	2.43	0.05

The log-normal standard deviation is equal to 0.05 because, for each SD level, the analysis was carried out by considering only the load variability.

4.7. *Scour Risk Assessment*

Scour risk was evaluated by integrating the product of scour hazard (SD probability) and scour fragility as given in Equation (3). The scour hazard denoted by $f_{SD}(x)$ was found in Figure 12. It presents the probability density of SD. The fragility denoted by $F_k(x)$

was found in Figure 16 according to SF. SF of 1.0 is the most critical case. Multiplying the scour hazard with the fragility and then integrating them over possible scour depth results in scour risk. Scour risk is expressed as probability of failure, P_f. For convenience, the probability of failure is converted into a reliability index as follows.

$$\beta = -\Phi^{-1}\left(P_f\right) \tag{14}$$

Reliability indices are listed in Table 3 and plotted in Figure 17. The scour risk was 1.919×10^{-7}~0.718 and the reliability index was 5.708~0.578, corresponding to the SF from 1.0 to 2.0. The level of target reliability index (β_t) can be referred from some design standards. DNV GL [29] proposes target failure probability of 10^{-4}, corresponding to β_t of 3.719, while IEC 61400-1 [30] proposes β_t of 3.3. Since the DNV guideline is for offshore wind turbine design, the reliability index was set higher than the IEC standard for onshore wind turbines. Furthermore, compared with the design standards, the reliability index evaluated for the PSB in Gunsan test bed seems higher than those standards if SF is below 1.5.

Table 3. Scour risk and reliability index.

SF	Sour Risk	Reliability Index
1.00	1.919×10^{-7}	5.078
1.25	6.178×10^{-7}	4.850
1.50	6.560×10^{-4}	3.213
1.75	0.058	1.570
2.00	0.718	0.578

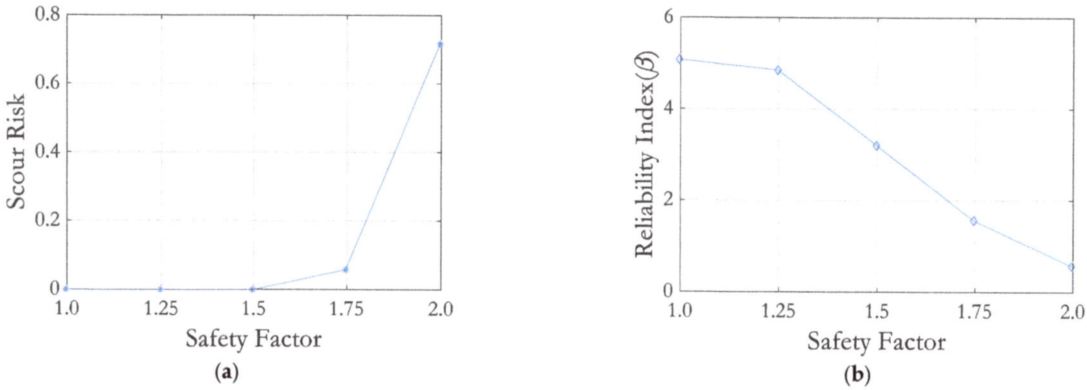

Figure 17. Result of scour risk and reliability index: (**a**) scour risk; (**b**) reliability index.

5. Conclusions

In this study, a scour risk assessment procedure was proposed. Scour hazard was calculated by giving variability to the variables of the empirical formula and was expressed as the probability of the scouring depth. Scour fragility was calculated as the probability that the response of the structure to the environmental load would exceed the limit state. Scour risk was obtained by integrating the product of the scour probability and the fragility for all possible scour depths. The scour risk assessment procedure was applied to PSB supporting 5.5 MW OWT installed on the Gunsan coast in Korea. From the numerical analysis, the reliability index was 4.85 if SF of 1.25 is applied. This level of reliability can be accepted enough when some OWT design standards are used.

Author Contributions: Conceptualization, Y.-J.K. and D.-H.K.; methodology, Y.-J.K. and D.-H.K.; software, Y.-J.K.; validation, D.-H.K.; formal analysis, Y.-J.K.; data curation, Y.-J.K. and D.-V.N.; writing—original draft preparation, Y.-J.K. and D.-V.N.; writing—review and editing, D.-H.K.; supervision, J.-H.L. and D.-H.K. All authors have read and agreed to the published version of the manuscript.

Funding: This research was supported by Basic Science Research Program through the National Research Foundation of Korea (NRF) funded by the Ministry of Education (No. 2020R1F1A1076884).

Institutional Review Board Statement: Not applicable.

Informed Consent Statement: Not applicable.

Data Availability Statement: Not applicable.

Conflicts of Interest: The authors declare no conflict of interest.

References

1. Oh, M.H.; Kwon, O.S.; Kim, K.S.; Jang, I.S. Economic Feasibility of Bucket Foundation for Offshore Wind Farm. *J. Korea Acad.-Ind. Coop. Soc.* **2012**, *13*, 1908–1914. [CrossRef]
2. Latini, C.; Zania, V. Dynamic lateral response of suction caissons. *Soil Dyn. Earthq. Eng.* **2017**, *100*, 59–71. [CrossRef]
3. Achmus, M.; Akdag, C.; Thieken, K. Load-bearing behavior of suction bucket foundations in sand. *Appl. Ocean Res.* **2013**, *43*, 157–165. [CrossRef]
4. Ding, H.; Liu, Y.; Zhang, P.; Le, C. Model tests on the bearing capacity of wide-shallow composite bucket foundations for offshore wind turbines in clay. *Ocean Eng.* **2015**, *103*, 114–122. [CrossRef]
5. Thieken, K.; Achmus, M.; Schröder, C. On the behavior of suction buckets in sand under tensile loads. *Comput. Geotech.* **2014**, *60*, 88–100. [CrossRef]
6. Ngo, D.-V.; Kim, Y.-J.; Kim, D.-H. Seismic Fragility Assessment of a Novel Suction Bucket Foundation for Offshore Wind Turbine under Scour Condition. *Energies* **2022**, *15*, 499. [CrossRef]
7. Yang, Q.; Yu, P.; Liu, Y.; Liu, H.; Zhang, P.; Wang, Q. Scour characteristics of an offshore umbrella suction anchor foundation under the combined actions of waves and currents. *Ocean Eng.* **2020**, *202*, 106701. [CrossRef]
8. Yang, T.; Qi, M.; Wang, X.; Li, J. Experimental study of scour around pile groups in steady flows. *Ocean Eng.* **2020**, *195*, 106651. [CrossRef]
9. Yu, T.; Zhang, Y.; Zhang, S.; Shi, Z.; Chen, X.; Xu, Y.; Tang, Y. Experimental study on scour around a composite bucket foundation due to waves and current. *Ocean Eng.* **2019**, *189*, 106302. [CrossRef]
10. Hu, R.; Wang, X.; Liu, H.; Lu, Y. Experimental Study of Local Scour around Tripod Foundation in Combined Collinear Waves-Current Conditions. *J. Mar. Sci. Eng.* **2021**, *9*, 1373. [CrossRef]
11. Hu, R.; Liu, H.; Leng, H.; Yu, P.; Wang, X. Scour Characteristics and Equilibrium Scour Depth Prediction around Umbrella Suction Anchor Foundation under Random Waves. *J. Mar. Sci. Eng.* **2021**, *9*, 886. [CrossRef]
12. Esteban, M.D.; Lopez-Gutierrez, J.S.; Negro, V.; Sanz, L. Riprap scour protection for monopiles in offshore wind farms. *J. Mar. Sci. Eng.* **2019**, *7*, 440. [CrossRef]
13. Lee, J.H.; Han, T.H.; Kim, S.B.; Yoon, G.L. Dynamic Instability Analysis of Wind Turbines with Mono Pile Foundation Considering Foundation Scour. In Proceedings of the Korean Society of Civil Engineers Conference 2013, Jeongseon-gun, Korea, 23–25 October 2013; pp. 892–896.
14. Ma, H.; Yang, J.; Chen, L. Effect of scour on the structural response of an offshore wind turbine supported on a tripod foundation. *Appl. Ocean Res.* **2018**, *73*, 179–189. [CrossRef]
15. Yanmaz, A.M.; Calamak, M. Evaluation of scour risk at foundations of river bridges. *Tek. Dergi/Tech. J. Turk. Chamb. Civ. Eng* **2016**, *27*, 7533–7549.
16. Khalid, M.; Muzzammil, M.; Alam, J. Reliability analysis of local scour at bridge pier in clay-sand mixed sediments. *Aquademia* **2018**, *2*, 1. [CrossRef]
17. Homaei, F.; Najafzadeh, M. A reliability-based probabilistic evaluation of the wave-induced scour depth around marine structure piles. *Ocean Eng.* **2020**, *196*, 106818. [CrossRef]
18. Kim, Y.J.; Lee, D.Y.; Kim, D.H. Risk Assessment of Offshore Wind Turbine Support Structures Considering Scouring. *J. Korean Soc. Coast. Ocean Eng.* **2020**, *32*, 524–530. [CrossRef]
19. Sumer, B.M.; Fredsøe, J. *The Mechanics of Scour in the Marine Environment*; World Scientific: Singapore, 2002. [CrossRef]
20. Shinozuka, M.; Hwang, H.; Reich, M. Reliability Assessment of Reinforced Concrete Containment Structures. *J. Nucl. Eng. Des.* **1984**, *80*, 247–267. [CrossRef]
21. *Bladed Multibody Dynamics User Manual (Ver. 4.4)*; Garrad Hassan & Partners Ltd.: Bristol, UK, 2013.
22. Abaqus, in Dassault Systemes Simulia Corporation. 2020. Available online: https://abaqus-docs.mit.edu/2017/English/SIMACAEEXCRefMap/simaexc-c-docproc.htm (accessed on 3 March 2022).
23. Papadrakakis, M.; Sapountzakis, E.J. *Matrix Methods for Advanced Structural Analysis*, 1st ed.; Elsevier: Amsterdam, The Netherlands, 2018. [CrossRef]

24. Morison, J.R.; Johnson, J.W.; Schaaf, S.A. The force exerted by surface waves on piles. *J. Pet. Technol.* **1950**, *2*, 149–154. [CrossRef]
25. *Report on the Estimation of Deep Sea Design Wave in All Seas II*; Korea Ocean Research & Development Institute: Busan, Korea, 2005.
26. Goda, Y. Revisiting Wilson's formulas for simplified wind wave prediction. *J. Waterw. Port Coast. Ocean Eng.* **2003**, *129*, 93–95. [CrossRef]
27. Available online: http://www.khoa.go.kr/oceangrid/gis/category/observe/observeSearch.do?type=TIDALCURRENT (accessed on 31 December 2019).
28. Bang, S.; Cho, Y. Use of Suction Piles for Mooring of Mobile Offshore Bases. 2001. Available online: https://apps.dtic.mil/sti/pdfs/ADA372819.pdf (accessed on 19 January 2022).
29. Det Norske Veritas. *Design of Offshore Wind Turbine Structures*; Offshore Standard DNV-OS-J101; Det Norske Veritas: Høvik, Norway, 2007.
30. *Wind Turbines Part I: Design Requirements*, 3rd ed.; International Standards IEC 61400-1; International Electrotechnical Commission: Geneva, Switzerland, 2005.

energies

MDPI

Article

On the Development of Offshore Wind Turbine Technology: An Assessment of Reliability Rates and Fault Detection Methods in a Changing Market

Alan Turnbull [1,*,†], Conor McKinnon [1,2,†], James Carrol [1] and Alasdair McDonald [2]

[1] Institute of Energy and Environment, Electronic and Electrical Engineering, University of Strathclyde, Glasgow G1 1XQ, UK; conor.mckinnon@strath.ac.uk (C.M.); j.carroll@strath.ac.uk (J.C.)
[2] Institute for Energy Systems, School of Engineering, University of Edinburgh, Edinburgh EH9 3DW, UK; alasdair.mcdonald@ed.ac.uk
* Correspondence: a.turnbull@strath.ac.uk
† These authors contributed equally to this work.

Abstract: Offshore wind turbine drive train technology is evolving as developers increase size, aim to maximise availability and adapt to changing electricity grid requirements. This work first of all explores offshore technology market trends observed in Europe, providing a comprehensive overview of installed and planned capacity, showing a clear shift from smaller high-speed geared machines to larger direct-drive machines. To examine the implications of this shift in technology on reliability, stop rates for direct-drive and gear-driven turbines are compared between 39 farms across Europe and South America. This showed several key similarities between configurations, with the electrical system contributing to largest amount of turbine downtime in either case. When considering overall downtime across all components, the direct-drive machine had the highest value, which could be mainly attributed to comparatively higher downtime associated with the electrical, generator and control systems. For this study, downtime related to the gearbox and generator of the gear-driven turbine was calculated at approximately half of that of the direct-drive generator downtime. Finally, from a perspective of both reliability and fault diagnostics at component level, it is important to understand the key similarities and differences that would allow lessons learned on older technology to be adapted and transferred to newer models. This work presents a framework for assessing diagnostic models published in the literature, more specifically whether a particular failure mode and required input data will transfer well between geared and direct-drive machines. Results from 35 models found in the literature shows that the most transferable diagnostic models relate to the hydraulic, pitch and yaw systems, while the least transferable models relate to the gearbox. Faults associated with the generator, shafts and bearings are failure mode specific, but generally require some level of modification to adapt features to available data.

Keywords: wind energy; offshore; reliability; fault detection; geared; direct drive; transfer learning

Citation: Turnbull, A.; McKinnon, C.; Carrol, J.; McDonald, A. On the Development of Offshore Wind Turbine Technology: An Assessment of Reliability Rates and Fault Detection Methods in a Changing Market. *Energies* **2022**, *15*, 3180. https://doi.org/10.3390/en15093180

Academic Editor: Paweł Ligęza

Received: 8 March 2022
Accepted: 21 April 2022
Published: 27 April 2022

1. Introduction

Wind turbine generator and drive train technology has developed rapidly over the last decade as utility-scale wind turbines have increased in size and contribute to a greater share of the electricity market [1–3]. This fundamental shift in the energy mix requires wind turbines to cope with greater flexibility in generation, with wind farms now operating more like traditional power plants to reach increased demand that meets current electricity grid conditions. As the generator and wider drive train configuration has adapted to meet changing grid requirements, wind turbine developers and operators have also been challenged to lower the overall LCOE by reducing the OPEX. The response to this challenge has been to find opportunities to maximise availability, increase system reliability, decrease the cost of repairs, reduce downtime and minimise lost production over the lifetime of a site due to unplanned maintenance activities [4]. In recent years, system reliability issues have also been addressed by simplifying and reducing the number of potential points of failure, with some OEMs taking the strategic decision to remove the gearbox and focus on direct-drive technology.

1.1. Problem Statement

In the offshore environment, where turbines are now increasing to 14 MW rated power, not only is access at a premium, but lost production is also very expensive to operators, with research recently suggesting that O&M could contribute up to 30–40% of the total LCOE [5]. This scenario provides developers with extra motivation to increase overall wind turbine reliability. This has lead some OEMs to focus on eradicating components that have conventionally caused expensive repairs and high amounts of downtime relative to other elements of the drive train. Examples of this behaviour would be in removing the gearbox for a direct-drive machine, or removing the high-speed stage of the gearbox for a medium-speed machine. For wind energy to continue to be financially viable, the wind industry must continue to adapt and improve technology; however, it must also ensure knowledge gained from older systems is understood and transferred where possible to ensure any lessons learned are appropriately recorded and applied.

When a new technology or wind turbine model is deployed, there is little or no operational data and maintenance records to understand reliability at a system-wide level. The transition to larger turbines with direct-drive train technology therefore poses an interesting hurdle to asset owners and operators that are looking to scale and optimise maintenance activities; in the context of wind turbine reliability and condition monitoring, how much insight can be drawn from data gathered on older technology and applied to modern direct-drive machines?

1.2. Motivation, Paper Structure and Novelty

Literature surrounding wind farm O&M commonly falls into several distinct categories; reliability analysis, performance optimisation, fault diagnostics, failure prognostics and maintenance optimisation. To date, reliability analysis has been used to determine critical components in order to focus efforts on fault detection, failure prediction and maintenance optimisation to minimise both OPEX and lost production, a process which has been identified as a key driver to achieve higher wind farm availability. Papers published to date have proposed a range of fault diagnostic methods primarily focusing on using SCADA data to detect anomalies across the wind turbine gearbox [6–8], blades [9], generator [10,11], pitch [12] and yaw [13] systems and main bearing [14]. Earlier approaches such as [15] used a linear auto-regressive model model to detect generator bearing failure by modelling bearing temperature and [16], which developed higher-order polynomial models of drive train temperatures. More recently, nonlinear auto-regressive neural networks with exogenous inputs (NARX) models have been used in [17,18] to detect gearbox issues. Several review papers [19–22] have also been published over the last several years providing a comprehensive overview. That being said, no study to date has attempted to review and evaluate previous work with an emphasis on highlighting the potential of transfer learning between direct-drive and geared machines.

Section 2 of this paper aims to provide a comprehensive overview of geared and direct-drive train taxonomy, with an emphasis on highlighting the key differences and similarities at assembly and component level that can be used to gather and transfer information. In addition, this section will take a look at real-world SCADA data to evaluate what information is typically gathered across different monitoring systems for each drive train configuration. This initial assessment is a vital step to determine which failure modes are likely to be common across both configurations and what associated monitoring data are readily available.

Using standardised research methods observed in the literature, Section 3 will present new results from a reliability study of 617 wind turbines across Europe and South America with a combined total of 217 operational years. Wind turbine stoppage rates and downtime related to a range of components are assessed, with the study encompassing both direct-drive and gear-driven systems all under 3.2 MW rated power. Initial results are then compared to results from other reliability studies found in the literature.

Section 4 builds on previous sections and presents a framework to assess existing literature in the area of fault diagnostics and prognostics. The aim of this review is to provide an overview of existing techniques and case studies that are most applicable to direct-drive machines, making use of existing datasets made up of mainly gear-driven wind turbines. Not only does this allow for some immediate insight into the direct-drive machines, this study also opens up opportunities in areas such as transfer learning and reinforcement learning to adapt models as more data and information are made available for modern, larger, direct-drive machines. Finally, drive train components and failure rates common to both configurations are brought together to assess which components are most critical for future research into direct-drive diagnostics and prognostics. In the context of existing literature in this area, the contribution of this work is to:

- Provide an overview of the current technology trends observed in the European offshore wind sector.
- Present results from a new reliability study of over 617 wind turbines, directly comparing down times associated with direct-drive and geared machines.
- Introduce a framework for evaluating how transferable previous diagnostic models published in the literature using older technology are when considering newer large-scale direct-drive generators.
- Deliver insight into key components that must be considered a priority for large-scale direct-drive generators with regards to diagnostic and prognostic modelling considering both reliability and previous research.

2. Drive Train Configuration Trends

When describing a wind turbine drive train configuration, it is typically expressed as a series of assemblies and components required to convert the kinetic energy in the rotor to electrical energy needed for a stable grid connection. In modern utility-scale wind turbines, there are four major categories as described in [4,23]. Note configuration type A and B from [4] have been excluded due to a focus on current utility-scale technology applicable to the offshore environment that meets modern grid requirements [24,25].

Full details of each configuration type along with schematic diagrams can be found in [4,23]; however, a brief overview of the important configurations used in this work will be provided. Configuration one is the doubly-fed induction generator (DFIG). A partial power converter is used to control the electrical current in the generator's rotor. Configuration two has a full-power converter which enables the decoupling of the generator and grid frequency. This means that the frequency on the generator side can be fully controlled allowing for enhanced grid services and the use of a gearbox can be avoided. A synchronous electrical generator (which can be either an electrically excited synchronous generator (EESG) or a permanent magnet synchronous generator (PMSG)) is directly coupled to the main shaft of the rotor. Configuration three is a gearbox-equipped wind turbine with a full-power converter and medium/high-speed synchronous generator, which can be EESG or PMSG. In this arrangement, it is possible to choose between a relatively small gearbox (with moderate gear ratios) at the expense of using a large medium-speed synchronous generator. On the other hand, it is possible to assemble a gearbox with a higher gear ratio in order to reduce the size of the generator (high-speed configuration with synchronous generator). Configuration four is a gearbox-equipped wind turbine with a full-power converter; however, it has a high-speed asynchronous generator. As the full-power converter enables the speed to be controlled by modifying the operating frequency, a squirrel cage induction generator (SCIG) is generally employed in this configuration. In the context of offshore wind energy more broadly, configuration three corresponds to a geared multistage high-speed wind turbine, configuration two is a direct-drive machine, while type three and four are hybrid models. In relation to the work completed in this paper, configuration two makes up the direct-drive wind turbine category, while configurations one, three and four are grouped together into the gear-driven wind turbine category.

According to the JRC Wind Energy Database, in terms of market share, geared turbines have dominated the global onshore market, with the vast proportion of these turbines onshore made up of a DFIG arrangement below 3 MW rated power output. This is particularly true across Europe, Asia and North America. Further analysis presented in [1] shows the evolution of configuration types with geographical location, with configuration four more prevalent in North America and configurations two and three having more market share in Europe and Asia. If we look offshore across Europe, DFIG models dominated the early market predominately close to shore. This has vastly changed over the last 5–7 years, with direct-drive and hybrid models now making up a significant proportion. PMSGs have seen an explosion in the Asian market in particular, while EESGs are typically more common in European waters. Conversely, PMSGs have been gaining more traction in Europe as turbines increase beyond 5–6 MW [2,23,26–29]. The technological shift towards direct-drive PMSGs over other types of generators is predominately due to the perceived increase in reliability that can be achieved with fewer components and importantly, no gearbox. Whether the reliability of the system as a whole does in fact increase is still up for debate, with further evidence required in order to conclusively state either way. This topic will be discussed in greater detail throughout the reliability analysis section of this paper.

Looking offshore, this shift is even more apparent, with direct-drive machines starting to dominate the UK market over the last 5 years. In fact, from the 1725 wind turbines currently installed or under development in UK waters since 2016, 70.1% (or 1221) have direct-drive PMSG technology. This accounts for an installed capacity of just over 11.1 GW since 2016, 73.4% of total capacity either installed or under development.

Figure 1 shows the technology shift in the UK and wider European market from the first offshore wind farm to all current wind farms either operational or currently in development (due to be commissioned by 2026). These plots were developed by the authors using open source information found in [30]. In Figure 1a, each circle represents a wind farm, with the size of each circle scaled with the number of turbines that make up the site. Direct-drive configurations are shown in red, while gear-driven wind turbines are represented in blue. The y-axis displays individual wind turbine rated power of each site which, as shown on the plot, has increased significantly over the last 20 years along with the average site size. Figure 1b shows these same technology trends but split into each European country for comparison. Observed trends are similar, with turbine rating and total site capacity getting larger, with a large number of direct-drive machines entering the market over the last 5 years. Looking more closely at wind turbine manufacturers, Figure 2 shows a breakdown of OEM market share across Europe. SGRE currently has the largest share, with over 58% of installed capacity, followed by Vestas (28%) and GE Renewable Energy (11%).

Each wind turbine drive train configuration can be broken down into a common series of major components, each assigned a unique set of failure categories, which will be discussed extensively throughout the next section. Looking specifically at SCADA data related to the generator and gearbox, Figure 3 shows the average number of data channels recorded for both direct-drive and geared machines. Three examples of each configuration type were used, each ranging from 1 to 3 MW rated power. This information has been included to showcase the key differences when it comes to which sensors are available to create features for fault detection. This approach will later be used to assess model transferability. Based on these examples, direct-drive models have on average fewer channels and sensors than their geared counterparts. With regards to the generator specifically, temperature readings are typically available for the bearings, stator and rotor across both configurations, along with generator shaft speed and electrical current, voltage and power measurements.

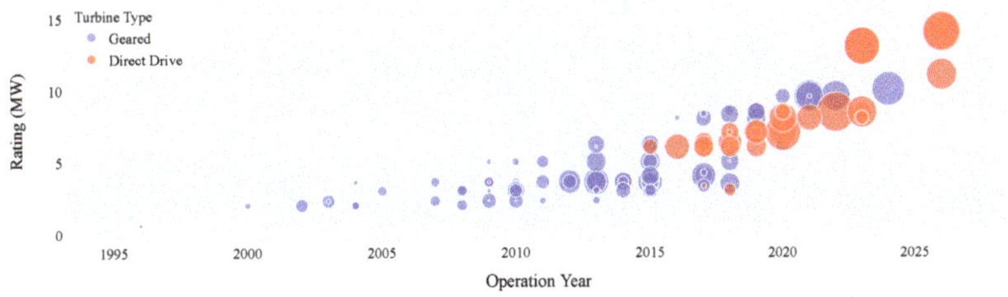

(a)

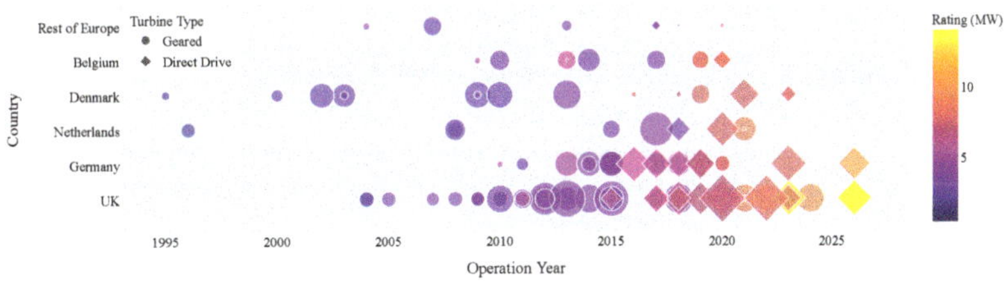

(b)

Figure 1. Overview of offshore wind turbines drive train technology trends in Europe. (**a**) Comparison of geared and direct-drive offshore wind turbine installed capacity in Europe; (**b**) installed capacity of different drives by country.

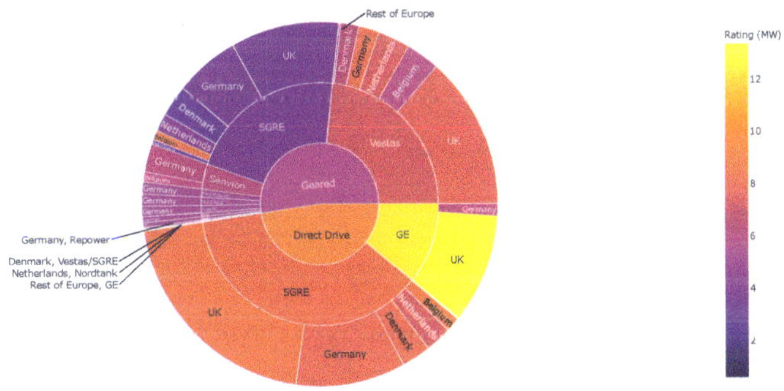

Figure 2. European offshore wind installed (or planned) capacity by drive train type and OEM.

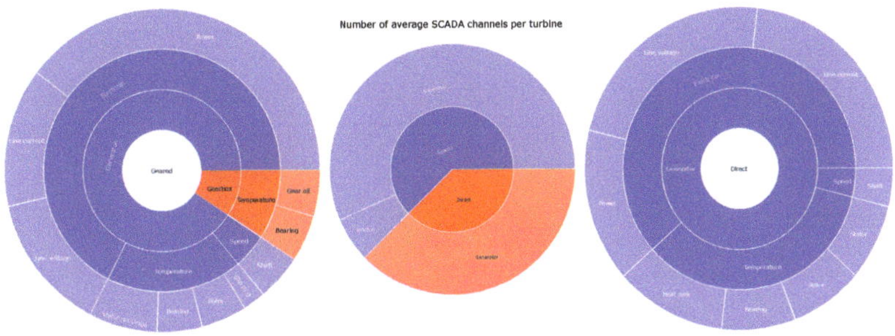

Figure 3. Number of SCADA channels assigned to each drive train category and component (showing only generator and gearbox).

3. Wind Turbine Stop Rate Analysis

Wind turbine reliability rates cover a range of metrics regarding wind turbine reliability. These can be wind turbine failure rates, stop rates, downtimes, or possibly lost production. For the purposes of this paper, the reliability rates examined are the stop rates and downtimes for wind farms in Europe and South America.

Wind turbine stoppages can occur for a range of reasons, from manual stops relayed from the owners or operators of the turbine to faults in the turbine components. Different metrics have been used in the past to assess the reliability of wind turbines, and their components. These metrics provide different levels of information, such as turbine availability percentage, stop rates, or failure rates. Wind turbine failures differ from wind turbine stops, and therefore cannot be directly compared. However, the general trends may still be examined and discussed, as stop rates will also include any failures, or faults, that caused the turbine to stop.

Previous studies have investigated the reliability of wind turbines, typically focusing on failure rates of wind turbine components. These are usually presented as number of failures per turbine per year, and are presented alongside the number of hours of downtime per turbine per year.

Several studies have collated reliability results from papers that have investigated wind turbine datasets. These studies typically present failures by component group, and these are usually similar with some slight variations in category definitions. The results presented in this section will be compared against the various past studies. Several of these studies also present the downtime per turbine per year in hours for the datasets. The studies cover wind turbines from across Europe, Asia, and the USA. The majority of these studies are focused on onshore turbines, and typically gear-driven turbines with some direct drives. In particular, a study from S. Ozturk et al. [31] has compared sub-1 MW geared and direct-drive wind turbines from the WMEP dataset in Germany. The papers collated in these reviews are described in Table 1.

The data provided for this study consisted of 39 farms, located in either South America or Europe. The majority of turbines assessed were located in South America, approximately 470, and the others, approximately 150, located in Europe. The European turbines are from either around the Mediterranean sea, or the British and Irish Isles.

Three of the farms investigated consisted of direct-drive turbines, approximately 50 turbines in total, with all of these located in Europe. The turbines rated between 1.5 and 3.5 MW, and aged between 2 and 13 years. The majority of turbines were operating for less than 7 years, with a small minority in operation longer. The data provided cover the life of the turbines investigated from when they were first operational until the beginning of 2020. A factor that is of interest in the wind energy community is how wind turbine reliability rates, such as stoppages, scale with turbine size. This link is unfortunately beyond the scope of this paper, as the dataset is limited to turbines of similar rating. Future work, utilising a dataset with more varied turbine rating, could examine the change in reliability rates between small and large rated turbines.

Table 1. Failure rate studies featured in the reviews previously discussed.

Author	Title	Dataset	Turbine Number	Years Collected	Country	Top 3 Failures	Top 3 Downtimes
M. Reder [32]	Wind Turbine Failures—Tackling current Problems in Failure Data Analysis	AWESOME	4300	-	Europe	<1 MW: (Gearbox, Blades, other blade brake)	<1 MW: (Gearbox, Generator, Blades)
						>1 MW: (Gearbox, Controller, Pitch)	>1 MW: (Gearbox, Generator, Blades)
						DD: (Controller, Met Station, Yaw)	DD: (Generator, Blades, Controller)
G. Wilson [33]	Assessing wind farm reliability using weather dependent failure rates	Blacklaw and Whitelee	Over 250	-	Scotland	Control, Drivetrain, Yaw	-
V. Hines [34]	Continuous Reliability Enhancement for Wind (CREW) Database: Wind Plant Reliability Benchmark	CREW	800–900	2013	USA	Rotor, Generator, Controls	Yaw, Brakes, Controls
Y. Lin [35]	Fault analysis of wind turbines in China	CWEA	111	2010	China	Pitch, Frequency Converter, Generator	-
			560	2011		Frequency Converter, Generator, Pitch	
			640	2012		Frequency Converter, Generator, Pitch	
C. Crabtree [5]	Wind Energy: UK experiences and offshore operational challenges	Egmond aan Zee	36	3 years	Netherlands	Control, Yaw, Scheduled, Pitch	Gearbox, Generator, Blades
I. Dinwoodie [36]	Analysis of offshore wind turbine operation & maintenance using a novel time domain meteo-ocean modeling approach	Egmond aan Zee	36	3 years	Netherlands	Control, Yaw, Scheduled, Pitch	Gearbox, Generator, Control
J. Ribrant [37]	Survey of failures in wind power systems with focus on Swedish wind power plants during 1997–2005	Elforsk	786	2000–2004	Sweden	Electric, Sensors, Blades/Pitch	Gearbox, Control, Electric
J. Ribrant [38]	Reliability performance and maintenance—A survey of failures in wind power systems	Elforsk	786	2000–2004	Sweden	Electric, Sensors, Blades/Pitch	Gearbox, Control, Electric
		VTT	92	2000–2004	Finland	Hydraulics, Blades/Pitch, Gearbox	Gearbox, Blades/Pitch, Hydraulics
		WMEP	650	2003–2004	Germany	Electric, Control, Sensors/Hydraulics	Generator, Gearbox, Drivetrain
Z. Ma [39]	A Study of Fault Statistical Analysis and Maintenance Policy of Wind Turbine System	Huadian	1313	2015	China	Transformer, Generator, Pitch	Transducer, Generator, Control

Table 1. *Cont.*

Author	Title	Dataset	Turbine Number	Years Collected	Country	Top 3 Failures	Top 3 Downtimes
C. Su [40]	Failures analysis of wind turbines: Case study of a Chinese wind farm	Jiangsu 1	61	2009–2017	China	Control, Pitch, Electrics	Control, Pitch/Blade, Electrics
		Jiangsu 2	47	2011–2017	China	Pitch, Control, Electrics	Pitch/Blades, Control, Electrics
G. Van Bussel [41]	Reliability, Availability and Maintenance aspects of large-scale offshore wind farms, a concepts study	LWK	643	1995–1999	Germany	Control, Inverter, Gearbox	-
G. Herbert [42]	Performance, reliability and failure analysis of wind farm in a developing Country	Muppandal	15	2000–2004	India	Blades, Gearbox, Hydraulics	-
M. Wilkinson [43]	Measuring wind turbine reliability: results of the Reliawind project	Reliawind	Around 350	-	Europe	Electrics, Rotor, Control	Electrics, Rotor, Control
R. Bi [44]	A survey of failures in wind turbine generator systems with focus on a wind farm in China	SUZHOU	134	2011	China	Pitch, Control, Sensors	Cables, Pitch, Control
F. Spinato [45]	Reliability of wind turbine subassemblies	Windstats Denmark (WSDK)	2345–851	-	Denmark	Converter, Yaw, Generator	-
		Windstats Germany (WSD), Schleswig Holstein (LWK)	1295–4285, 158–643	-	Germany	Electrical, Converter, Rotor	-
P. Tavner [46]	Reliability analysis for wind turbines	Windstats Germany (WSD)	up to 4500	1994–2004	Germany	Grid/Electrical, Yaw, Pitch Control	-
		Windstats Denmark (WSDK)	up to 2500	1994–2004	Denmark	Yaw, Hydraulic, Generator	-
S. Ozturk [31]	Failure Modes, Effects and Criticality Analysis for Wind Turbines Considering Climatic Regions and Comparing Geared and Direct Drive Wind Turbines	WMEP—DD 500 kW	1500	1989–2006	Germany	Control, Electric, Generator/Hub	Rotor Blades, Parts/Housing, Drive Train
		WMEP—GD 200 kW	1500	1989–2006	Germany	Control, Electric, Hydraulic	Gearbox, Electric, Rotor Blades/Control/Parts/Housing
		WMEP—GD 300 kW	1500	1989–2006	Germany	Electric, Control, Hydraulic	Gearbox, Generator, Rotor Blades
		WMEP—GD 500 kW	1500	1989–2006	Germany	Electric, Control, Yaw	Generator, Control, Electric
S. Faulstich [47]	Wind turbine downtime and its importance for offshore deployment	WMEP	1500	1989–2006	Germany	Electrical system, Electrical Control, Sensors	Gearbox, Drivetrain, Generator
B. Hahn [48]	Reliability of Wind Turbines: Experiences of 15 years with 1500 WTs	WMEP	1500	1991–2006	Germany	Electrical, Plant Control, Sensors	Generator, Gearbox, Drivetrain

The categories used here were defined based on a combination of two reliability studies [5,49], which was done to allow for ease of comparison with previous work. The first set of categories were taken from Figure 8 in [49], with the addition of Grid from [5], and finally Nacelle, Shafts and Bearings categories were added.

Stop rate was calculated by taking the number of stops in each category and dividing through by the number of turbines and years of operation in each farm, this gave a number of stops per turbine per year.

$$S = \frac{N_s}{N_T * T} \quad (1)$$

where S is the stop rate, N_s is the number of stoppages recorded for that category, N_T is the number of turbines in that farm, and T is the years in operation for that farm. Downtime was calculated, for each farm, by dividing the total days of downtime per category by the number of turbines in the farm and the number of years in operation.

$$DT = \frac{T_D}{N_T * T} \quad (2)$$

where DT is the downtime, T_D is the total days of downtime for recorded, N_T is the number of turbines in the farm, and T is the years in operation for that farm. These two values allow for fair comparison across all farms and turbines, as it negates the effect of farm size or age.

Figure 4 shows the average stop rates and downtimes for the direct-drive and gear-driven farms respectively. As can be seen, the generator has roughly double the stop rate in direct-drive turbines, and there is also a much greater average downtime for direct-drive generators. The top three stoppage and downtime categories for each turbine type are shown in Table 2. There are some similarities between both turbine types; however, from Figure 4, it can be seen that direct-drive turbines seem to have higher overall downtimes. Even for components that are assumed to be similar between turbine configurations, there are quite large differences in stop rate and downtimes. For example, the sensors and pitch systems both have differences in stop rate and downtime, whilst being components that should not differ too much between configuration. It is possible that the pitch system could be hydraulic or electric, and this could bring about some of the change. There may also be a difference in turbine operation that could account for these differences.

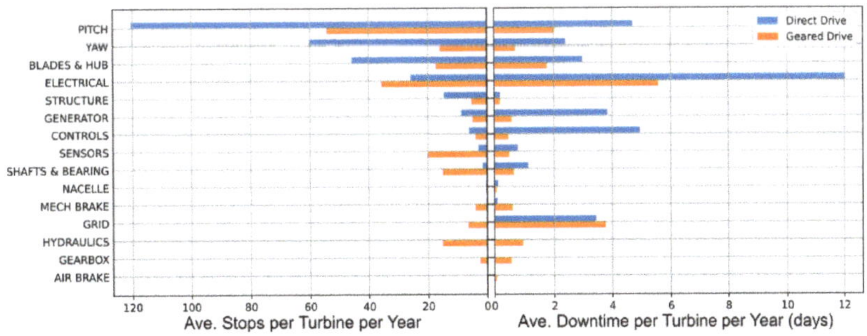

Figure 4. Comparison of the Geared and Direct Drive turbines within the dataset based on average stop rates and downtimes for each configuration.

Before any comparisons are made between the results presented here and in other studies, it is important to note that stoppages are not the same as failures, therefore a direct comparison cannot be made. Stoppages can include failures; however, they also include stoppages due to alarms. These alarms can be for any reason, such as temporary overheating of a component. It is possible that these stops are indicators of failure; however, developing a model to represent this link is beyond the scope of this paper.

Table 2. Top three stoppage and downtime categories per turbine type.

Turbine Type	Top 3 Stoppage Categories	Top 3 Downtime Categories
Direct-Drive Turbines	Pitch, Yaw, Blades and Hub	Electrical, Controls, Pitch
Gear-Driven Turbines	Pitch, Electrical, Sensors	Electrical, Grid, Pitch

Comparison with Previous Studies

Table 1 presents the details of previous failure rate and downtime studies examined. The table outlines the different databases used by each paper, the number of turbines contained, the time period examined, and the country of origin for each database. The last two columns of the database present the top three turbine components by failure rate, and downtime per failure respectively. From this table, two studies examined the reliability data for direct-drive turbines in Europe. The first, from M. Reder et al. [50], found that the top three, in descending order, components by failure rate were the Controller, Met Station, and Yaw systems, and by downtime were the Generator, Blades, and Controller. The second study, from S. Ozturk et al. [31] found that the top three components by failure rate were Controls, Electric Systems, then the Generator and Hub were tied for third. The top three by downtime were the Rotor Blades, Parts/Housing, and then the Drivetrain. When this is compared against Figure 4, it can be seen that there are some differences. The top three categories are shown in Table 2; however, these are for stoppages rather than failures. So it may not be appropriate to make a direct comparison.

The frequency of which each component is featured within the top 3 failures or downtimes is plotted in Figure 5. This bar chart plots the number of times each component was featured in the top three of either metric in Table 1. This chart can then be used to find out the overall top three components for both failure rates and downtimes. For failure rates, the top three components were the Controller, Pitch, and Generator, with the Controller, Gearbox, and Generator coming in at top three for downtime.

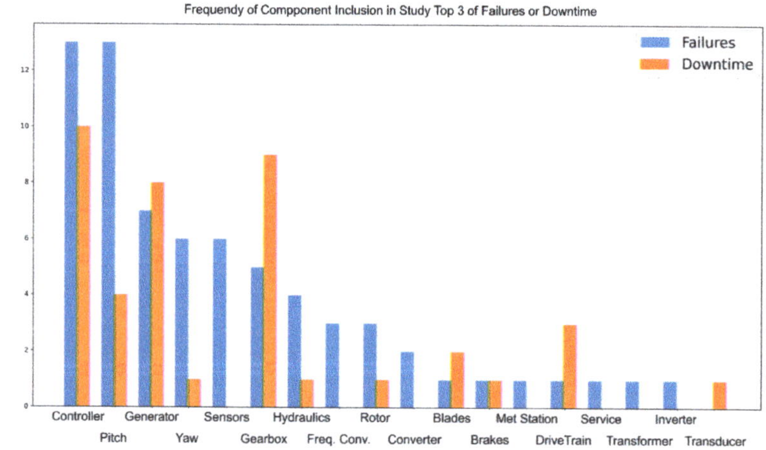

Figure 5. Frequency of each component being within studies top three failures or downtimes from Table 1.

One review from Crabtree et al. [5] presented the stop rate for the Egmond aan Zee offshore wind farm situated in the Netherlands. This farm consisted of 36 turbines with 3 years of operational data. These were geared 3 MW turbines, and unlike other studies presented stop figures instead of fault data. The control system, yaw system, and service stop categories were the top three, with the Gearbox, Generator, and Blades stops being the top three categories for downtime. This review is of particular interest as it presents stop rates, which can be directly compared against the results presented here.

Compared with Figure 4, the geared turbines from this study have relatively low stops due to the yaw and control systems, and low service stops on average with quite extreme outliers. For the downtime, the Generator, Gearbox, and Blades were relatively low.

Stop rates are explicitly different from failure rates in several ways—for example stop rates are caused by an wind turbine stoppage, whereas failure rates are due to an unscheduled, or unplanned, failure of the turbine due to some fault or malfunction. Therefore, several categories are found in stop data that would not be found for failure data, such as scheduled service as these are known in advance, or grid failures which are outwith the operator control. Stoppages are also typically more frequent and should have lower downtimes on average per stoppage as they are usually less severe than turbine failures. Within the stop data there will be failure examples, as these are examples of wind turbine stoppages; however, they will be less frequent.

4. Framework for Assessing the Transferability of Diagnostic Techniques between Drive Trains

4.1. Framework

In the previous section, reliability rates (wind turbine stop rates and downtimes) for both direct and geared wind turbines were presented, where several key differences were highlighted. Building on these results, a framework for assessing the transferability of failure modes and associated sensors will now be presented. The aim for this framework is to help determine how well a particular fault could be diagnosed in modern direct-drive wind turbines using data and diagnostic models demonstrated on geared wind turbines. This framework does not make any attempt to predict future reliability rates of larger direct-drive machines. Stop rates and downtimes presented in Section 3 will be used in conjunction with the framework to assess which components need to be the focus of future diagnostic research. An example of how this framework can be used is presented later in this section, which examines previous fault detection papers by assessing their failure mode and input data to examine how transferable the fault case from each paper was. By doing this, we can see which components in particular are suitable to assessment in the future with direct-drive machines, and whether the data inputs used previously would be suitable or would require some level of processing. The transferability of each paper was assessed over two dimensions to examine overall transferability. The first dimension examined how transferable the sensors, or data channels, used by each paper to predict, or diagnose, the fault. The second dimension assessed how well the specific failure mode transferred between geared and direct-drive turbines based on fundamental understanding of the physics of failure.

Two small-scale decision trees were drawn up to assess each paper examined for their transferability, one for each dimension. The first tree, Figure 6, is used to assess the transferability of the sensors, or channels, used in each paper. So this assesses the paper's selection of features for modelling a particular failure mode—and how well these features, and required sensors, transfer from a geared to a direct-drive machine. The first question, in the top diamond, asks if the direct-drive would have all the sensors required for the features used by the paper. The second question, in the middle two diamonds, asks if the majority of the sensors used in the paper would be in the same location within the turbine. The third question, the bottom four diamonds, asks if these sensors are of the same specification as those on an arbitrary direct-drive machine. This question is essentially used to assess if the general expected data range you would get from this sensor would be the same as one on a geared machine. An example could be bearing temperature, you could consider if a similar thermistor could be used to cover the expected range and resolution of recorded temperatures. This helps to assess scale of component as well, for example a generator bearing in a direct-drive machine will be much larger than one in a similarly rated geared machine. For all of these questions, it was assumed that the direct-drive and geared machines were of the same arbitrary turbine model; however, the only difference being that the the direct-drive machine had no gearbox and the generator was of an appropriate size for the same power rating as the geared machine.

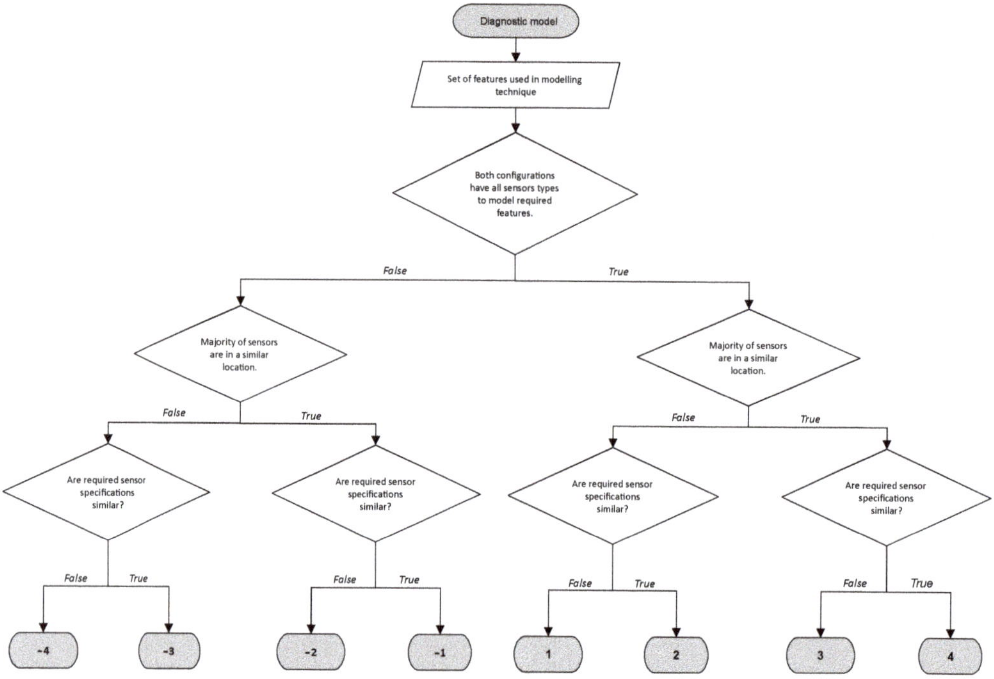

Figure 6. Decision tree for sensor/channel group transferability.

The second decision tree, Figure 7, assesses how transferable the physics of failure mode of each paper is. This tree examines the particular component that failed, and any failure mode information provided by each paper. First the papers are split by if the fault, or damage, can be found anywhere across either turbine configuration. Next it is split by whether it can be found on a specific component—if it can then it asks if the fault progresses in the same physical manner, and if not then it asks if the model corresponds to a specific failure mode. If it cannot be found on a specific component, then the questions determine if it is on an assembly within the turbine, and then if it follows the same physical manner, and lastly if it is specific to the failure mode examined. Again these questions were answered under the assumption that the turbines would be of similar models, with the exception of the gearbox and generator.

To test these decision trees, a database of past fault detection papers were collected and their fault and input data were collated. Table 3 shows all the past papers examined, and their scores based on the decision trees referenced earlier. The majority of these faults occurred on the drivetrain, with some in the blades. SCADA data were the focus of this study; however, some papers were assessed that used images as their input data. The input data that each paper utilised were assessed with the first decision tree, and the fault itself was assessed using the second decision tree. The data ranged both in terms of years assessed, but also in the turbine rating. All the turbines within the dataset of papers examined were geared turbines.

Table 3. Database of papers examined for their transferability.

Author	Year	Turbine Rating	Fault Examined	Data	Sensor Score	Component Score
Dhiman [6]	2021	Sub 1 MW	Gearbox	SCADA	−4	−5
L. Yang [7]	2021	Unknown	Gearbox	SCADA	−4	−5
X. Yang [51]	2021	Unknown	Blade Damage	Images	4	5
Turnbull [10]	2020	0.5–1 MW	High Speed Shaft	SCADA	−2	−2
		2–4 MW	Generator Bearing		3	2
W. Chen [52]	2021	1.5 MW	Blade Ice Accretion	SCADA	4	5
S. Moreno [53]	2020	2 MW	Load and Wind Sensor Failure	SCADA	4	4
X. J. Zeng [54]	2018	1.5 MW	Gearbox Oil Temperature Over Limit Fault	SCADA	−3	−3
M. Beretta [11]	2020	2 MW	Bearing HSS Replacement	SCADA	3	1
			Generator Brushes			
			Generator Non-Drive End Bearing			
J. Chen [55]	2020	1.6 MW	Overheating Generator Bearing	SCADA	−2	2
Rezamand [9]	2020	~2.5 MW	Blade Fault	SCADA	4	4
X. Liu [56]	2020	Unknown	Gearbox and Generator	SCADA	−2	−5
McKinnon [57]	2020	Unknown	High Speed Shaft Faults	SCADA	3	2
Y. Wang [58]	2019	Unknown	Blade Damage	Images	4	5
J. Carroll [59]	2019	2–4 MW	Gearbox Bearing	SCADA and Vibration	−4	−2
			Gear Tooth Fault		−2	−5
McKinnon [57]	2020	2–4 MW	Intermediate Gear Fault	SCADA	−4	−5
H. Yun [60]	2019	Unknown	Ice Detection	SCADA	4	5
C. Yang [61]	2019	Unknown	Pitch Limit Switch and Angle Encoder	SCADA	3	5
L. Wei [12]	2018	2 MW	Pitch System	SCADA	4	5
R. Pandit [13]	2018	2.3 MW	Yaw Error	SCADA	4	5
H. Zhao [62]	2018	1.5 MW	Gearbox	SCADA	−4	−5
			Generator Rear Bearing		3	2
			Inverter Failure		3	5
Y. Zhao [63]	2017	1.5 MW	Generator Fault	SCADA	3	3
Y. Zhao [64]	2016	Unknown	Generator Fault	SCADA	3	3
M. Beretta [14]	2021	2 MW	Main Bearing	SCADA	−2	5
McKinnon [65]	2021	1.8 MW	Pitch System Bearing	SCADA	4	5
M. Cardoni [66]	2021	Unknown	Oil leaks between HSS and Generator	Images	4	5
P. Mucchielli [67]	2021	Unknown	A Range	SCADA	4	1
X. Liu [68]	2021	Unknown	Gearbox Planetary Bearing	SCADA	−2	−2
			Gearbox HSS Bearing		−2	1
			Gearbox		−4	−5
A. Heydari [69]	2021	2 MW	Gearbox Bearing Fault	SCADA	−2	−2
L. Xiang [70]	2022	750 kW	Gearbox Gear Failure	SCADA	−2	−5

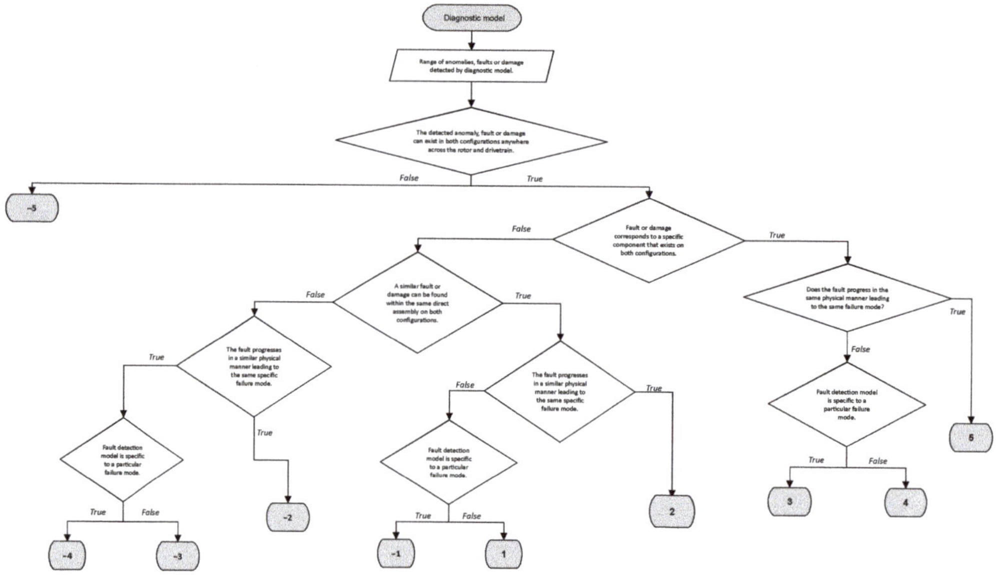

Figure 7. Decision tree for component failure mode transferability.

This technique could potentially be expanded to papers that utilised high-frequency conditon monitoring systems (CMS) data, which is commonly used in the literature. SCADA data, which is 10 min aggregate data, is what is examined here. High-frequency CMS data may be of interest in transfer learning for two reasons, the first being the lack of direct-drive data for the larger machines, but also because CMS data are much more difficult to acquire. SCADA systems are commonly installed on wind turbines; however, CMS sensors are less so (particularly on older machines). Typically, CMS data are specifically used to target an area of interest; however, it is becoming cheaper to install relative to power output and therefore more common.

4.2. Framework Application Results

Results are presented based on the framework explained in the above section for the literature stated in Table 3. Figure 8 shows the component tranferability scores on the x-axis and the sensor tranferability scores on the y-axis. Each single point represents a diagnostic model presented in the literature (see Table 3) with the fault grouped into the corresponding component or wider assembly. The mean value per component for each axis was calculated and plotted as a larger cross on Figure 8. Note that this does not correspond to any particular model, but simply represents the average transferability of that component or assembly. All 35 models from Table 3 are shown, although some points have identical coordinates and cannot be easily distinguished. The most transferable diagnostic models relate to the hydraulic, pitch and yaw systems, which makes sense due to them being mostly universal to different wind turbine drive train configurations. For this same reason, sensor faults and electrical issues also scored highly. The least transferable models were, as expected given the drive train topology, associated with the gearbox. The components with the most variability in scoring were the the generator, shafts and bearings. For these, the overall transferability was largely dependent on both the sensors required to create model features and specific failure mode being diagnosed.

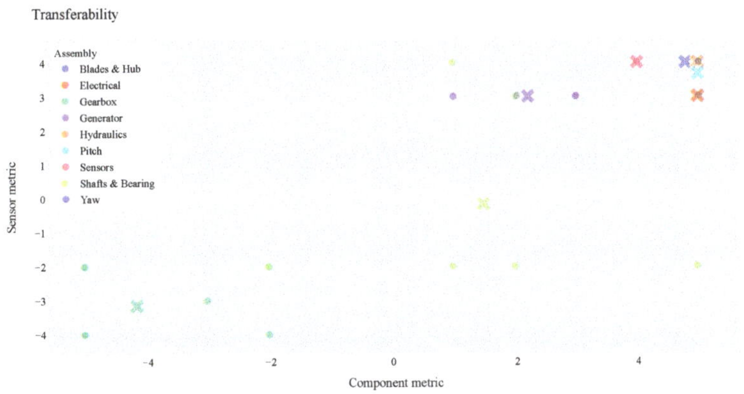

Figure 8. Results showing transferability of diagnostic model.

4.3. Discussion

The framework presented is built upon several questions determined by the authors to distinguish diagnostic models by two key metrics. The first decision tree (Figure 6) is designed to split models and features by which channels and sensors are similar across both turbine configurations. This decision is of course influenced by the selection of features used in each paper, which can lead to some variation in what channels were deemed relevant. The papers chosen for this study have been peer reviewed; therefore, it is assumed that the channels were selected either through some expert domain knowledge, or through a data-driven technique to select variables relevant to the component examined. For this reason it can also be assumed that these features would prove to be related to the component and how transferable these features are will help decide how transferable each particular failure mode is.

The second decision tree (Figure 7) is designed to compare faults across different components, as well as the manner in which a particular fault is likely to develop through time. Although the decision tree itself is designed to be as general as possible, the specific failure modes related to the diagnostic models presented here do not represent all possible failure modes, or even components, within the wind turbine. In Figure 7, faults to be diagnosed are first split at a component level, as this is the fundamental aspect of how transferable a fault is. From here, faults are split by the manner they are likely to progress and present themselves, which is more specific to the fault and failure mode itself. This requires some expert knowledge of wind turbines and mechanical or electrical systems depending on the individual fault.

Some papers presented in Table 3 were less specific about the fault than others. For example, Dhiman [6] presents a gearbox fault, whereas Zeng [54] presents a gearbox oil temperature over limit fault. The more specific a fault, the easier it can be scored using the decision trees; however, it is more sensitive. Conversely, as a more general fault is harder to score without further information about the failure mode from the authors, more general faults will typically score lower than a specific fault. Similarly, those papers which utilise a wider range of SCADA channels, compared to ones that use channels picked for their relevance to the fault, will typically score lower. By using a wider range of data, it is less likely that all of the features used will still be transferable to a direct-drive machine. For example, while a paper may examine a generator fault which is fairly transferable, the authors may incorporate many features from the gearbox. Obviously, most SCADA channels in the gearbox will not be transferable to those in a direct-drive turbine.

The transferability of each diagnostic model is shown in Figure 8. Each quadrant relates to a different level of transferability. The bottom left quadrant is the least transferable section, with both the required sensors and physics of failure relating to a particular component fault being unsuitable. The top left quadrant is where the sensors required to create the model features are available, and hence more transferable, but the fault is not relevant. An example of this could be a gearbox failure sensed through other means such as power curve analysis. The bottom right quandrant represents models which scored a high score on the component metric, but a low score on sensor metric. An example of this could be in detecting a high-speed bearing fault using features primarily created from sensors associated with the gearbox. Although the sensors are not available across both configurations, high-speed bearing faults could still be relevant to a lower-speed shafts on the direct-drive machine. The top right quandrant consists of the most transferable fault cases, with both the sensors and components being relevant and transferable.

The most transferable components in this upper right quadrant appear to be the components more ubiquitous across turbine configurations, such as pitch and yaw systems. The channels relevant to these faults are usually quite ubiquitous as well. The generator is also a fairly transferable component; however, due to its size compared to that of a direct-drive turbine, some processing may be required to the channels to engineer more appropriate features. As expected, the least transferable component is the gearbox; however, as can be seen, some failure modes and sensors are slightly more transferable to a direct drive. For example, a gearbox bearing may exhibit similar behaviour to other bearings within a direct-drive turbine. One interesting component group is general shafts and bearings of the turbine. These are typically transferable faults; however, the choice of sensors may need examined. Many of these failures focus on the high-speed shaft, which is not present in a direct-drive turbine, but could be transferred with some feature engineering.

5. Conclusions

This paper presents a comprehensive look at the future trends in European offshore wind energy. First, it was shown that offshore wind in Europe is currently moving more towards direct-drive turbines, with each individual installed turbine having a higher rated power. Overall capacity of new sites is also increasing on average. To examine the implications of this change, the stop rates for direct-drive and gear-driven turbines were compared between 39 farms across Europe and South America. It was found that there was some differences, and in particular the top components by stop rate and downtime were different from previous papers that had examined failure rates of wind turbines. In the future, it would be of value to develop a way of mapping stop rates to failure rates, thereby allowing a more direct comparison to be made.

Finally, this paper presented a framework for analysing how well published fault detection models transfer between geared and direct-drive turbines. For this, two decision trees were created to enable a quantitative score to be placed on both the required input channels and failure mode respectively. Overall transferability could then be assessed by considering both metrics together. It was found that components, or assemblies, that were ubiquitous across turbine configurations, such as the pitch system, were more transferable. Whereas, as expected, the gearbox was the least transferable component. The generator, shafts and bearings were somewhat transferable; however, in general, these would require some level of feature engineering to improve the potential performance. While this paper has examined over 25 papers in testing of the proposed framework, not all turbine components or fault conditions were presented. Further study may be needed to apply this framework to these components. Additionally, this paper has focused mostly on SCADA data from wind turbines; however, this technique could be applied to papers that have utilised high-frequency CMS data.

Based on market trends, it is important for researchers to focus their efforts on developing fault detection techniques for the most critical components related to large direct-drive technology. Since no such reliability study exists in the literature on large direct-drive wind turbines, for the time being, previously observed reliability rates must be used and adapted where appropriate. Based on these studies, the most common critical components stated with extended downtime are related to the controller, pitch system, generator and gearbox. Based on the transferability scores presented in this paper, the components that would require further work are those related to the direct-drive generator. Additional work is required in this area in order to utilise, adapt and improve existing fault detection methods.

Author Contributions: Conceptualisation, A.T. and C.M.; methodology, A.T. and C.M.; validation, A.T., C.M. and J.C.; formal analysis, A.T. and C.M.; data curation, A.T. and C.M.; writing—original draft preparation, A.T. and C.M.; writing—review and editing, J.C. and A.M.; visualisation, A.T. and C.M.; funding acquisition, J.C. and A.M. All authors have read and agreed to the published version of the manuscript.

Funding: This research was funded by EPSRC grant number EP/T031549/1 and Supergen ORE hub 2018 grant number EP/S000747/1.

Institutional Review Board Statement: Not applicable.

Informed Consent Statement: Not applicable.

Conflicts of Interest: The authors declare no conflict of interest.

Abbreviations

The following abbreviations are used in this manuscript:

CMS	Conditon monitoring systems
DFIG	Doubly-fed induction generator
EESG	Electrically excited synchronous generator
LCOE	Levelised cost of energy
OPEX	Operational expenditure
O&M	Operations and maintenance
OEM	Original equipment manufacturer
PMSG	Permanent magnet synchronous generator
SCADA	Supervisory control and data acquisition
SCIG	Squirrel cage induction generator

References

1. Hernandez, C.V.; Telsnig, T.; Pradas, A.V. *JRC Wind Energy Status Report: 2016 Edition | EU Science Hub*; Publications Office of the European Union: Luxembourg, 2017.
2. Ackermann, T.; Söder, L. An overview of wind energy-status 2002. *Renew. Sustain. Energy Rev.* **2002**, *6*, 67–127. [CrossRef]
3. Hansen, A.D.; Hansen, L.H. Wind turbine concept market penetration over 10 years (1995–2004). *Wind Energy* **2007**, *10*, 81–97. [CrossRef]
4. Hansen, A.D.; Iov, F.; Blaabjerg, F.; Hansen, L.H. Review of Contemporary Wind Turbine Concepts and Their Market Penetration. *Wind Eng.* **2004**, *28*, 247–263. [CrossRef]
5. Crabtree, C.J.; Zappalá, D.; Hogg, S.I. Wind energy: UK experiences and offshore operational challenges. *Proc. Inst. Mech. Eng. Part A J. Power Energy* **2015**, *229*, 727–746. [CrossRef]
6. Dhiman, H.S.; Deb, D.; Muyeen, S.M.; Kamwa, I. Wind Turbine Gearbox Anomaly Detection based on Adaptive Threshold and Twin Support Vector Machines. *IEEE Trans. Energy Convers.* **2021**, *36*, 3462–3469. [CrossRef]
7. Yang, L.; Zhang, Z. Wind Turbine Gearbox Failure Detection Based on SCADA Data: A Deep Learning-Based Approach. *IEEE Trans. Instrum. Meas.* **2021**, *70*, 3507911. [CrossRef]
8. Liu, Z.; Zhang, L. A review of failure modes, condition monitoring and fault diagnosis methods for large-scale wind turbine bearings. *Measurement* **2020**, *149*, 107002. [CrossRef]
9. Rezamand, M.; Kordestani, M.; Carriveau, R.; Ting, D.S.; Saif, M. A New Hybrid Fault Detection Method for Wind Turbine Blades Using Recursive PCA and Wavelet-Based PDF. *IEEE Sens. J.* **2020**, *20*, 2023–2033. [CrossRef]
10. Turnbull, A.; Carroll, J.; McDonald, A. Combining SCADA and vibration data into a single anomaly detection model to predict wind turbine component failure. *Wind Energy* **2021**, *24*, 197–211. [CrossRef]
11. Beretta, M.; Cárdenas, J.J.; Koch, C.; Cusidó, J. Wind fleet generator fault detection via SCADA alarms and autoencoders. *Appl. Sci.* **2020**, *10*, 8649. [CrossRef]
12. Wei, L.; Qian, Z.; Yang, C.; Pei, Y. Wind turbine pitch system condition monitoring based on performance curves in multiple states. In Proceedings of the 2018 9th International Renewable Energy Congress, IREC 2018, Hammamet, Tunisia, 20–22 March 2018; pp. 1–6. [CrossRef]
13. Pandit, R.K.; Infield, D. SCADA-based wind turbine anomaly detection using Gaussian process models for wind turbine condition monitoring purposes. *IET Renew. Power Gener.* **2018**, *12*, 1249–1255. [CrossRef]

14. Beretta, M.; Vidal, Y.; Sepulveda, J.; Porro, O.; Cusidó, J. Improved ensemble learning for wind turbine main bearing fault diagnosis. *Appl. Sci.* **2021**, *11*, 7523. [CrossRef]
15. Garlick, W.; Dixon, R.; Watson, S. A Model-Based Approach to Wind Turbine Condition Monitoring Using SCADA Data. In Proceedings of the 20th International Conference on Systems Engineering, Coventry, UK, 8–10 September 2009.
16. Wilkinson, M.; Darnell, B.; Delft, T.V.; Harman, K. Comparison of methods for wind turbine condition monitoring with SCADA data. *IET Renew. Power Gener.* **2014**, *8*, 390–397. [CrossRef]
17. Bangalore, P.; Letzgus, S.; Karlsson, D.; Patriksson, M. An artificial neural network-based condition monitoring method for wind turbines, with application to the monitoring of the gearbox. *Wind Energy* **2017**, *20*, 1421–1438. [CrossRef]
18. Cui, Y.; Bangalore, P.; Tjernberg, L.B. An anomaly detection approach based on machine learning and scada data for condition monitoring of wind turbines. In Proceedings of the 2018 International Conference on Probabilistic Methods Applied to Power Systems, PMAPS 2018, Boise, ID, USA, 24–28 June 2018.
19. Tsui, K.L.; Chen, N.; Zhou, Q.; Hai, Y.; Wang, W. Prognostics and health management: A review on data driven approaches. *Math. Probl. Eng.* **2015**, *2015*, 793161. [CrossRef]
20. Liu, R.; Yang, B.; Zio, E.; Chen, X. Artificial intelligence for fault diagnosis of rotating machinery: A review. *Mech. Syst. Signal Process.* **2018**, *108*, 33–47. [CrossRef]
21. Tautz-Weinert, J.; Watson, S. Using SCADA data for wind turbine condition monitoring—A review. *IET Renew. Power Gener.* **2017**, *11*, 382–394. [CrossRef]
22. Stetco, A.; Dinmohammadi, F.; Zhao, X.; Robu, V.; Flynn, D.; Barnes, M.; Keane, J.; Nenadic, G. Machine learning methods for wind turbine condition monitoring: A review. *Renew. Energy* **2019**, *133*, 620–635. [CrossRef]
23. Serrano-González, J.; Lacal-Arántegui, R. Technological evolution of onshore wind turbines—A market-based analysis. *Wind Energy* **2016**, *19*, 2171–2187. [CrossRef]
24. Li, H.; Chen, Z. Overview of different wind generator systems and their comparisons. *IET Renew. Power Gener.* **2008**, *2*, 123–138.:20070044. [CrossRef]
25. Baroudi, J.A.; Dinavahi, V.; Knight, A.M. A review of power converter topologies for wind generators. *Renew. Energy* **2007**, *32*, 2369–2385. [CrossRef]
26. Chen, Z. Wind turbine drive train systems. In *Wind Energy Systems: Optimising Design and Construction for Safe and Reliable Operation*; Elsevier Ltd.: Amsterdam, The Netherlands, 2010; pp. 208–246. [CrossRef]
27. Llorente Iglesias, R.; Lacal Arantegui, R.; Aguado Alonso, M. Power electronics evolution in wind turbines—A market-based analysis. *Renew. Sustain. Energy Rev.* **2011**, *15*, 4982–4993. [CrossRef]
28. Wagner, H.J.; Mathur, J. Introduction to Wind Energy Systems. In *Green Energy and Technology*; Springer: Berlin/Heidelberg, Germany, 2013. [CrossRef]
29. Joselin Herbert, G.M.; Iniyan, S.; Sreevalsan, E.; Rajapandian, S. A review of wind energy technologies. *Renew. Sustain. Energy Rev.* **2007**, *11*, 1117–1145. [CrossRef]
30. Wind Europe Intelligence Platform—Interactive Data and Maps. 2021. Available online: https://windeurope.org/intelligence-platform/product/european-offshore-wind-farms-map-public/ (accessed on 3 December 2021).
31. Ozturk, S.; Fthenakis, V.; Faulstich, S. Failure modes, effects and criticality analysis for wind turbines considering climatic regions and comparing geared and direct drive wind turbines. *Energies* **2018**, *11*, 2317. [CrossRef]
32. Reder, M.; Yürüşen, N.Y.; Melero, J.J. Data-driven learning framework for associating weather conditions and wind turbine failures. *Reliab. Eng. Syst. Saf.* **2018**, *169*, 554–569. [CrossRef]
33. Wilson, G.; McMillan, D. Assessing wind farm reliability using weather dependent failure rates. *J. Phys. Conf. Ser.* **2014**, *524*, 012181. [CrossRef]
34. Hines, V.A.; Ogilvie, A.B.; Bond, C.R. *Continuous Reliability Enhancement for Wind (CREW) Database: Wind Plant Reliability Benchmark*; Technical Report; Sandia National Laboratory: Albuquerque, NM, USA, 2013.
35. Lin, Y.; Tu, L.; Liu, H.; Li, W. Fault analysis of wind turbines in China. *Renew. Sustain. Energy Rev.* **2016**, *55*, 482–490. [CrossRef]
36. Dinwoodie, I.; Quail, F.; McMillan, D. Analysis of offshore wind turbine operation & maintenance using a novel time domain meteo-ocean modeling approach. In Proceedings of the ASME Turbo Expo, Copenhagen, Denmark, 11–15 June 2012; Volume 6, pp. 847–857. [CrossRef]
37. Ribrant, J.; Bertling, L. Survey of failures in wind power systems with focus on Swedish wind power plants during 1997–2005. In Proceedings of the 2007 IEEE Power Engineering Society General Meeting, Tampa, FL, USA, 24–28 June 2007; pp. 1–8. [CrossRef]
38. Ribrant, J. *Reliability Performance and Maintenance—A Survey of Failures in Wind Power Systems*; Technical Report; KTH School of Electrical Engineering: Stockholm, Sweden, 2006.
39. Ma, Z.; An, G.; Sun, X.; Chai, J. A Study of Fault Statistical Analysis and Maintenance Policy of Wind Turbine System. In Proceedings of the International Conference on Renewable Power Generation, Beijing, China, 17–18 October 2015; pp. 1–4.
40. Su, C.; Yang, Y.; Wang, X.; Hu, Z. Failures analysis of wind turbines: Case study of a Chinese wind farm. In Proceedings of the 2016 Prognostics and System Health Management Conference, PHM-Chengdu 2016, Chengdu, China, 19–21 October 2017; pp. 1–6. [CrossRef]
41. Van Bussel, G.J.W.; Zaaijer, M.B. Reliability, Availability and Maintenance aspects of large-scale offshore wind farms, a concepts study. In Proceedings of the MAREC 2001 Marine Renewable Energies Conference, Newcastle, UK, 27–28 March 2001; Volume 113, pp. 119–126.

42. Herbert, G.M.; Iniyan, S.; Goic, R. Performance, reliability and failure analysis of wind farm in a developing Country. *Renew. Energy* **2010**, *35*, 2739–2751. [CrossRef]
43. Wilkinson, M.; Harman, K.; Hendriks, B. Measuring wind turbine reliability: Results of the Reliawind project. *Wind Energy* **2011**, *35*, 102–109. [CrossRef]
44. Bi, R.; Qian, K.; Zhou, C.; Hepburn, D.M.; Rong, J. A survey of failures in wind turbine generator systems with focus on a wind farm in China. *Int. J. Smart Grid Clean Energy* **2014**, *3*, 366–373. [CrossRef]
45. Spinato, F.; Tavner, P.J.; Van Bussel, G.J.; Koutoulakos, E. Reliability of wind turbine subassemblies. *IET Renew. Power Gener.* **2009**, *3*, 387–401. [CrossRef]
46. Tavner, P.J.; Xiang, J.; Spinato, F. Reliability analysis for wind turbines. *Wind Energy* **2007**, *10*, 1–18. [CrossRef]
47. Faulstich, S.; Hahn, B.; Tavner, P. Wind Turbine Downtime and its Importance for Offshore Deployment. *Wind Energy* **2011**, *14*, 327–337. [CrossRef]
48. Hahn, B.; Durstewitz, M.; Rohrig, K. Reliability of Wind Turbines: Experiences of 15 years with 1500 WTs. In *Wind Energy: Proceedings of the Euromech Colloquium*; Springer: Berlin, Germany, 2007; pp. 329–332.
49. Artigao, E.; Martin-Martinez, S.; Ceña, A.; Honrubia-Escribano, A.; Gomez-Lazaro, E. Failure rate and downtime survey of wind turbines located in Spain. *IET Renew. Power Gener.* **2021**, *15*, 225–236. [CrossRef]
50. Reder, M.D.; Gonzalez, E.; Melero, J.J. Wind Turbine Failures-Tackling current Problems in Failure Data Analysis. *J. Phys. Conf. Ser.* **2016**, *753*, 072027. [CrossRef]
51. Yang, X.; Zhang, Y.; Lv, W.; Wang, D. Image recognition of wind turbine blade damage based on a deep learning model with transfer learning and an ensemble learning classifier. *Renew. Energy* **2021**, *163*, 386–397. [CrossRef]
52. Chen, H.; Zuo, Y.; Chau, K.T.; Zhao, W.; Lee, C.H. Modern electric machines and drives for wind power generation: A review of opportunities and challenges. *IET Renew. Power Gener.* **2021**, *15*, 1864–1887. [CrossRef]
53. Moreno, S.R.; Coelho, L.D.S.; Ayala, H.V.; Mariani, V.C. Wind turbines anomaly detection based on power curves and ensemble learning. *IET Renew. Power Gener.* **2020**, *14*, 4086–4093. [CrossRef]
54. Zeng, X.J.; Yang, M.; Bo, Y.F. Gearbox oil temperature anomaly detection for wind turbine based on sparse Bayesian probability estimation. *Int. J. Electr. Power Energy Syst.* **2020**, *123*, 106233. [CrossRef]
55. Chen, H.; Ma, H.; Chu, X.; Xue, D. Anomaly detection and critical attributes identification for products with multiple operating conditions based on isolation forest. *Adv. Eng. Inform.* **2020**, *46*, 101139. [CrossRef]
56. Liu, X.; Lu, S.; Ren, Y.; Wu, Z. Wind turbine anomaly detection based on SCADA data mining. *Electronics* **2020**, *9*, 751. [CrossRef]
57. McKinnon, C.; Carroll, J.; McDonald, A.; Koukoura, S.; Infield, D.; Soraghan, C. Comparison of new anomaly detection technique for wind turbine condition monitoring using gearbox SCADA data. *Energies* **2020**, *13*, 5152. [CrossRef]
58. Wang, Y.; Yoshihashi, R.; Kawakami, R.; You, S.; Harano, T.; Ito, M.; Komagome, K.; Iida, M.; Naemura, T. Unsupervised anomaly detection with compact deep features for wind turbine blade images taken by a drone. *IPSJ Trans. Comput. Vis. Appl.* **2019**, *11*, 3. [CrossRef]
59. Carroll, J.; Koukoura, S.; McDonald, A.; Charalambous, A.; Weiss, S.; McArthur, S. Wind turbine gearbox failure and remaining useful life prediction using machine learning techniques. *Wind Energy* **2019**, *22*, 360–375. [CrossRef]
60. Yun, H.; Zhang, C.; Hou, C.; Liu, Z. An Adaptive Approach for Ice Detection in Wind Turbine with Inductive Transfer Learning. *IEEE Access* **2019**, *7*, 122205–122213. [CrossRef]
61. Yang, C.; Qian, Z.; Pei, Y. Condition Monitoring for Wind Turbine Pitch System Using Multi-parameter Health Indicator. In Proceedings of the 2018 International Conference on Power System Technology, POWERCON 2018, Guangzhou, China, 6–8 November 2019; pp. 4820–4825. [CrossRef]
62. Zhao, H.; Liu, H.; Hu, W.; Yan, X. Anomaly detection and fault analysis of wind turbine components based on deep learning network. *Renew. Energy* **2018**, *127*, 825–834. [CrossRef]
63. Zhao, Y.; Li, D.; Dong, A.; Kang, D.; Lv, Q.; Shang, L. Fault prediction and diagnosis of wind turbine generators using SCADA data. *Energies* **2017**, *10*, 1210. [CrossRef]
64. Zhao, Y.; Li, D.; Dong, A.; Lin, J.; Kang, D.; Shang, L. Fault prognosis of wind turbine generator using SCADA data. In Proceedings of the NAPS 2016—48th North American Power Symposium, Denver, CO, USA, 18–20 September 2016; pp. 1–6. [CrossRef]
65. Mckinnon, C.; Carroll, J.; Mcdonald, A.; Koukoura, S.; Plumley, C. Investigation of anomaly detection technique for wind turbine pitch systems. In Proceedings of the 9th Renewable Power Generation Conference (RPG Dublin Online 2021), Online, 1–2 March 2021.
66. Cardoni, M.; Pau, D.P.; Falaschetti, L.; Turchetti, C.; Lattuada, M. Online learning of oil leak anomalies in wind turbines with block-based binary reservoir. *Electronics* **2021**, *10*, 2836. [CrossRef]
67. Mucchielli, P.; Bhowmik, B.; Ghosh, B.; Pakrashi, V. Real-time accurate detection of wind turbine downtime—An Irish perspective. *Renew. Energy* **2021**, *179*, 1969–1989. [CrossRef]
68. Liu, X.; Teng, W.; Wu, S.; Wu, X.; Liu, Y.; Ma, Z. Sparse dictionary learning based adversarial variational auto-encoders for fault identification of wind turbines. *Meas. J. Int. Meas. Confed.* **2021**, *183*, 109810. [CrossRef]

69. Heydari, A.; Garcia, D.A.; Fekih, A.; Keynia, F.; Tjernberg, L.B.; De Santoli, L. A Hybrid Intelligent Model for the Condition Monitoring and Diagnostics of Wind Turbines Gearbox. *IEEE Access* **2021**, *9*, 89878–89890. [CrossRef]
70. Xiang, L.; Yang, X.; Hu, A.; Su, H.; Wang, P. Condition monitoring and anomaly detection of wind turbine based on cascaded and bidirectional deep learning networks. *Appl. Energy* **2022**, *305*, 117925. [CrossRef]

energies

MDPI

Article

Reliability Testing of Wind Power Plant Devices with the Use of an Intelligent Diagnostic System

Stanislaw Duer [1,*], Jacek Paś [2], Marek Stawowy [3], Aneta Hapka [4], Radosław Duer [1], Arkadiusz Ostrowski [5] and Marek Woźniak [5]

1 Department of Energy, Faculty of Mechanical Engineering, Technical University of Koszalin, 15-17 Raclawicka St., 75-620 Koszalin, Poland; radoslaw.duer@wpl.pl
2 Faculty of Electronic, Military University of Technology of Warsaw, 2 Urbanowicza St., 00-908 Warsaw, Poland; jacek.pas@wat.edu.pl
3 Department of Transport Telecommunication, Faculty of Transport, Warsaw University of Technology, Koszykowa St. 75, 00-662 Warsaw, Poland; marek.stawowy@pw.edu.pl
4 Faculty of Electronic and informatics, Technical University of Koszalin, 2 Sniadeckich St., 75-620 Koszalin, Poland; aneta.hapka@wp.pl
5 Doctoral School, Technical University of Koszalin, 2 Sniadeckich St., 75-620 Koszalin, Poland; arkadiusz.ostrowski@s.tu.koszalin.pl (A.O.); marek.wozniak@s.tu.koszalin.pl (M.W.)
* Correspondence: stanislaw.duer@tu.koszalin.pl; Tel.: +48-943-478-262

Abstract: This paper introduces the issue of reliability simulation studies of wind farm equipment in the process of an operation. By the improvement, retrofitting and insertion of new (optimal) solutions to change the quality and terms of the use of wind farm equipment, an evaluation of their impact on reliability under real conditions can be carried out over a long period of time. Over a brief period, testing the reliability of a technical facility is only possible in a simulation. The aspect of evaluating the reliability of wind farm equipment after the application of intelligent systems, including the Wind Power Plant Expert System (WPPES), can be tested in the manner of a simulation. It was accepted in this article that the operation of the wind farm equipment is detailed based on Markov processes. The results of such research activities are the development of reliable and appropriate strategies and an exploitation policy of PE facilities. The above-mentioned issues in such a comprehensive approach have not been fully presented in the literature. The process of exploitation of complex technical objects such as PE devices is a complex random technical and technological process.

Keywords: simulation testing; reliability; intelligent systems; servicing process; Markov processes; expert system

Citation: Duer, S.; Paś, J.; Stawowy, M.; Hapka, A.; Duer, R.; Ostrowski, A.; Woźniak, M. Reliability Testing of Wind Power Plant Devices with the Use of an Intelligent Diagnostic System. *Energies* **2022**, *15*, 3583. https://doi.org/10.3390/en15103583

Academic Editors: Alessandro Cannavale and Paweł Ligęza

Received: 20 March 2022
Accepted: 9 May 2022
Published: 13 May 2022

Publisher's Note: MDPI stays neutral with regard to jurisdictional claims in published maps and institutional affiliations.

1. Introduction

1.1. Motivation and Incitement

This paper presents the problems of the simulation study of the reliability of wind farm (WF) equipment in its operation process. The problem of investigating the process of operation of complex technical objects, including wind farm (WF) equipment and wind turbines (TWG), is an important cognitive issue. This problem is particularly important for the owners and users of WF. It must answer the question of how to organize the organizational and technical activities in the technical maintenance system of WF facilities. It is only a properly organized system of WF renewal that may optimize well the use of these facilities. The results of such research activities are the development of reliable and appropriate strategies and the operation policy of PE facilities. The above-mentioned issues in such a comprehensive approach have not been fully presented in the literature. The process of the operation of complex technical objects such as PE devices is a multifaceted random technical and technological process. The process of the operation of technical objects and equipment is defined as a random set of the state of the use and the state of the technical operation of an object. The intricacy of this process results from the complexity of the events (elements) that describe and occur during this process. The set of those elements that enter the process of WF equipment operation, which require cognition, description and examination, include: WF equipment, the process of diagnosing (examining the state) of WF during its use and the process of the technical service of a given object in which it is renewed. The issues concerning the description and testing of individual elements describing the process of the operation of technical objects have been well-presented in publications. However, there is a lack of studies that holistically present the issue of testing and organizing the operation process of complex technical objects. Hence, this article undertakes the task of a simulation study of the reliability of the wind farm (WF) equipment in its operation process. The following research questions must be solved for this work. The initial issue is to comprehend and describe the WF equipment diagnostics problem. Another issue is understanding and describing the process of wind farm equipment operation (usage and maintenance). Understanding and describing the concerns of the organization of the technical service system in the operation process of the object under consideration is an essential topic covered in this article. The major goal of this article's production was to comprehend and describe the challenges surrounding the arrangement of an object's operating process in order to model it. Three models (A, B and C) of the operation process of wind farm equipment were developed and described in this article for the purpose of the organization of the simulation studies. Another issue addressed in this article is the problem of a reliability assessment of WF equipment after the application of intelligent systems, including the Wind Power Plant Expert System (WPPES), which supports its efficient operation; this is the main research objective of this article. An important objective of this article was to develop a simulation study plan; for this purpose, appropriate quantities that describe WF reliability were selected. Based on the literature and the authors' own work, the input data for the study was prepared. The quantities adopted describe the functional properties of the object investigated: the time of the object's use T—the time of the object remaining in a fit state, T_{NA}—the time of removing the unfitness of the object and T_{NP}—the time required to perform a planned repair. In article [1], the author introduced, inter alia, the idea and methodologies of the development of the models of the operation process of complex technical objects. This includes such models that describe the use and maintenance of systems using trivalent logic diagnostics and artificial neural network information. The problem of the modeling systems and the operation processes of complex technical objects is complex, and it involves many scientific fields, such as mathematics, the theory of operation, the reliability theory, technical diagnostics, the systems modeling theory, computer science, etc. Each of the directions is currently being developed quite intensively. The problem of performance modeling and simulation studies of the efficiency of wind farm equipment presented in this paper offers an innovative and future-proof solution.

1.2. Literature of Review

An important element in modeling the operation process of a complex technical facility is the development of a model of the renewal process of an intelligent maintenance system. These issues have been presented in publications, including those by Buchannan et al. and Duer et al. [2–6]. In his work, the author presented issues related to the definition of system maintenance models. The system designed for an automatic regeneration of object properties provides a basis for an optimization of costs related to preventive measures. This system fully minimizes the costs associated with organizing the maintenance system. The regeneration of the facility is carried out when required. This is ensured by an intelligent object diagnostic system built based on an artificial neural network—in particular, a network that reliably identifies the object's states for which preventive measures need to be performed [7–10]. The system eliminates the costs related to the regeneration of those elements of the facility that do not require it and that are in working order. The intelligent maintenance system designed (including the intelligent diagnostic system) for the facility ensures the regeneration of those internal (structural) elements that need it and are in a state of incomplete efficiency {1} or unfitness {0}.

The study by Kacalak et al. [11–15] contained a description of the effective measurement path, which constitutes an important element in the design of the diagnostic system. Furthermore, the theoretical basis for the system's design is presented for designing a measurement system using a computerized measurement card, the aim of which is to create a measurement base for the diagnostic system.

The issues of modeling the process of the operation of technical facilities were presented in the publications by Nakagawa and Pokoradi et al. [16–21]. This research presented a mathematical approach to modeling this process. This article addresses the problem of a quality assessment of such an organized maintenance process. To that end, this paper presents a simulation test program. The research program includes a characterization of the models of the operation processes of technical facilities and the establishment of test input data, which provides the quantities of the operation time of a technical object being the summary duration time of the regeneration (repairs) and the use of the object and the determination of the indexes of a qualitative assessment of the regeneration of the object in the operation process. The results of the research are justified in the example of simulation studies on the effects of the operation process with the reclamation of a technical facility in an intelligent system incorporating an artificial neural network.

The studies authored by Dyduch, Epstein et al. and Siergiejczyk et al. contained a description of the reliability, and an operational analysis is necessary [22–29]. In their lectures, they presented a reliability and operational analysis of complex technical and other facilities, such as power systems in transport telematics systems (PSSs in TTDs). The article discussed PSS in a TTD from both the primary and backup sources, as well as alternative solutions utilized in power supply systems. Bulletin outlines, among other things, are the solutions employed in power systems. This enables the determination of the relationships that indicate the plausibility of the other system in a fully functional state, a security emergency and a safety emergency. A qualitative examination of PSS in TTD, as well as the quality ratio for supply continuity, was conducted and evaluated. This index demonstrates the CQoPS reliance on numerous quality metrics, not only to indicate dependability but also to indicate the quality of the continuity of supply. The example illustrates the computation of the CQoPS factor both for the main and backup supply using three observations, each affecting the quality. The considerations presented for qualitative and reliability operational modeling PSS can be used in other public facilities as well (including those classified as crucial infrastructure).

Zajkowski [30] presented the organization, execution and analysis of the simulations held to assess the quality of the maintenance system of WPP wind farm equipment. An important aspect for the reader is the presentation of the models of WPP wind farm equipment operation processes. The reader will find the problems of the construction and organization of the operation process of complex technical objects [31]. Three models of WPP in the simulations are used the processes of wind turbine equipment operation. The prime model is Model A, i.e., an operation process of a wind power plant that uses an intelligent maintenance system. The latter model is Model B, i.e., an information-intensive operation process in bivalent logic: the operating process is a model with the maintenance system organized by planning its optimal preventive actions. The third is Model C, i.e., an operation process of a wind power plant with a maintenance system that is organized classically without any condition survey in the evaluation process: the maintenance strategy is based on the manual planning of preventive measures and an arbitrary selection of its scope by the operator.

The article by Duer [32] depicted an organization, implementation and an analysis of the enforcement carried out to assess the quality of the maintenance system for the wind farm equipment WPP. A significant aspect for the reader is the presentation of the models of wind farm equipment WPP operation processes. The reader will become acquainted with the issues of the construction and organization of the operation process of complex technical facilities in [33]. Three models for wind farm usage operation processes WPP have been used in simulations. The first model is Model A: an operation process of a wind power plant that uses an intelligent maintenance system with an artificial neural network. The other model is Model B: the process of the operation of an object that uses information in bivalent logic: a model with an upkeep system organized by the scheduling of its optimal preventive actions. The third one is Model C: a classically organized operation process of a wind power plant with a maintenance system without any condition testing in the assessment process: the maintenance strategy is based on a manual planning of preventive actions and a free choice of their range by the operator.

1.3. Contribution and Paper Organization

The operation process of the wind farm equipment, together with the technical maintenance system, is based on the diagnostic information concerning the test information and identifying the technical condition of the object tested [34]. The diagnostics of the wind farm equipment uses state recognition in binary or ternary logic [35]. Diagnostics using trivalent logic is of the greatest practical importance in the organization of the operation process and renewing of the technical object. The artificial neural network used in this model points out that the assessment system used in this WPP operating process is intelligent. A characterization of this diagnostic process using an artificial neural network, and an expert system WPPES is not the target and subject of the issues addressed in this article. Different models of the WPPS usage process are known, and models are described that appear in the studies included in the bibliography. Simulation studies of the process of the operation of technical objects shall be carried out based on the input data. The aforementioned data refers to:

- Fault-free operation of a WPP;
- Operating time between WPP wind farm damages;
- Period of (WPP) Wind Power Plants repair;
- The duration of ineffective (WPP) Wind Power Plants shutdown.

The article raises the problem of simulation testing of the dependability of wind farm equipment in the operating process. The main objective of this article is the concept of assessing the dependability of WF appliances after the use of intelligent systems, along with the Wind Power Plant Expert System (WPPES), which supports its effective use. The Kolmogorov–Chapman equations were adopted in an analytical description of the models developed by the wind farm operational process. In a simulation analysis of the reliability of wind farm equipment, the reliability value known in the literature [36] will be determined in the form of the availability factor Kg(t). This article covers the issues of a qualitative evaluation of the renewal process of operational properties in complex technical objects in an intelligent maintenance system. The concerns surrounding the study of the operating process of complicated technological items are not covered in depth in the literature. A complex technical object operation is a complex and random technical and technological procedure. A random collection of the states of use and maintenance of a technical object is described as the process of operating technical objects and equipment. The complexity of this process is the result of the complexity of the events (elements) that describe and occur during it. The following are the aspects that occurred throughout the use of the object and require description and testing: the technical object, the process of diagnosing the object in its state of use and the process of the maintenance of a given object. The issues of the description and research of individual elements that characterize the process of the operation of technical objects are well-presented in previous publications. However, no research comprehensively presents the issues of studying the operation process of complex technical objects.

In most publications, researchers of the issues of the reliability and quality of the operation of systems and complex technical objects use research methods based on the use of the Kolmogorov–Chapman equation. Such methods were presented in [36]. In these works, researchers use a qualitative assessment of the operation process in the form of the availability coefficient ($K_g(t)$). Another practical approach used in this assessment may be to convert the determined value of the coefficient or readiness function $K_g(t)$ into a form of probability that the object under test will be in a fit or operational state. The innovation of our article and the subject presented in it is the use of new functions developed by the authors for an assessment of reliability issues and the operation process of systems and complex technical objects in the form of ($F_C(t)$): the quality of a qualitative assessment of the operation process of the facility and ($F_{ch}(t)$): the absorption function of the operation process. These two values are only used in this publication when assessing the reliability of technical systems and facilities.

The problems of a simulation study of the reliability of wind farm (WF) devices in the process of the operation presented in the article will be solved as follows. The second part of the article will cover the problems of understanding and a description of the operation processes (use and operation) of the wind farm devices. To organize the simulation studies in the article, three models developed (A, B and C) of the operation process of the wind farm devices were instroduced and described. Another issue addressed in this part of the article is the problem of a reliability assessment of the WF equipment after the application of intelligent systems, including WPPES, which supports its efficient operation, as the main research objective of this article. The third part of this article covers the planning and organization of the simulation study. For this purpose, relevant quantities describing the reliability of the WF have been selected. Based on the literature and the authors' own analytical work, the input data for the study was prepared. The quantities adopted describe the functional properties of the object investigated: T—the time of the object remaining in the state of fitness, T_{NA}—the time of removing the unfitness of the object and T_{NP}—the time of performing the planned repair. In the fourth part of the article, the results of the simulation tests are presented for their prepared input data.

2. Methodology of Building Models of the Wind Farm Equipment Operation Process

The operation process of a complex technical facility: the Wind Power Equipment with expert assistance WPPES systems is a stochastic process $S(t)$, the elements of which (S_i) belong to the subsets of the states of an object {S}: operation and maintenance. Analyzing possible operating situations in which the facility could be found after any number of runs, the states of the object in operation can be determined, which creates a set of the states of the object $S(t)$. Each of the possible states of the object in the process of operation that may occur is determined during the process of diagnosing the object. The set of the object's state in the process of the object's operation {S}, depending on the accepted valence of the valuation of states, is divided into two or three subsets {S_1, S_{01} and S_{10}}, where: S_0—the subset of operating states, S_1—the subset of non-operational states and S_{10}—the subset of insufficient operational states. The subset of the states of the operation of an operation process is a single-element process S_0 of the object's states in the operation process.

The condition in which the facility is used in the operation process is the state of the operation process Z_1 in which the facility performs the required functions by its use. If the facility is no longer in use because it is not performing its required function, it must be repaired during maintenance.

The distinct states in the diagnosis process when the object is not in use belong to the subset of the object's handling states {S_1}. This subset is a multi-element collection that includes the following states: {S_1, S_{01} and S_{10}} where: S_1—scheduled maintenance, S_{01}—unscheduled surveys and S_{10}—inefficient use (the shutdown state). Concerning the research needs expressed in the literature [1–3,33], different methods are used to represent the process of the operation of a technical facility. Most frequently, the operating model of an object is presented graphically. A process graph is a graphical presentation of the realization of the object's operation process. A questionnaire analysis offers another method of presenting the implementation of the facility operation process.

2.1. Indices That Characterize the Process of Operating a Technical Facility

From the set illustrated in the literature [32] of these indicators characterizing the process of the operation of a technical object, the value best reflecting the operation process is the availability indicator K_g and the accessibility function $K_g(t)$. The process for calculating the availability feature $K_g(t)$ is generally simplified when calculated for a limit value at $t \to \infty$. The size is closely related to the stationary characteristics of the damage and maintenance process. Due to this, the availability rate K_g is the most appropriate measure to set out the efficiency of the operation process, which links both the utility and economic characteristics of the facility. The accessibility factor K_g of the object is the likelihood of the event that the object is operational after a sufficiently long period of operation ($\to\infty$). The accessibility factor K_g determines the average proportion of the technical object's service life in the total service life, as represented by the following relationship:

$$K_g = \lim_{t\to\infty} K_g(t) = \lim_{t\to\infty} K_{gsr}(t) \quad (1)$$

where $K_g(t)$ is the medium value of the accessibility factor K_g.

The determination of the accessibility factor K_g requires that the operation process of a given object is known exactly. While having the determined quantity that expresses the effectiveness of the operation process of an object of any class, one can determine the quality function of the object's operation process.

The quality function of the object's operation process F_c is a dependence that characterizes the object's operation process concerning the effectiveness and the quantity of the input used during the total object's operation period, which is presented in the form of the following dependence:

$$F_c = \frac{K_g}{N_e} \quad (2)$$

The quality function of the facility's operation process expresses the effects achieved in the facility operation process in the form of the accessibility factor K_g versus the operational input N_e that incurred during a specific period (t) of the facility's functioning.

In the quality assessment of the dependability of complex engineering objects, the volume that the author referred to as the absorption function of the object operation process (F_{ch}) can be of a high research relevance interest. The uptake function of the facility operation process F_{ch} is a benchmark indicating the amount of expenditure (costs) that incurred in a given process of the facility's operation so that the quality function of the facility's operation process (F_c) would then have the paramount value. The absorption function of the facility's operation process (F_{ch}) shall be determined based on the following relationship:

$$F_{ch} = 1 - F_c \quad (3)$$

where F_{ch} is the absorption function of the facility's operation process.

2.2. Two-State Model of the Wind Farm Equipment Operation Process (Model C)

The characterization of a simple model of the operation process of a technical facility consists of a sequence of the random times of use and repair that occur alternately in this process. In the article, the terms: "technical facility" and "wind farm equipment" are equivalent to each other. The new facility is being put into operation, and it performs its required function. At the moment of failure after the time (τ_1), the object goes to the emergency repair state S_1. In the state of emergency repair S_1, the process of restoring the object to its full operational properties over time (T_{NA}) takes place. As soon as the damage is removed, the object with the intensity μ, where: $\mu = 1/T_{NA}$ [1/h], returns to the fit state S_0, where it performs its task. Since there is no possibility of recognizing the state of incomplete fitness (inoperability) in the facility, the facility is used as long as it is subject to further damage. As soon as a failure occurs, the object goes into the S_1 emergency repair state. In the S_1 state, the object renewal process takes place, which consists of restoring complete functional properties to the object. As soon as the damage over time (T_{NA}) is removed, the object goes back to the fit state S_0 with intensity μ.

The relations between the states in model C mean the following (Figure 1):

- λ—indicates the intensity of the transition of the system from state S_1;
- μ—informs about the intensity of the system transition from state S_1 to state S_0. The graphic structure of the two-state model is as follows (Figure 1).

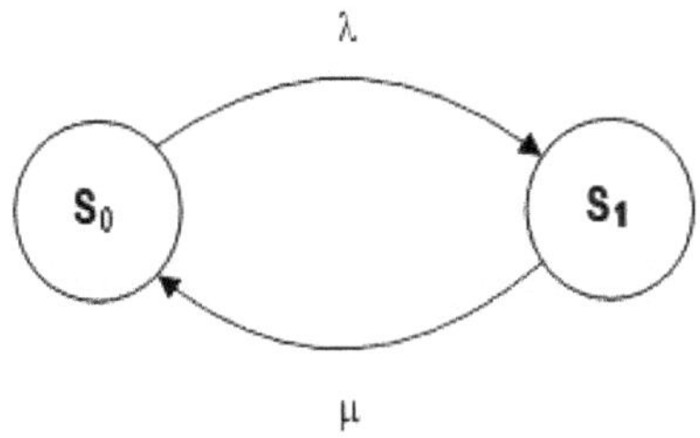

Figure 1. Diagram of a simple process for handling wind farm equipment, model C.

In order to determine the likelihoods of the system in the particular states of an interest to us, the operation process model of the wind farm equipment presented in Figure 2 should include the following equations:

$$\begin{cases} -\lambda \cdot P_0 + \mu \cdot P_1 = 0 \\ \lambda \cdot P_0 - \mu \cdot P_1 = 0 \end{cases} \quad (4)$$

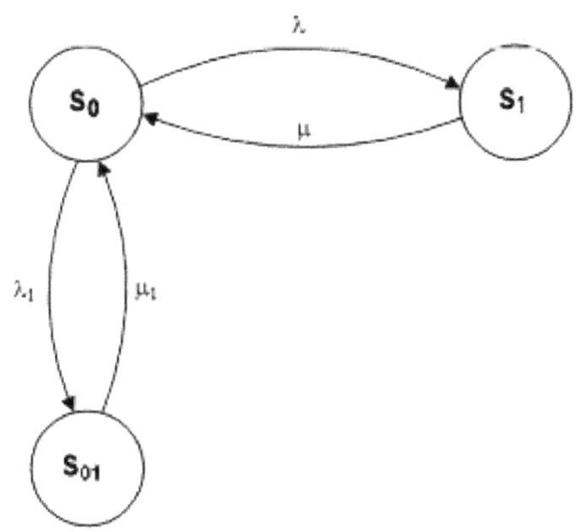

Figure 2. Diagram of the process of operating wind farm equipment with intelligent WPPES systems and an artificial neural network decision support concerning safety in its use (model A).

In the matrix notation, Relationship (4) can be presented as follows:

$$\begin{bmatrix} -\lambda & \mu \\ \lambda & -\mu \end{bmatrix} \cdot \begin{bmatrix} P_0 \\ P_1 \end{bmatrix} = \begin{bmatrix} 0 \\ 0 \end{bmatrix} \quad (5)$$

By transforming Equation (5), the following dependencies can be determined:

$$P_1 = \frac{\lambda}{\mu} \cdot P_0 \quad (6)$$

Obviously enough, it is known that the relationship is correct:

$$P_0 + P_1 = 1 \tag{7}$$

Therefore:

$$P_0 \cdot \left(1 + \frac{\lambda}{\mu}\right) = 1 \tag{8}$$

$$K_{g1} = P_0 = \frac{1}{\left(1 + \frac{\lambda}{\mu}\right)} \tag{9}$$

$$K_{g1} = P_0 = \frac{\mu}{\mu + \lambda} \tag{10}$$

The symbols in the equations above mean the following:

- λ—indicates the intensity of the system transition from state S_0 to state S_1;
- μ—informs about the intensity of the system transition from state S_1 to state S_0;
- P_0—the probability of the system will remain in S_0;
- P_1—the probability of the system will remain in S_1;
- $K_g(t)$—readiness factor.

Using Expression (10), it is possible to determine the value of the probability that the wind farm equipment system is in working condition. It is numerically equal to the readiness ratio $K_g(t)$. We assume that the modeling of the operation process consists of determining the probabilities of the system of the wind farm equipment in particular states $\{S_0, S_1\}$, and therefore, the following needs to be determined:

- the probability function of the system remaining in S_0;
- the probability function of the system remaining in S_1.

2.3. Three-State Operational Process Model of the Wind Farm Equipment (Model A)

A subset of the service condition and a subset of the service state make up each process of any complicated technical entity. The task of describing the relationship of transitions between the stages emphasized in the object's operating process under investigation is tough. The selection of data for simulation testing necessitates a detailed analysis and understanding of the wind farm equipment's real functioning process. The most difficult work in any facility's operation process is to develop a model of the structure of the facility's operation process in the form chosen (analytical or graphic). With the constructed process of the facility's operation tested and knowing the relations between the operating states interpreted, it is possible to define selected values of reliability that characterize this model.

An analysis of the ways of changing the states of the object operated according to the model described (Figure 2) demonstrates that the object can be in one of the following conditions:

- S_0—use of the technical object;
- S_1—scheduled maintenance—NP preventive repair;
- S_{01}—unscheduled maintenance: the fit state S_0 with intensity μ.

The relations between the states in model A mean:

- λ—interpretation of the intensity of the transition of the system from state S_0 to state S_1;
- μ—interpretation of the intensity of the transition of the system from state S_1 to state S_0, only an interpretation of the intensity of the transition of the system from state S_0 to state S_{01} and only an interpretation of the intensity of the transition of the system from state S_{01} to state S_0.

In the publications by numerous authors [36], the three-state model of operation is called the prophylaxis (servicing) model by age. This statement may be correct only when the model developed by the operating process concerns a technical object in which its diagnostics is carried out, along with the possibility of assessing the states. Therefore, in a situation where a technical facility is diagnosed, especially in the three-valued state assessment [2–4,33], and a set of wind farm devices, it is equipped with intelligent WPPES advisory systems and an artificial neural network decision support concerning safety in its use. Model A is an operational process model. The system of the WF is equipped with intelligent WPPES advisory systems support for decision-making concerning the safety of its use (Figure 2).

The transition between the states in this model is as follows: an early moment in time, the object is turned on for work; its use or being ready for use (on-call) begins. When using a technical facility, based on a forecast, with high reliability, it is possible to determine the moment in time Δt in which the facility will be in the condition of planned repair. Therefore, for known times Δt, the planned repairing of the object is scheduled. Hence, after time, at the moment (t1 = t0 + Θ) of operation, the object is in a subset of service states. The object passes from state S0 to state S_1 with the intensity λ, where: λ = 1/Θ [1/h]. In the S_1 condition, the planned repair is carried out during T_{NP}. At the moment of performing the scheduled repair, the object with intensity μ, where: μ = 1/T_{NP} [1/h], goes to state S_0 use. The technical object in use in state S_0 may be damaged. Then, the technical object with damage intensity λ_1 goes to state S_{01}. Removal of the failure of the object in the state of unscheduled repair of S_{01} during T_{NA} causes the object to move from unplanned repair intensity μ_1 to the use state S_0.

To determine the likelihood of the presence of wind farm devices in individual states, the model of its working process set out in Figure 2 is to be described by the following equations:

$$\begin{cases} -\lambda \cdot P_0 + \mu \cdot P_1 - \lambda_1 \cdot P_0 + \mu_1 \cdot P_{01} = 0 \\ \lambda \cdot P_0 - \mu \cdot P_1 = 0 \\ \lambda_1 \cdot P_0 - \mu_1 \cdot P_{01} = 0 \end{cases} \tag{11}$$

In the matrix notation, Relationship (11) can be presented as follows:

$$\begin{bmatrix} -(\lambda+\lambda_1) & \mu & \mu_1 \\ \lambda & -\mu & 0 \\ \lambda_1 & 0 & -\mu_1 \end{bmatrix} \cdot \begin{bmatrix} P_0 \\ P_1 \\ P_{01} \end{bmatrix} = \begin{bmatrix} 0 \\ 0 \\ 0 \end{bmatrix} \tag{12}$$

By transforming Equation (12), the following relationships were obtained:

$$P_1 = \frac{\lambda}{\mu} \cdot P_0 P_{01} = \frac{\lambda_1}{\mu_1} \cdot P_0 \tag{13}$$

Obviously enough, it is known that the equation is correct:

$$P_0 + P_1 + P_{01} = 1 \tag{14}$$

Therefore:

$$P_0 \cdot \left(1 + \frac{\lambda}{\mu} + \frac{\lambda_1}{\mu_1}\right) = 1 \tag{15}$$

$$K_{g2} = P_0 = \frac{1}{\left(1 + \frac{\lambda}{\mu} + \frac{\lambda_1}{\mu_1}\right)} \tag{16}$$

$$K_{g2} = P_0 = \frac{\mu \cdot \mu_1}{\mu \cdot \mu_1 + \lambda \cdot \mu_1 + \lambda_1 \cdot \mu} \tag{17}$$

The symbols in the equations above mean the following:

- Λ—interpretation of the intensity of the transition of the system from state S_0 to state S_1;

- μ—interpretation of the intensity of the transition of the system from state S_1 to state S_0, only an interpretation of the intensity of the transition of the system from state S_0 to state S_{01} and only an interpretation of the intensity of the transition of the system from state S_{01} to state S_0;
- P_0—the likelihood that the system will remain in S_0;
- P_1—the likelihood that the system will remain in S_1;
- P_{01}—the likelihood that the system will remain in S_{01}.

Using Expressions (17), it is possible to determine the value of the probability of a wind farm system remaining in the operational state. It is numerically equal to the readiness index. If we assume that the modeling of the operation process consists of determining the probabilities of a wind farm system remaining in particular states {S_0, S_1 and S_{01}}, then the following values need to be determined:

- The probability function of the system remaining in state S_0;
- The probability function of the system remaining in state S_1;
- The probability function of the system remaining in state S_{01}.

2.4. Four-State Model of the Operation Process of Wind Farm Equipment (Model B)

A model of an actual facility's operating process using a typical usage and maintenance system (model B). In this model, it was assumed that the wind farm's equipment did not include any sophisticated WPPES decision support system for ensuring the safety of its operation. Undetected faults and inefficiencies in this class of things may present themselves automatically when they arise. Then, there is no way of knowing when the facility in use will fail. That model accurately depicts the operation of this type of facility. As a result, the condition of ineffective usage S_{10} was identified in the operating process for these undetected malfunctions. Using this type of model, the operational properties of facilities are examined, e.g., those stored (over a long period of time) or those facilities in which there are no diagnostic systems and, therefore, no fault signaling (fault signaling factor $\alpha = 0$). The problem related to the description of this type of operation process mod was presented in previous studies [2,32]. The graphic form of this model is shown in Figure 3.

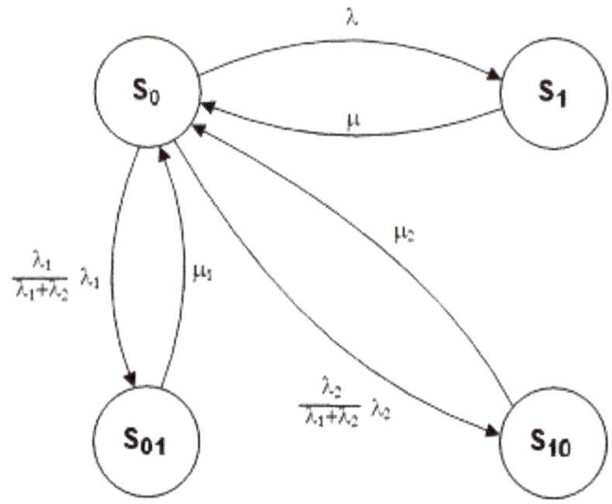

Figure 3. Diagram of the operation process of a wind equipment farm that is not provided with any intelligent WPPES safety decision support systems (model B).

The analysis of the model of the operation process of the wind farm equipment presented in Figure 3 demonstrates that the facility may be in one of the following states:

- S_0—effective use of the facility;
- S_1—scheduled maintenance—preventive NP;

- S_{01}—unscheduled maintenance;
- S_{10}—ineffective use of the facility.

The transitions between the states in model B mean the following:

- λ—has an interpretation of the intensity of the system's transition from state S_0 to state S_1;
- μ—has an interpretation of the system's transition from state S_1 to state S_0;
- $\frac{\lambda_1}{\lambda_1+\lambda_2}\lambda_1$—has an interpretation of the intensity of the transition of the system from state S_0 to state S_{01};
- μ_1—has an interpretation of the intensity of the transition of the system from state S_{01} to state S_0;

1. $\frac{\lambda_2}{\lambda_1+\lambda_2}\lambda_2-$ —has an interpretation of the intensity of the transition of the system from state S_0 to state S_{10};

- μ_2—has an interpretation of the intensity of the transition of the system from state S_{10} to state S_0.

A technical facility used without the WPPES expert system is damaged, and it modifies its place in the operating process. The object moves with intensity $\frac{\lambda_2}{\lambda_1+\lambda_2}\lambda_2$ from state S_0 to the state of ineffective use S_{10}. The time of the technical object remaining in the state of ineffective use S_{10} is determined by a random variable τ_{NA}. The value of the useful life of an ineffective technical facility results from the wind farm operating conditions or from changes in the weather conditions. The technical object, once the unfitness has been located and removed, that determines the cause of its ineffective use is again transferred to the S_0 use state with an intensity of μ_1. To identify the probability of the system remaining in a particular state, the network of crossings presented in Figure 4 shall be described with the following equations:

$$\begin{cases} -\lambda\cdot P_0+\mu\cdot P_1-\lambda_1\cdot\frac{\lambda_1}{\lambda_1+\lambda_2}\cdot P_0+\mu_1\cdot P_{01}-\lambda_2\cdot\frac{\lambda_2}{\lambda_1+\lambda_2}\cdot P_0+\mu_2\cdot P_{10}=0 \\ \lambda\cdot P_0-\mu\cdot P_1=0 \\ \lambda_1\cdot\frac{\lambda_1}{\lambda_1+\lambda_2}\cdot P_0+\mu_1\cdot P_{01}=0 \\ \lambda_2\cdot\frac{\lambda_2}{\lambda_1+\lambda_2}\cdot P_0+\mu_2\cdot P_{10}=0 \end{cases} \quad (18)$$

In the matrix notation, Relationship (18) can be presented as follows:

$$\begin{bmatrix} -\left(\lambda+\frac{\lambda_1}{\lambda_1+\lambda_2}\cdot\lambda_1+\frac{\lambda_2}{\lambda_1+\lambda_2}\cdot\lambda_2\right) & \mu & \mu_1 & \mu_2 \\ \lambda & -\mu & 0 & 0 \\ \frac{\lambda_1}{\lambda_1+\lambda_2}\cdot\lambda_1 & 0 & -\mu_1 & 0 \\ \frac{\lambda_2}{\lambda_1+\lambda_2}\cdot\lambda_2 & 0 & 0 & -\mu_2 \end{bmatrix}\cdot\begin{bmatrix} P_0 \\ P_1 \\ P_{01} \\ P_{10} \end{bmatrix}=\begin{bmatrix} 0 \\ 0 \\ 0 \\ 0 \end{bmatrix} \quad (19)$$

By transforming Equation (19), the following relationships were obtained:

$$\begin{aligned} P_1 &= \tfrac{\lambda}{\mu}\cdot P_0 \\ P_{01} &= \tfrac{\lambda_1}{\lambda_1+\lambda_2}\cdot\tfrac{\lambda_1}{\mu_1}\cdot P_0 \\ P_{10} &= \tfrac{\lambda_2}{\lambda_1+\lambda_2}\cdot\tfrac{\lambda_2}{\mu_2}\cdot P_0 \end{aligned} \quad (20)$$

Obviously enough, it is known that the relationship is correct:

$$P_0+P_1+P_{01}+P_{10}=1 \quad (21)$$

Therefore:

$$P_0\cdot\left(1+\frac{\lambda}{\mu}+\frac{\lambda_1}{\lambda_1+\lambda_2}\cdot\frac{\lambda_1}{\mu_1}+\frac{\lambda_2}{\lambda_1+\lambda_2}\cdot\frac{\lambda_2}{\mu_2}\right)=1 \quad (22)$$

$$K_{g7}=K_0=\frac{1}{\left(1+\frac{\lambda}{\mu}+\frac{\lambda_1}{\lambda_1+\lambda_2}\cdot\frac{\lambda_1}{\mu_1}+\frac{\lambda_2}{\lambda_1+\lambda_2}\cdot\frac{\lambda_2}{\mu_2}\right)} \quad (23)$$

$$K_{g7} = K_0 = \frac{(\lambda_1 + \lambda_2)\cdot\mu\cdot\mu_1\cdot\mu_2}{(\lambda_1 + \lambda_2)\cdot\mu\cdot\mu_1\cdot\mu_2 + \lambda(\lambda_1 + \lambda_2)\cdot\mu_1\cdot\mu_2 + \lambda_1^2\cdot\mu\cdot\mu_2 + \lambda_2^2\cdot\mu\cdot\mu_1} \quad (24)$$

Using Expression (23), it is possible to determine the values that are of interest and related to the probabilities of a wind farm system remaining in its various operating states. If the assumption is made that the modeling of the operation process consists of determining the probabilities of a wind farm system remaining in individual states $\{S_0, S_1, S_{01}$ and $S_{10}\}$, then the following values need to be determined:

- The likelihood function of the system being in state S_0;
- The likelihood function of the system being in state S_1;
- The likelihood function of the system being in state S_{01};
- The likelihood function of the system being in state S_{10}.

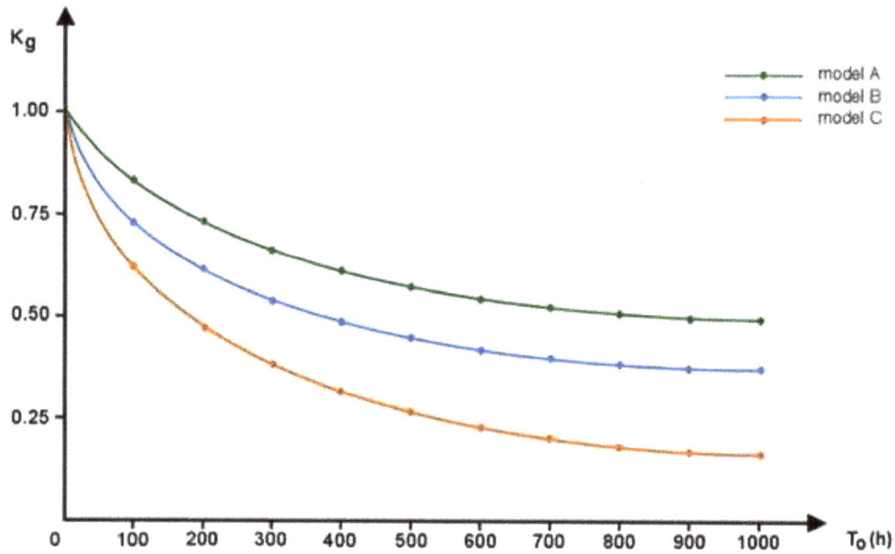

Figure 4. Graph of $K_g(t)$ readiness function for the models tested (A, B and C) of the wind farm equipment in the operation process, where T_o is the time of the operation process.

3. Results of the Research into Reliability Testing in the Operation of the Wind Farm Equipment and Discussion

A simulation test method may be used to perform reliability testing on wind farm equipment that has an intelligent WPPES system to support its successful operation. The genuine values of the operational (use) and nonoperational (damage) periods that characterize the usage of a set of wind farm devices equipped with an intelligent WPPES system in the actual operation process are required for this distribution of the facility's reliability tests. The input data gathered in this manner serves as the foundation for a simulation study for the operational process models generated.

The source of the aforementioned input data for the research may consist of the values obtained from the observation of the actual process of the facility's operation tested, and it may include a properly prepared and implemented simulation experiment. Simulation tests have been carried out for ongoing expenditure on the operation of the object tested over the entire lifetime (simulation test time of the object—T_o).

As the assessment of the actual operation process is too time-consuming, three models of the operation process are designed for the simulation tests:

1. Model A: this is the operation process of wind farm facilities equipped with intelligent decision support systems concerning safety in its use (a model including the operating system with WPPES);
2. Model B: the process of operating wind turbine equipment without any intelligent safety decision support systems (a model with the operating system including WPPES);
3. Model C: a simple process of operating wind farm equipment.

The same input data is used in the reliability simulation testing of the wind farm equipment operation process. The input data adopted for the simulation is as follows:

- λ—damage intensity = 0.00022831 (1/h);
- μ—repair intensity = 0.5 (1/h);
- λ_1—type I inspection intensity = 0.0004566 (1/h);,
- μ_1—type I operational maintenance intensity = 0.45 (1/h);
- λ_2—intensity of type II inspections = 0.0001141 (1/h);
- μ_2—type II operational maintenance intensity = 0.25 (1/h).

The reliability of the facility after renovation in the maintenance regime was tested for three models of the facility's operation processes (models: A, B and C). The results obtained are presented graphically in Figures 4 and 5.

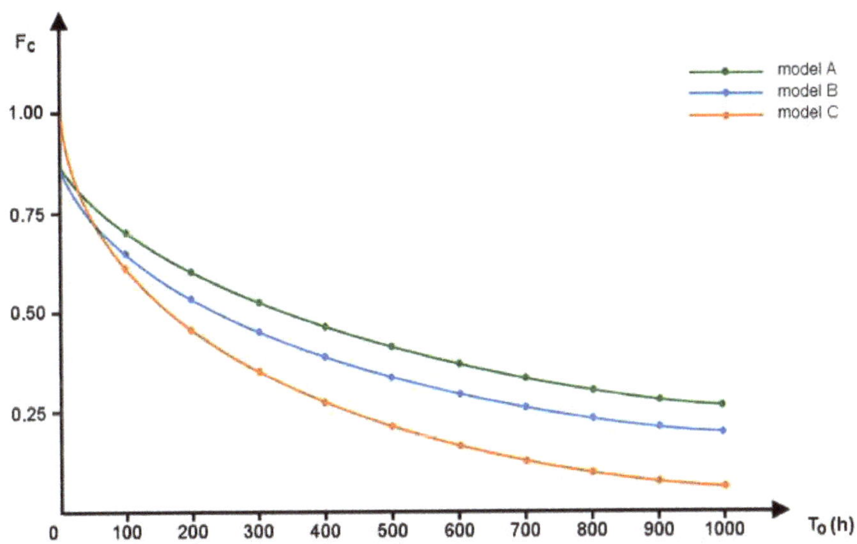

Figure 5. Graph of the use quality function Fc for the models tested (A, B and C) of the wind farm devices in the operation process, where: To—is the time of the operational process.

In the simulation studies, the indicators characterizing the reliability in the process of the operation of wind farm equipment using the WPPES expert system and the artificial neural network (model A) and FW without the WPPES system (models B and C) were examined. In the simulation analysis, the following values of the reliability assessment of the object were investigated: availability factor K_g and unavailability factor F_{ch}.

The test results obtained are presented in the charts (Figures 4 and 5).

The reliability values (K_g) and (F_{ch}) are determined and presented in Figures 4 and 5, and they are expressed as a function of the test time (T_o), of the operation process.

The following requests can be obtained from the analysis of the reliability values (K_g) and (F_{ch}) presented in Figure 4:

1. The results of the size (K_g) test for the model (A and B) are descending, according to the corresponding theoretical characteristics [5,10]. The highest value (K_g) for

model A is K_{gA} = 0.7508. However, for other models, this value is: K_{gB} = 0.4931 and K_{gC} = 0.2332, respectively.

2. The value (K_g) tested in the simulation process has also some practicalities: it speaks of the efficiency of the organization of the system related to the application and maintenance of wind farm equipment. From the definition of this value, expressed as Formula (14) (K_g) [1,2,5], it follows that the availability factor (the basic reliability value tested) determines the ability (readiness) to use the wind farm equipment by its time of use (T).
   That is why the system of the organization of the operation process presented in a model (A) is the most effective, because the value K_{gA} = 0.7508 is the utmost value. For this value, we may conclude that, during a given lifetime (t = T) for the time (t = 3/4 T) of the wind farm equipment, it is technically fit and capable of performing its functions (i.e., the features required).
   On the other hand, for the three models tested (A, B and C), the least effective organization of the wind farm equipment operation process is found in the model (C), while, for model C, the values tested are K_{gC} = 0.2332. Based on this figure (K_{gC}), we may conclude that during the use (t = 1 − 0.2332 T), the wind farm equipment is unable to perform its tasks (this is its ineffective application—standstill S_{10}) (Figure 5).

3. The quality function investigated in the operational process (F_C) of the wind farm devices in models (A, B and C), which is presented graphically in Figures 5 and 6, shall be expressed in the simulation time (T_o) adopted. From the definition of the size (F_C), Dependence (2) shows that this value is largely influenced by the expenditure incurred in the organization of the use (operation process) of the wind farm equipment possessing the WPPES expert system in a given model. To test the dependability of the wind farm equipment, the value of fixed inputs for the models (A, B and C) expressed as a linear function was adopted. The analysis (Figure 6) shows that the highest quality of the operational process (F_C) is for a model (A), where: F_{CA} = 0.2239, while, for models B and C, these values are, respectively, lower, and they amount to F_{CB} = 0.1979; F_{CC} = 0.0959. It may be assumed that the organization of the use and maintenance system of the wind farm equipment in a model (A) is more expensive. It is a more high-tech process of operating wind farm equipment.
4. The reliability test of the wind farm equipment in the operation process (F_{ch}) exhibits practicalities (Figure 6). This informs the researcher about the ineffectiveness of the organization (structure) of the proposed process of operating the wind farm equipment under examination. This value represents information on the average share of downtime (damage and maintenance) over the entire period of testing (use and maintenance) of the wind farm equipment.
   It follows from the definition of this quantity tested that the nonavailability coefficient (F_{ch}) of the wind farm equipment, which is presented in the form of Dependence (16) [36], determines the inability (unfitness, nonavailability, etc.) to use the wind farm equipment.
   Therefore, we may conclude that the higher the value (F_{ch}) is, the greater the reliability of the devices of the wind farm tested is. Based on an analysis of the research results presented graphically, it was established that the most ineffective system for the organization of the wind farm equipment operation process is presented in Model C. The value (F_{ch}) tested accepts the value (F_{chC} = 0.9750), and it is the maximum value for the three models tested. For model A, this value is (F_{chA} = 0.7508), and it is the smallest value. We can also state that, in the operation phase (t = T) tested in time (t = 3/4 T) in model C, the set of the wind farm equipment was technically incapable of carrying out its tasks (the features required).
   On the other hand, for the other models (A and B) tested, the smallest ineffectiveness of the organization of the service system in model A for this value was F_{chA} = 0.233. Based on one of these values (F_{chA}), it can be concluded that the reliability of the wind farm equipment in model A was high during operation (t = 1 − 0.2332 T). Therefore,

the FW can carry out its tasks (this is an ineffective use: a state of standstill). This state was probably caused by damage to the wind farm equipment in model B or its being in the maintenance system (regeneration time).

5. The study of the quantities that describe the reliability (durability) of the wind farm equipment in the operational process is an important (basic) tool in taking actions to optimize the efficiency of the operation process of the wind farm equipment.
6. The research conducted confirms that one of the criteria (methods) for assessing the quality of the model of the wind farm equipment operation process may be a reliability assessment.

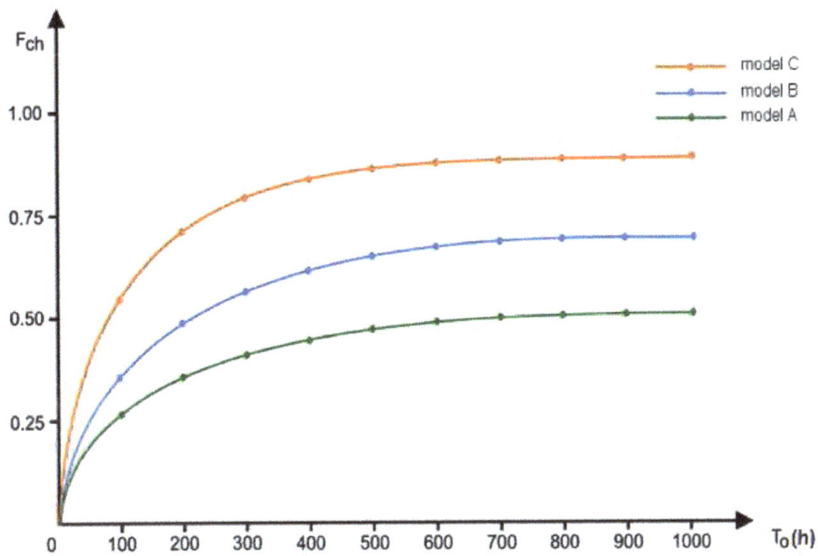

Figure 6. Graph of changes in the facility's nonavailability coefficient F_{ch} in the facility's operation process for the operating system models tested (models: A, B and C), where: T_o is the time of operation process.

The results of the research conducted serve to confirm the correctness of the concept adopted related to the organization of the system for the use and maintenance of wind farm equipment possessing the WPPES system.

4. Discussion

The article presents a method for testing and evaluating the reliability properties of wind farm devices in the operation process. Two new quantities were additionally proposed for the reliability tests of the object tested: the facility operation process quality function (Fc(t)) and the operation process absorption (non-readiness) function ($F_{ch}(t)$). These two quantities, which feature the reliability of the working process of wind farm equipment, are original values that have not yet been published in the literature.

The study, analysis and monitoring of the real operational process to identify the input data for the testing formed the foundation for the organization of the wind farm device simulation tests. The literature on the subject of dependability testing of complicated technical devices is constantly evolving. Three models of the wind farm operating process are presented in this study for simulation investigations. The first model performed was model A, which describes the operation process of wind farm installations equipped with intelligent systems that support decision-making regarding the safety of their use. The next model developed was model B, which describes the operation process of wind farm equipment without any intelligent WPPES safety decision support systems. Another model was model C, which describes a simple (theoretical) operational process of wind farm equipment. To organize simulation tests, this article covers and describes the following issues:

1. Development of the models (A, B and C) of the operation process of the wind farm equipment.
2. Elaboration of a test plan.
3. Arrangement: input data for the tests describing the functional properties of the object tested, for example, the object's use time T—the object's working lifetime, T_{NA}—time to remove the unfitness of the object and T_{NP}—scheduled repair time.
4. Understanding and describing the operation process (use and maintenance) of the wind farm equipment.

The problem of testing the reliability properties of the wind farm equipment during its operation presented in the article is a difficult task. The difficulty of this is due to the acquisition of input data for the research. Numerical data describing the operation of the wind farm equipment can be obtained over a long period of time. The observation time (a measurement of the downtime and service life) in the actual operation process of the wind farm equipment is practically expressed in years during its "life". The only rational (sensible) attitude towards this type of study is a simulation test. This type of research requires the knowledge and a description of the actual operation process of the wind farm equipment and the determination of reliable input data for the research. At the core of each research, there is a good testing plan (how and in what manner to test) of the wind farm equipment. The grounds for the simulation study of the operation process of wind farm devices are formed by the models developed of the operation process organization.

The following reliability values were examined in the simulation tests to establish the credibility of the wind farm device in the operational process:

- The value of coefficient (K_g) for the models of (A, B and C). In the model, the value tested is $K_{gA} = 0.7508$. On the other hand, for the remaining models, this value is: $K_{gB} = 0.4931$ and ($K_{gC} = 0.2332$, respectively. Thus, the most effective system is the organization of the use of wind farm equipment in the operation process in model A;
- The operation process quality (F_C) function of wind farm devices in the models (A, B and C): the highest quality of the use of the wind farm equipment in the operation process (F_C) is for model A, where: $F_{CA} = 0.2239$. However, for models B and C, these values are respectively lower, and they amount to $F_{CB} = 0.1979$; $F_{CC} = 0.0959$.
- Unreliability of wind farm equipment in the operation process (F_{ch}): the value (F_{ch}) examined for model C is $F_{chC} = 0.9750$, and this is the maximum value. For model A, this value is $F_{chA} = 0.7508$, and this is the smallest value. We can also conclude that, in the lifetime investigated (t = T) for time (t = 3/4 T) in model C, the farm's equipment set is capable of work.

What forms the basis for the reliability of any simulation study results is the study plan developed, the input data and the quality of the models developed of a given process. The first problem is fundamental. Reliable input data can be determined only based on the study of appropriate time quantities occurring in the real study of the operation process of wind farm equipment. In the simulation study, the time data from the study of the technical documentation analysis of the operation process of the equipment of one wind farm located in the north of Poland was adopted. Models of the operation process of wind farm equipment are theoretical models presented in the literature-based Kolmogorov–Chapman equations. The models are developed properly and correctly. A good software tool: MaTlabon, was used in the simulation studies. On this basis, it can be assumed that the results obtained of the simulation studies are correct and reliable.

5. Conclusions

This paper presents the problem of simulation studies on the reliability of wind farm equipment during its operation using input data from WF equipment. The parameters of these studies describe the actual operation of the wind farm equipment. We found that reliability studies of the real process of any technical object can be carried out over a long period of time. Observation, measurement time: periods of unfitness and fitfulness in the real process of operation of the wind farm equipment are practically expressed in years during its "life". Therefore, the current acquisition of data on the reliability of the studied object can be obtained through simulation studies reflecting the actual operation of the WF equipment. This type of research requires learning and descripting the real process of the wind farm equipment operation and determining reliable input data for the research by the investigators. The second task that must be released in this type of research is to develop models of the exploitation process of the technical object under study. In this paper, for the purpose of simulation studies, three models (A, B and C) of the exploitation process of wind farm equipment were developed. The expected result of the simulation study of the exploitation process of the wind farm equipment was the reliability, which determined the extent to which reliability indicators influence new solutions for equipping wind farms with new intelligent decision support systems for their safe operation. This article is the only one in the literature on this topic that fully represents the expected reliability results.

Author Contributions: Conceptualization, resources, methodology, software and validation, A.O. and J.P.; formal analysis, investigation and data curation, M.S. and M.W.; writing—original draft preparation, writing—review and editing and visualization, S.D. and R.D. and supervision, project administration and funding acquisition, A.H. All authors have read and agreed to the published version of the manuscript.

Funding: This research was funded by the Faculty of Electronics and Informatics, Technical University of Koszalin, Sniadeckich Str. 2, 75-620 Koszalin, Poland.

Institutional Review Board Statement: Not applicable.

Informed Consent Statement: Not applicable.

Data Availability Statement: The data presented in this article are available at the request of the corresponding author.

Conflicts of Interest: The authors declare no conflict of interest.

Symbols and Acronyms

$X(e_{i,j})$	diagnostic signal in the jth element of the ith set
$X(w)(e_{i,j})$	model signal for $X(e_{i,j})$ signal
F_C max	max. value of the function of the use of the object
$W(\varepsilon\ (e_{i,j}))$	value of state assessment logics for jth element within ith
{2, 1, 0}	module (from the set of the accepted three-value logic of the state assessment)
$K_g(t)$ or K_g	the average value of availability function or factor K_g
F_c	the quality function of the object's operation process
F_{ch}	function of the object operation process
λ	damage intensity
T_o	the time of operation process
μ	repair intensity
λ_1	intensity of type I inspections
μ_1	type I operational maintenance intensity
λ_2	intensity of type II inspections
μ_2	type II operational maintenance intensity
P_0	probability of the system being in state S_0
P_1	probability of the system being in state S_1
P_{01}	probability of the system being in state S_{01}
P_{10}	probability of the system being in state S_{10}
WPPES	Wind Power Plant Expert System

References

1. Buchannan, B.; Shortliffe, E. *Rule Based Expert Systems*; Addison—Wesley Publishing Company: London, UK; Amsterdam, The Netherlands; Don Mills, ON, Canada; Sydney, Australia, 1985; p. 387.
2. Duer, S. Modelling of the operation process of repairable technical objects with the use of information from an artificial neural network. *Expert Syst. Appl.* **2011**, *38*, 5867–5878. [CrossRef]
3. Duer, S. Artificial neural network in the control process of object's states basis for organization of a servicing system of a technical objects. *Neural Comput. Appl.* **2011**, *21*, 153–160. [CrossRef]
4. Duer, S. Examination of the reliability of a technical object after its regeneration in a maintenance system with an artificial neural network. *Neural Comput. Appl.* **2012**, *21*, 523–534. [CrossRef]
5. Duer, S.; Bernatowicz, D.; Wrzesień, P.; Duer, R. The Diagnostic System with an Artificial Neural Network for Identifying States in Multi-Valued Logic of a Device Wind Power. In *Communications in Computer and Information Science*; Springer: Berlin/Heidelberg, Germany, 2018; Volume 928, pp. 442–454.
6. Duer, S. Assessment of the Operation Process of Wind Power Plant's Equipment with the Use of an Artificial Neural Network. *Energies* **2020**, *13*, 2437. [CrossRef]
7. Dempster, A.P. Upper and lower probabilities induced by a multi-valued mapping. *Ann. Math. Stat.* **1967**, *38*, 325–339. [CrossRef]
8. Dhillon, B.S. *Applied Reliability and Quality, Fundamentals, Methods, and Procedures*; Springer: London, UK, 2006; p. 186.
9. Hayer-Roth, F.; Waterman, D.; Lenat, D. *Building Expert Systems*; Addison—Wesley Publishing Company: Boston, MA, USA, 1983; p. 321.
10. Hojjat, A.; Shih, L.H. *Machine Learning, Neural Networks, Genetic Algorithms, and Fuzzy Systems*; John Wiley & Sons, Inc.: Hoboken, NJ, USA, 1995; p. 398.
11. Kacalak, W.; Majewski, M. New Intelligent Interactive Automated Systems for Design of Machine Elements and Assem-blies. In Lecture Notes in Computer Science 7666, Part IV, Proceedings of the 19th International Conference on Neural Information Processing—ICONIP 2012, Doha, Qatar, 12–15 November 2012; Springer: Berlin/Heidelberg, Germany, 2012; pp. 115–122.
12. Kunjumuhammed, L.P.; Pal, B.C.; Oates, C.; Dyke, K.J. Electrical Oscillations in Wind Farm Systems: Analysis and Insight Based on Detailed Modeling. *IEEE Trans. Sustain. Energy* **2015**, *7*, 51–62. [CrossRef]
13. Lipinski, D.; Majewski, M. System for Monitoring and Optimization of Micro- and Nano-Machining Processes Using Intelligent Voice and Visual Communication. In Lecture Notes in Computer Science, Proceedings of the 14th International Conference on Intelligent Data Engineering and Automated Learning—IDEAL 2013, Anhui, China, 20–23 October 2013; Springer: Berlin/Heidelberg, Germany; Volume 8206, pp. 16–23.
14. Mathirajan, M.; Chandru, V.; Sivakumar, A.I. Heuristic algorithms for scheduling heat-treatment furnaces of steel casting industries. *Sadhana* **2007**, *32*, 111–119. [CrossRef]
15. Majewski, M.; Kacalak, W. Building Innovative Speech Interfaces Using Patterns and Antipatterns of Commands for Controlling Loader Cranes. In Proceedings of the CSCI 2016, Las Vegas, NV, USA, 15–17 December 2016; IEEE Computer Society; IEEE Xplore Digital Library: Las Vegas, NV, USA, 2016; pp. 525–530.
16. Nakagawa, T. *Maintenance Theory of Reliability*; Springer: London, UK, 2005.
17. Ito, K.; Nakagawa, T. Optimal inspection policies for a storage system with degradation at periodic tests. *Math. Comput. Model.* **2000**, *31*, 191–195. [CrossRef]
18. Palkova, Z.; Okenka, I. *Programovanie*; Slovak University of Agriculture: Nitra, Slovakia, 2007.
19. Pogaku, N.; Prodanovic, M.; Green, T.C. Modeling, Analysis and Testing of Autonomous Operation of an Inverter-Based Microgrid. *IEEE Trans. Power Electron.* **2007**, *22*, 613–625. [CrossRef]
20. Pokoradi, L. Logical Tree of Mathematical Modeling. *Theory Appl. Math. Comput. Sci.* **2015**, *5*, 20–28.
21. Shahanaghi, K.; Babaei, H.; Bakhsha, A. A Chance Constrained Model for a Two Units Series Critical System Suffering from Continuous Deterioration. *Int. J. Ind. Eng. Prod. Res.* **2009**, *20*, 69–75.
22. Siergiejczyk, M.; Paś, J.; Rosiński, A. Issue of reliability–operation evaluation of electronic transport systems used in the railway environment with consideration of electromagnetic interference. *IET Intell. Transp. Syst.* **2016**, *10*, 587–593. [CrossRef]
23. Siergiejczyk, M.; Rosiński, A. Analysis of Power Supply Maintenance in Transport Telematics System. *Solid State Phenom.* **2014**, *210*, 14–19. [CrossRef]
24. Siergiejczyk, M.; Stawowy, M. Modelling of uncertainty for continuity quality of Power supply. In Risk, Reliability, and Safety: Innovating Theory and Practice: Modelling of Uncertainty for Continuity Quality of Power Supply, Proceedings of the ESREL 2016, Glasgow, Scotland, 25–29 September 2016; Walls, L., Revie, M., Bedford, T., Eds.; CRC Press/Balkema: London, UK, 2017; pp. 667–671.
25. Dyduch, J.; Paś, J.; Rosiński, A. *The Basic of the Operation of Transport Electronic Systems*; Publishing House of Radom University of Technology: Radom, Poland, 2011.
26. Epstein, B.; Weissman, I. *Mathematical Models for Systems Reliability*; CRC Press/Taylor & Francis Group: Boca Raton, FL, USA, 2008.
27. Stawowy, M.; Olchowik, W.; Rosiński, A.; Dąbrowski, T. The Analysis and Modelling of the Quality of Information Acquired from Weather Station Sensors. *Remote Sens.* **2021**, *13*, 693. [CrossRef]
28. Sun, J. Impedance-Based Stability Criterion for Grid-Connected Inverters. *IEEE Trans. Power Electron.* **2011**, *26*, 3075–3078. [CrossRef]

29. Wiliams, J.M.; Zipser, D. A learning Algorithm for Continually Running Fully Recurrent Neural Networks. *Neural Comput.* **1989**, *1*, 270–280. [CrossRef]
30. Zajkowski, K. The method of solution of equations with coefficients that contain measurement errors, using artificial neural network. *Neural Comput. Appl.* **2014**, *24*, 431–439. [CrossRef] [PubMed]
31. Duer, S. Investigation of the operation process of a repairable technical object in an expert servicing system with an arti-ficial neural network. *Neural Comput. Appl.* **2010**, *19*, 767–774. [CrossRef]
32. Duer, S.; Zajkowski, K.; Harničárová, M.; Charun, H.; Bernatowicz, D. Examination of Multivalent Diagnoses Developed by a Diagnostic Program with an Artificial Neural Network for Devices in the Electric Hybrid Power Supply System "House on Water". *Energies* **2021**, *14*, 2153. [CrossRef]
33. Bernatowicz, D.; Duer, S.; Wrzesień, P. Expert System Supporting the Diagnosis of the Wind Farm Equ. In *Communica-tions in Computer and Information Science;* Springer: Poznan, Poland, 2018; Volume 928, pp. 432–441.
34. Duer, S.; Zajkowski, K.; Harničárová, M.; Charun, H.; Bernatowicz, D. Neural Networks in the Diagnostics Process of Low-Power Solar Plant Devices. *Energies* **2021**, *14*, 2719. [CrossRef]
35. Bedkowski, L.; Dabrowski, T. *Basic of the Maintenance Theory*; WAT: Warsaw, Poland, 2006; p. 187.
36. Badrzadeh, B.; Gupta, M.; Singh, N.; Petersson, A.; Max, L.; Høgdahl, M. Power system harmonic analysis in wind pow-er plants—Part I: Study methodology and techniques. In Proceedings of the 2012 IEEE Industry Applications Society Annual Meeting, Las Vegas, NV, USA, 7–11 October 2012; pp. 1–11.

energies

Article

Dynamic Error Correction Method in Tachometric Anemometers for Measurements of Wind Energy

Paweł Ligęza

Strata Mechanics Research Institute, Polish Academy of Sciences, Reymonta 27, 30-059 Krakow, Poland; ligeza@imgpan.pl

Abstract: Measurements of air flow velocity are essential at every stage of the design, construction and operation of wind turbines. One of the basic measurement tools in this area is the tachometric anemometer, which is based on the simple physical phenomenon of the air kinetic energy exchange with a rotating measuring element. Tachometric anemometers have favorable operational features and good static metrological parameters. However, in the case of fast-changing flows, the measurement is burdened with a significant dynamic error, and the measured average value of the velocity is overestimated. This article presents the concept and results of pilot studies of a dynamic error correction method of tachometric anemometers. The correction consists of the precise measurement of the rotor's rotational velocity and determination of the measured air velocity, taking into account the dynamics of the instrument. The developed method can be used in tachometric anemometers intended for laboratory, technical and industrial measurements in time-varying flows. One of the important application areas is the measurement of wind energy.

Keywords: wind energy; air velocity; tachometric anemometer; dynamic error; measurement correction

Citation: Ligęza, P. Dynamic Error Correction Method in Tachometric Anemometers for Measurements of Wind Energy. *Energies* **2022**, *15*, 4132. https://doi.org/10.3390/en15114132

Academic Editors: João Carlos de Campos Henriques and Maria Grazia De Giorgi

Received: 11 April 2022
Accepted: 2 June 2022
Published: 4 June 2022

1. Introduction

Measurements of the air flow velocity are of key importance at each stage of the implementation of wind energy projects [1]. Anemometric measurements are used in laboratory tests in the process of modeling and optimization of devices [2]. When selecting the location of turbines and wind farms, comprehensive measurements of the characteristics of local air flows are necessary [3]. This applies to daily and annual characteristics, flow stability, turbulence intensity, spatial profiles, wind shear and other parameters [4]. Anemometric measurements are also important for making long-term forecasts of the energy efficiency of wind farms [5]. At the stage of turbine operation, it is necessary to constantly monitor wind parameters, both in terms of optimizing energy efficiency and predicting threats and failures, and in the process of maintenance and repairs [6,7]. Measurements of the flow velocity are also extremely important in the field of ecology and research on the impact of wind turbines on the environment, animals and humans [8].

Instruments used in wind energy for anemometric measurements are based on the use of various physical phenomena. Depending on the metrological requirements, mechanical anemometers, pressure tubes, thermal anemometers, ultrasonic anemometers, and optical anemometers are used [9]. The basic type of anemometer used in technical and industrial measurements is the mechanical tachometric anemometer.

The name "tachometric anemometers" was introduced by the author to refer to the whole group of mechanical anemometers having a rotating measuring element. These anemometers have various structures, but the common element is the mechanical rotor rotating under the influence of air flow. An overview of the design of tachometric anemometers and their applications in the mining industry, which is interesting from the point of view of development, is presented in [10]. The flow velocity is measured indirectly by measuring the rotational speed of the rotor. Depending on the purpose, the rotors of tachometric anemometers have different shapes, and use different methods of reading the rotor rotational speed and different systems of signal processing, linearization, filtration and averaging.

In addition, regarding wind energy, tachometric anemometers are one of the basic types of measuring instruments. The work [11] analyzes the operation of the cup-anemometer group over long time intervals for the purpose of determining the operating conditions of wind energy. The signals of anemometers placed at different heights in one of the meteorological stations in Saudi Arabia were examined. The analysis of the signals from cup anemometers allowed for the formulation of conclusions concerning the influence of the height of the anemometers' placement on the tower on the measurement results. The influence of the shading of the measuring tower was investigated. Annual mean, median, and standard deviation values of the wind velocity were found to be substantially the same for sensors placed at the same location at all altitudes. Based on the analysis of the measurement results, specific, optimal types of wind turbines were selected for the tested location. The article [11] proves the thesis that metrological issues and the properties of measuring instruments constitute an important element of wind energy optimization.

Tachometric anemometers are based on the exchange of kinetic energy between the tested gas stream and a rotating measuring element. The most commonly used measuring elements are a vane, cup, turbine or propeller rotor. Other constructions of the measuring element can also be encountered. The angular velocity of rotation of this element is a measure of the flow velocity. Tachometric anemometers are often used due to their simple operating principle and good static parameters. The main advantages of these instruments are: simple and durable structure; linear dependence of rotor velocity on flow rate; low sensitivity to other flow parameters, such as temperature, pressure and humidity; their averaging properties; and low price of the devices.

One of the disadvantages of tachometric anemometers is that they do not allow the determination of the wind direction. Additional mechanical sensors or other measurement methods, such as ultrasonic anemometers [12], must then be used. However, this type of anemometer is currently less commonly used than mechanical tachometric anemometers in wind energy measurement. One of the relevant factors here is certainly the relatively high price of ultrasonic anemometers, which, however, due to studies such as those presented in [12], may soon change.

However, the main disadvantage of mechanical tachometric anemometers seems to be their poor dynamic properties. The dynamic properties of mechanical tachometric anemometers are an important issue analyzed in many scientific publications. The work [13] is dedicated to the issues of construction and dynamics of cup anemometers. The anemometer with three cups, which is the author's original design, is tested here. The construction of the equation of motion for the rotor of the vane anemometer and the study of the effect of overestimating the average velocity measurement is included in the article [14], and is the basis for considerations in subsequent publications. In article [15], the analysis covers the transient states of the turbine anemometer response to the test excitations. The article [16] is devoted to measurements in fast-changing and pulse flows with the use of an anemometer with a turbine. An overview of the applications of anemometers in research on turbulent flow issues is presented in [17]. In many articles analyzing dynamic processes, selected detailed metrological issues are considered. The work [18] presents the key issues related to the measurement of atmospheric turbulent flows. The use of mechanical anemometers for the measurement of time-varying flows in liquids is discussed in [19]. The work [20], by comparison, is devoted to the issue of measurements in pulsed fluid flows with the use of miniature turbines. The influence of air density on the calibration of the anemometer is the subject of the work [21], while in the article [22], using CFD techniques, the influence of the medium viscosity on the work of the anemometer is analyzed. In the work [23], an attempt was made to apply artificial neural networks to model static and dynamic phenomena for the cup anemometer. The analysis of the anemometer response to acoustic fluctuations is discussed in the article [24].

The relatively large inertia of the measuring element implies a significant dynamic error in the measurements of fast-changing flows. This error overestimates the measured average velocity and limits the applicability of these anemometers in velocity fluctuation studies [25]. A method of improving the dynamic properties of anemometers through the use of active control in a feedback loop was previously developed [26]. However, this method requires a significant expansion of the hardware structure of the instrument. The author of this article also developed a method of software correction of the output signal from the thermal anemometer [27]. However, this article presents a new method of dynamic error correction for tachometric anemometers. It should be emphasized that, in this article, we deal with only one element concerning the assessment and improvement of the parameters of tachometric anemometers. The aim of the article is to present the assumptions of the new method, the possibility of its technical application, and the preliminary, pilot results of the research verifying the method. Comprehensive tests of the prototype anemometer in a wide range of changes in flow parameters and other metrological parameters were not carried out. The aim of the article is to present the theoretical foundations of the method and an exemplary application to the constructors of apparatus and measuring instruments in order to enable its application in the developed instruments intended for measuring specific issues. Each device, depending on the shape and parameters of the rotor and the signal processing algorithm used, will have specific metrological parameters. The determination of these parameters and the associated measurement uncertainty are an important element of a correct measurement process [28].

2. The Method of Dynamic Measurement Error Correction of Tachometric Anemometers

In order to analyze the proposed method, let us consider a simplified tachometric anemometer model using the rotor kinetic energy balance [29]. The change in kinetic energy E of the rotor over time $\mathrm{d}t$ is related to the transfer of kinetic energy by the gas mass part $\mathrm{d}m$ acting on the rotor in this time:

$$\frac{\mathrm{d}E}{\mathrm{d}t} = \frac{\mathrm{d}m(V^2 - v^2)}{2\mathrm{d}t} \tag{1}$$

Velocity V of the gas reaching the rotor is the flow velocity to be measured, and v is the velocity of the gas mass part $\mathrm{d}m$ leaving the rotor in the time $\mathrm{d}t$. In the presented model, it is assumed that the velocity v is a linear function of the angular velocity of the rotor ω, with the proportionality coefficient a depending on the rotor shape:

$$v = a\omega. \tag{2}$$

This assumption [29] results from the fact that the gas mass $\mathrm{d}m$ acting on the rotor is linearly controlled by the rotor angular velocity, similarly to the case of a fan. The mass of gas that exchanges kinetic energy with the rotor at time $\mathrm{d}t$ is given by the equation:

$$\mathrm{d}m = \rho s v \mathrm{d}t, \tag{3}$$

where:

ρ—gas density,

s—cross-section of the rotor in the plane perpendicular to the velocity V. Substituting (2) and (3) into (1), and denoting the moment of inertia of the rotor as I, we obtain the relationship:

$$I\frac{\mathrm{d}\omega}{\mathrm{d}t} = \frac{\rho s a}{2}(V^2 - a^2\omega^2). \tag{4}$$

This equation is a simplified dynamic mathematical model of the rotor of a tachometric anemometer. It shows the dependence of the rotor angular velocity ω on the measured flow velocity V. The parameters of Equation (6) I, s, a, are the quantities characterizing the mechanical properties of the rotor, while the density ρ characterizes the tested gas.

A typical measurement with a standard tachometric anemometer consists in placing the anemometer rotor in the tested flow and measuring the mean angular velocity of the rotor during a given period. The value v_1 of the measured flow velocity is determined in accordance with Model (4), assuming that the measurement is performed in the steady state:

$$v_1 = a\omega. \tag{5}$$

The parameter a of Equation (5) is most often determined during instrument calibration in a wind tunnel. In many cases, the static relationship (5) additionally takes into account the intercept, higher powers of angular velocity or complex algebraic expressions [30]. This allows the static error to be minimized.

Such a measurement in fast-changing flows is always burdened with a dynamic error. This error is caused by not taking into account the dynamic states described by Equation (4). This phenomenon overestimates the measured average velocity and limits the possibility of using tachometric anemometers to measure velocity fluctuations. The proposed method of correction of the dynamic measurement error of tachometric anemometers consists in determining the v_2 value of the measured flow momentary velocity from the dynamic Model (4) in accordance with the relationship:

$$v_2 = \sqrt{a^2\omega^2 + b^2\frac{\mathrm{d}\omega}{\mathrm{d}t}}, \tag{6}$$

where:

$$b = \sqrt{\frac{2I}{\rho s a}}. \tag{7}$$

In the process of calibrating the anemometer, both the static parameter a and the dynamic process parameter b should be determined. Note that both parameters have a length dimension. The momentary angular velocity of the rotor in the presented method must be measured with high accuracy and time resolution. This makes it possible to determine the time derivative of the angular velocity of the rotor with sufficient approximation for the required accuracy of the measurement.

Standard, commercial tachometric anemometers measure the angular velocity or rotor rotation frequency with a low time resolution [31]. The maximum number of counted pulses during one revolution is usually several dozen, and is typically the number of pulses related to the number of rotor segments, or the number of marks of the coding disc placed on its axis [32]. Therefore, it is not possible to apply the proposed method directly to standard instruments. However, magnetic, precision angular position sensors are now easily accessible and cheap. Their advantages are a resolution of 12 to 14 bits per revolution and a very short conversion time. The use of such sensors enables the implementation of the proposed method in tachometric anemometers.

3. Experimental Verification of the Method

The block diagram of the test stand for the verification of the method of correcting the dynamic error of the anemometer is shown in Figure 1.

The tested tachometric anemometer was a vane anemometer type. The rotor radius is about 45 mm, with a shape comprising eight blades with an angle of inclination of 45°. On the rotor axis there is a magnet co-operating with the angular position sensor with a resolution of 14 bits, model AS5147 from AMS. The sensor through the SPI interface and the Controller block sends information about the rotor position to the Computer. A sampling time of 5 ms was used. The rotor parameters of Equation (6) were experimentally determined in a calibration wind tunnel stand [33] with results of a = 34 mm, b = 281 mm. In order to carry out dynamic tests, the anemometer is placed in a test stream of air. The test flow is generated by a wind tunnel equipped with a fan, a honeycomb stream straightening system and an outlet nozzle. It is not an easy task to create a reference flow for the dynamic testing of anemometers. It is practically impossible to build an experimental stand that produces a variable in time flow with preset shape, amplitude and frequency parameters. Various methods are used to evaluate the dynamic parameters of anemometers. One of these is to develop a good anemometer model and evaluate the dynamic properties on the basis of simulation tests. However, this is an approximate method. Various modulation methods are used to generate the actual test stream. Depending on the required parameters, it may provide appropriate control of the tunnel drive motor, swinging shutters, vibrating elements (membranes) on the walls of the tunnel, or elements causing vortexes of the stream. In our tests, the flow is modulated by a rotating shutter. The shutter is made in the form of a metal mesh semicircle. For half the period of spinning the shutter, the air stream is free, and for the second half of the period, it is dumped with a mesh. The mesh lowers the stream velocity by about 15%. In this way, a periodic, quasi-rectangular air flow is generated. The air velocity and shutter frequency are set from the Computer via the Controller block. Additionally, the system is equipped with a hot-wire Constant-Temperature Anemometer [34], which is used to control the test flow. This anemometer allows for observation of the shape of the generated flow, and measurement of its amplitude and frequency parameters. A hot-wire sensor made of thin tungsten wire having a diameter of 5 μm and resistance at a reference room temperature of 6 Ω was used here. The sensor cooperates with the Constant-Temperature Anemometer with a filament overheating ratio of 1.8. This allows a 20 kHz bandwidth in the tested velocity range to be obtained. This is at least one thousand times more than the frequencies of interest for tachometric anemometers. The Constant-Temperature Anemometer was calibrated in a reference tunnel in the range of 0.5 to 20 m/s. The uncertainty of calibration is 2%. The signal from the anemometer is converted into a digital signal, which was linearized and filtered in the computer.

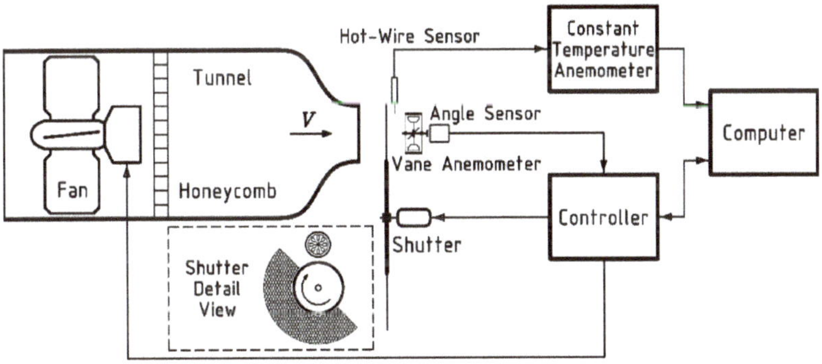

Figure 1. Diagram of the test stand.

Initial tests were carried out for two average velocity values: 2.5 and 12.5 m/s. Four shutter frequencies were used: 0.2, 1, 5 and 25 Hz. The change in the angular position of the anemometer rotor over time was recorded by the Controller and Computer. On this basis, the angular velocity of the rotor was calculated by dividing the difference in angular positions by the sampling time. Its derivative was calculated from two successive values of the rotor angular velocity. In order to minimize signal noise, the algorithm of inertial averaging of the derivative with a time constant of 10 ms was used. The measurements carried out in this way, together with the determined parameters, made it possible to calculate the measured flow velocity. The value of v_1 of the measured velocity was calculated for the standard method from the dependence in Equation (5). Using the proposed method of dynamic correction, the value of v_2 from Equation (6) was also calculated. The measurement results for the average flow velocity of 2.5 m/s are shown in Figure 2.

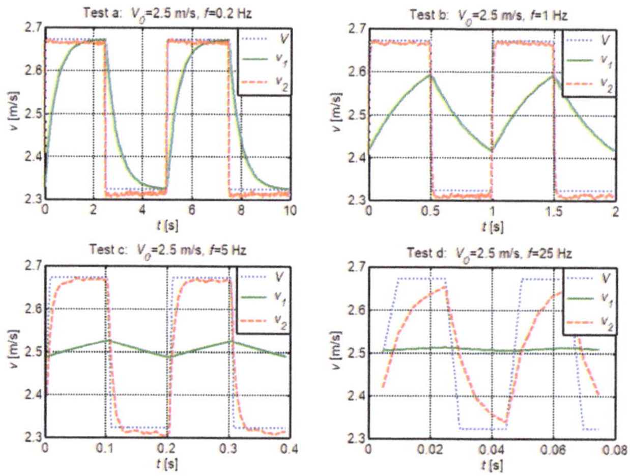

Figure 2. Measurement results for the average velocity of 2.5 m/s and the frequency of flow modulation: (**a**) 0.2 Hz, (**b**): 1 Hz, (**c**): 5 Hz, (**d**): 25 Hz.

For the standard measurement method, waveform v_1, a significant dynamic measurement error in relation to the shape of excitation V is visible. The error is caused by a relatively large rotor inertia. At 1 Hz, the waveform still roughly reaches the excitation value. For the frequency of 25 Hz, the variability of the waveform v_1 is practically invisible. In contrast, we observe an overestimation in the average velocity. Based on the response time to the rectangular excitation, the frequency bandwidth of the anemometer was estimated to be 0.07 Hz in this case.

It can be seen from the v_2 waveform that the application of the described correction method allows for a significant reduction in the dynamic error. The waveform for 1 Hz is practically in line with the excitation, and for 5 Hz the dynamic error is still small. The estimated frequency bandwidth here is around 2.75 Hz. It is therefore 40 times wider than that of the standard method. A small amount of noise is visible on waveforms calculated using the presented correction method. It is inseparably related to the principle of calculating the derivative from the discrete signal.

Similar tests were carried out for an average velocity of 12.5 m/s. The results are shown in Figure 3.

The dynamic error of the waveform v_1 for the standard measurement method is smaller than that previously found due to the rotor inertia. At 0.2 Hz, the waveform roughly reaches the excitation value. For the frequency of 25 Hz, the variability of the waveform is still well observable. Figure 3d clearly shows the overestimation of the average velocity above 12.5 m/s. Due to the small amplitude of the excitation changes, amounting to about 15%, the increase in the average velocity is not very large. Based on the response time to the rectangular excitation, the frequency bandwidth of the anemometer in this case was estimated to be 0.36 Hz.

The course of the measured velocity v_2 shows also a significant reduction in the dynamic error. However, the waveforms are similar to those obtained for the lower velocity. For frequencies up to 5 Hz, the waveform is practically compliant with the excitation. A slight overshoot is observed here. The estimated frequency bandwidth is approximately 3.1 Hz. Thus, it is almost 10 times greater than for the standard method. In the conducted tests, the change in the corrected anemometer bandwidth with the increase in f the flow velocity is rather low. This may be related to the applied method of inertial averaging of the derivative of the angular velocity of the rotor. Here, we deal with a necessary compromise between the bandwidth and the noise in the measured signal.

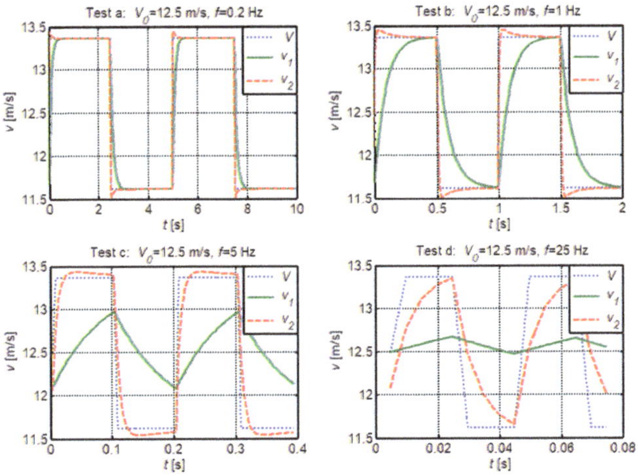

Figure 3. Measurement results for the average velocity of 12.5 m/s and the frequency of flow modulation: (**a**): 0.2 Hz, (**b**): 1 Hz, (**c**): 5 Hz, (**d**): 25 Hz.

This article presents the results of verification research for two velocity values and four flow pulsation frequencies. These results are representative of the whole conducted research and allow for the formulation of general, presented conclusions regarding the confirmation of the presented method. The values of velocities and frequencies were selected in the range that reflects the actual conditions of using tachometric anemometers in typical applications. It is necessary to emphasize once again the fundamental difficulties associated with testing the dynamic properties of anemometers. The author is aware that the presented experimental studies can be extended, both in terms of the range of parameters and the methodology of research and apparatus. However, according to the author, the results of the research are fully acceptable as confirmation of the effectiveness of the proposed method. Further research, if it is going to be conducted, should concern the application of the method in devices designed for, for example, serial production.

4. Conclusions

The article presents the concept of a method of correction of the dynamic measurement error of tachometric anemometers. The results of the pilot studies carried out for the vane anemometer confirmed the adopted assumptions and the effectiveness of the method.

Measurements of velocities fluctuations using tachometric anemometers are disturbed by a dynamic error, and the measured average velocity is overestimated. However, it is possible to correct these errors by processing the output signal. The correction algorithm based on the dynamic model of the anemometer enables the extension of the frequency bandwidth and reduction in the dynamic error. The algorithm requires measuring the angular velocity of the rotor with high accuracy and time resolution, and calculating the derivative of the angular velocity. The calculated derivative of the angular velocity of the rotor is used for the measurement correction. The quality of the estimation of this quantity affects the effectiveness of the method. It is important to optimize the derivative calculation algorithm to minimize the dynamic error, but also the noise of the measurement signal.

It is not a trivial issue to create a reference flow for testing the dynamic properties of anemometers. For the purposes of the research presented in this article, a rotating diaphragm modulating the air stream was used. To control the test flow, a Constant-Temperature Anemometer with a hot-wire sensor having a diameter of 5 μm was used. The tests showed a significant reduction in the dynamic error and the extension of the frequency bandwidth of the vane anemometer by about 40 times for the velocity of 2.5 m/s, and 10 times for the velocity of 12.5 m/s. The quantitative parameters of the improvement in the dynamic properties of the measurement are mainly related to the quality of the estimation of the derivative of the angular velocity of the rotor. The presented method and its possible variants can improve the dynamic parameters of tachometric anemometers used in the measurement of fast-changing flows. This requires the use of a high-performance angular velocity sensor and signal processing system in these anemometers.

The article concerns only one selected problem in the field of tachometric anemometers. It proposes an original method of minimizing the dynamic error in the measurement of flows that vary in time. The issues related to mechanical anemometers vary and cannot be discussed in detail in the case of the pilot verification tests presented here. Issues such as total measurement uncertainty, directional characteristics of the anemometer, influence of the medium parameters on the measurement, invasiveness of the sensor, and many others are considered in other articles on measurements with mechanical anemometers. The article is a preliminary conceptual work and presents pilot research confirming the assumptions of the method. The implementation of the presented method in actual, commercial measuring instruments intended for laboratory, technical and industrial tests [35] requires taking into account individual metrological requirements and parameters of the anemometer. These works should take into account the parameters of the tested flow, the required metrological properties, the type, structure and parameters of the rotor, and other individual features.

Funding: This research received no external funding.

Institutional Review Board Statement: Not applicable.

Informed Consent Statement: Not applicable.

Data Availability Statement: Data available via the author's email.

Conflicts of Interest: The author declare no conflict of interest.

References

1. Ligęza, P. Basic, Advanced, and Sophisticated Approaches to the Current and Forecast Challenges of Wind Energy. *Energies* **2021**, *14*, 8147. [CrossRef]
2. Scappatici, L.; Bartolini, N.; Castellani, F.; Astolfi, D.; Garinei, A.; Pennicchi, M. Optimizing the design of horizontal-axis small wind turbines: From the laboratory to market. *J. Wind. Eng. Ind. Aerodyn.* **2016**, *154*, 58–68. [CrossRef]
3. Azlan, F.; Kurnia, J.C.; Tan, B.T.; Ismadi, M.-Z. Review on optimization methods of wind farm array under three classical wind condition problems. *Renew. Sustain. Energy Rev.* **2021**, *135*, 110047. [CrossRef]
4. Kim, D.-Y.; Kim, Y.-H.; Kim, B.-S. Changes in wind turbine power characteristics and annual energy production due to atmospheric stability, turbulence intensity, and wind shear. *Energy* **2021**, *214*, 119051. [CrossRef]
5. Hanifi, S.; Liu, X.; Lin, Z.; Lotfian, S. A Critical Review of Wind Power Forecasting Methods—Past, Present and Future. *Energies* **2020**, *13*, 3764. [CrossRef]
6. Guerrero-Villar, F.; Dorado-Vicente, R.; Fike, M.; Torres-Jiménez, E. Influence of ambient conditions on wind speed measurement: Impact on the annual energy production assessment. *Energy Convers. Manag.* **2019**, *195*, 1111–1123. [CrossRef]
7. Du, Y.; Zhou, S.; Jing, X.; Peng, Y.; Wu, H.; Kwok, N. Damage detection techniques for wind turbine blades: A review. *Mech. Syst. Signal Process.* **2020**, *141*, 106445. [CrossRef]
8. Leung, D.Y.C.; Yang, Y. Wind energy development and its environmental impact: A review. *Renew. Sustain. Energy Rev.* **2012**, *16*, 1031–1039. [CrossRef]
9. Suomi, I.; Vihma, T. Wind Gust Measurement Techniques—From Traditional Anemometry to New Possibilities. *Sensors* **2018**, *18*, 1300. [CrossRef] [PubMed]
10. Unwin, I.D.; Phil, M. The Measurement of Air Flow in British Coal Mines: A Historical Review; B. SC. D.I.S. 2007. Available online: http://www.cmhrc.co.uk/cms/document/air_flow_2007.pdf (accessed on 1 June 2022).
11. Baseer, M.A.; Meyer, J.P.; Rehman, S.; Mahbub, A.M.; Al-Hadhrami, L.M.; Lashin, A. Performance evaluation of cup-anemometers and wind speed characteristics analysis. *Renew. Energy* **2016**, *86*, 733–744. [CrossRef]
12. Pugi, L.; Allotta, B.; Boni, E.; Guidi, F.; Montagni, M. Massai TommasoIntegrated Design and Testing of an Anemometer for Autonomous Sail Drones. *J. Dyn. Sys. Meas. Control* **2018**, *140*, 055001.
13. Patterson, J. The cup anemometer. *Trans. Roy. Soc. Can.* **1926**, *20*, 1–54.
14. Ower, E. On the response of a vane anemometer to an air-stream of pulsating speed. *Phil. Mag. Ser.* **1937**, *157*, 992–1005. [CrossRef]
15. Grey, J. Transient response of the turbine flowmeter. *Jet Propuls.* **1956**, *26*, 98–100.
16. Lee, W.F.Z.; Kirik, M.J.; Bonner, J.A. Gas Turbine Flowmeter Measurement of Pulsating Flow. *J. Eng. Power* **1975**, *97*, 531–539. [CrossRef]

17. Wyngaard, J.C. Cup, Propeller, Vane, and Sonic Anemometers in Turbulence Research. *Annu. Rev. Fluid Mech.* **1981**, *13*, 399–423. [CrossRef]
18. Kristensen, L. Cup anemometer behavior in turbulent environments. *J. Atmos. Ocean. Technol.* **1998**, *15*, 5–17. [CrossRef]
19. Lee, B.; Cheesewright, R.; Clark, C. The dynamic response of small turbine flowmeters in liquid flows. *Flow Meas. Instrum.* **2004**, *15*, 239–248. [CrossRef]
20. Lee, B. Dynamic Response of Small Turbine Flowmeters in Pulsating Liquid Flows. In *Submitted for the Degree of Doctor of Philosophy, Department of Systems Engineering*; Brunel University: Uxbridge, UK, 2002.
21. Pindado, S.; Sanz, A.; Wery, A. Deviation of Cup and Propeller Anemometer Calibration Results with Air Density. *Energies* **2012**, *5*, 683–701. [CrossRef]
22. Tegtmeier, C. CFD Analysis of Viscosity Effects on Turbine Flow Meter Performance and Calibration. Master's Thesis, University of Tennessee, Knoxville, TN, USA, 2015. Available online: http://trace.tennessee.edu/utk_gradthes/3415 (accessed on 1 June 2022).
23. Bégin-Drolet, A.; Lemay, J.; Ruel, J. Time domain modeling of cup anemometers using artificial neural networks. *Flow Meas. Instrum.* **2013**, *33*, 10–27. [CrossRef]
24. Stoltenkamp, P.W.; Bergervoet, J.T.M.; Willems, J.F.H.; van Uittert, F.M.R.; Hirschberg, A. Response of turbine flow meters to acoustic perturbations. *J. Sound Vib.* **2008**, *315*, 258–278. [CrossRef]
25. Jamróz, P. Effect of the Continuous Traverse Trajectory and Dynamic Error of the Vane Anemometer on the Accuracy of Average Velocity Measurements at the Cross-Section of the Mine Heading–Model-Based Testing. *Arch. Min. Sci.* **2014**, *59*, 1051–1060. [CrossRef]
26. Ligęza, P.; Jamróz, P.; Ostrogórski, P. Methods for dynamic behavior improvement of tachometric and thermal anemometers by active control. *Measurement* **2020**, *166*, 108147. [CrossRef]
27. Ligęza, P. Method of testing fast-changing and pulsating flows by means of a hot-wire anemometer with simultaneous measurement of voltage and current of the sensor. *Measurement* **2022**, *187*, 110291. [CrossRef]
28. Pindado, S.; Cubas, J.; Sorribes-Palmer, F. The Cup Anemometer, a Fundamental Meteorological Instrument for the Wind Energy Industry. Research at the IDR/UPM Institute. *Sensors* **2014**, *14*, 21418–21452. [CrossRef]
29. Ligęza, P. An alternative mathematical model of vane anemometers based on the balance of power. *Flow Meas. Instrum.* **2017**, *54*, 210–215. [CrossRef]
30. Guerrero-Villar, F.; Dorado-Vicente, R.; Medina-Sánchez, G.; Torres-Jiménez, E. Alternative Calibration of Cup Anemometers: A Way to Reduce the Uncertainty of Wind Power Density Estimation. *Sensors* **2019**, *19*, 2029. [CrossRef]
31. Ramos-Cenzano, A.; Ogueta-Gutierrez, M.; Pindado, S. On the signature of cup anemometers' opto-electronic output signal: Extraction based on Fourier analysis. *Measurement* **2019**, *145*, 495–499. [CrossRef]
32. Güçyetmez, M.; Keser, S.; Hayber, S.E. Wind speed measurement with a low-cost polymer optical fiber anemometer based on Fresnel reflection. *Sens. Actuators A Phys.* **2022**, *339*, 113509. [CrossRef]
33. Jamróz, P. Interaction between the Standard and the Measurement Instrument during the Flow Velocity Sensor Calibration Process. *Processes* **2021**, *9*, 1792. [CrossRef]
34. Ligęza, P. Constant-Temperature Anemometer Bandwidth Shape Determination for Energy Spectrum Study of Turbulent Flows. *Energies* **2021**, *14*, 4495. [CrossRef]
35. Skotniczny, P.; Ostrogórski, P. Three-dimensional air velocity distributions in the vicinity of a mine heading's sidewall. *Arch. Min. Sci.* **2018**, *63*, 335–352.

energies

MDPI

Communication

A Hot-Wire Anemometer with Automatically Adjusted Dynamic Properties for Wind Energy Spectrum Analysis

Paweł Ligęza * and Paweł Jamróz

Strata Mechanics Research Institute, Polish Academy of Sciences, Reymonta 27, 30-059 Krakow, Poland; jamroz@imgpan.pl
* Correspondence: ligeza@imgpan.pl

Abstract: Hot-wire anemometric measurements are often related to the determination of flow parameters in a high frequency range. Such knowledge is particularly important when analyzing the flow phenomena in the vicinity of wind turbines. The reliability of obtained results is determined by the knowledge of the properties of the system used for measurements. It concerns both the static and dynamic characteristics of individual measurement channels. In studies of hot-wire anemometric systems, a problem related to the unstable transmission bands of such systems and their high dependence on measurement conditions and the system configuration itself has been commonly indicated. This paper presents the results of an investigation of a new type of hot-wire anemometer, allowing for automatic adjustment of its dynamic characteristics under real working conditions. The presented system is dedicated to the analysis of the wind energy spectrum in experimental laboratory tests on reduced-scale models and to specialized in situ measurements.

Keywords: flow measurement; wind energy spectrum; hot-wire anemometer; transmission band adjustment

Citation: Ligeza, P.; Jamróz, P. A Hot-Wire Anemometer with Automatically Adjusted Dynamic Properties for Wind Energy Spectrum Analysis. *Energies* **2022**, *15*, 4618. https://doi.org/10.3390/en15134618

Academic Editor: Davide Astolfi

Received: 16 May 2022
Accepted: 22 June 2022
Published: 24 June 2022

1. Introduction

Among the many topics related to wind energy [1], there is a special group associated with precise measurements of flow parameters. Such topics include farm site assessment and wake interference among the wind turbines. During such research, various types of precise anemometers are used. One type is a hot-wire anemometric system dedicated to the measurement velocity signals in a wide frequency range. A reliable knowledge of the wind is significant for the wind farm location planning and implementation. Wind data acquired by the meteorological stations provide knowledge about the distribution of frequency, velocity, and energy in different directions. It is used for making preliminary estimates about the wind energy potential of a site [2].

A lot of research related to the wind turbines focuses on the model tests in a wind tunnels. In such research the scale effect must be taken into account. It requires the use of various types of measuring equipment, allowing for measurements in a wide spectrum of frequencies [3–6]. Many experimental investigations focus on the effects of various flow parameters, including the intensity of turbulence [7–9]. In presented cases the basic measurement instrument capable to measure velocity with frequencies of tens of kilohertz is a hot-wire anemometer.

Based on recorded velocity changes, it is possible to determine the characteristic parameters of analyzed flows, including the turbulence. This phenomenon occurs in flows of any kind, and it is characterized by its randomness in both time and space domains. With respect to its time randomness, the assessment of turbulence requires the measurement systems that enable measurements with the best possible dynamic parameters. Therefore, the design of hot-wire anemometric measurement systems comprises a complex process in which an appropriate selection of electronic components providing the fastest possible response of the system to changes occurring in the flow plays a crucial role.

While measuring the flow velocity with the use of such hot-wire anemometric techniques, the problems related to changes of characteristics of a hot-wire anemometric system, which are conditional based on the environment in which these measurements are performed, can often be encountered. Research on sensitivity of hot-wire anemometric measurement methods with respect to environmental conditions, as well as the configuration of measurement systems, concerns their static characteristics, allowing for precise determination of flow velocity. Comprehensive analysis of this problem resulted in the development of measurement systems allowing for compensation of the effect of variable environmental measurement conditions, such as the influence of variable flow temperature on characteristics of the measuring hot-wire anemometer [10]. Rarely, the discussion deals with the problem of the dynamic sensitivity of the measurement of hot-wire anemometric systems, which is closely related to the issue of the dynamic response of the hot-wire anemometric system. The presented system is dedicated to the analysis of the wind energy spectrum in experimental laboratory tests on reduced-scale models and to specialized in situ measurements.

2. Dynamic Properties of the Hot-Wire Anemometric Systems

The first elaborations concerning the transmission band of hot-wire anemometers contained detailed analyses of measurement probes as well as experimental investigations of the transmission band of hot-wire anemometric sensors. An analysis of the influence of flow velocity on the transmission band of hot-wire anemometric sensors operating in the constant-current operational mode can be found in [11]. The author describes the results of an experimental investigation of a transmission band of a platinum wire at different velocities of the flow, in which the wire was placed. A detailed analysis of the wire's dynamics under various flow conditions suggests its huge changes depending on flow velocity. Studies of dynamics of thermoresistant wire elements are still under way and obtained results allow for development of complex systems for correction of dynamic errors acquired during temperature measurements [12–15].

The dynamic response of the hot-wire anemometric systems doesn't depend only on the sensor properties themselves. It is also conditional on the electronic configuration of the measurement module itself, among other things. Therefore, the analysis of the transmission band of such systems must be examined with special consideration given to components deciding the rate of the system's reaction to fluctuating measurement signals, such as the properties of the electronic components from which these systems are made.

In the literature concerning this subject, a number of studies raising the problem of the dynamic response of hot-wire anemometric systems are available. In the study [16], a description of the constant-temperature hot-wire anemometric system in the bridge configuration by means of the third-order linear differential equation allowing for determination of its dynamic response was proposed. Testing of dynamics of such systems in real-life conditions was hindered, owing to the lack of possibility to generate precisely specified, controlled, harmonically variable fields of flow velocity. In order to verify the dynamic response of such systems, a generator of sine or rectangular voltage wave included in one branch of the bridge circuit of a hot-wire anemometer, simulating the step-change excitation of the velocity of flowing medium, was used. This method was used to determine the optimal dynamic response of the hot-wire anemometric system, which would ensure the system's maximum bandwidth—Figure 1.

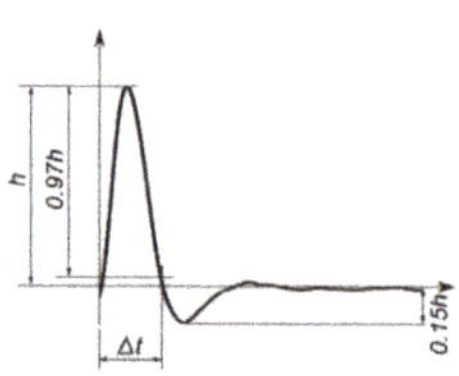

Figure 1. Optimal dynamic response of the hot-wire anemometer.

In such case, the system's bandwidth frequency f_c can be determined on the basis of measured time Δt, which is the time interval in which the system's response reaches 3% of its maximum value (1).

$$f_c = \frac{1}{1.3\Delta t} \tag{1}$$

Such a method for investigating the dynamics of hot-wire anemometers was adopted as the standard. Later on, a series of studies dealing with detailed analysis of dynamic properties of hot-wire anemometric measurement systems was performed. Studies [17–20] presented the results of model testing of hot-wire anemometric measurement systems, which consider additional factors influencing the response of hot-wire anemometers such as: the frequency characteristics of the amplifier used in real-life conditions, the impedance of electric circuit, the offset, type of measuring wire and the properties of the medium in which the measurements are performed.

3. The Hot-Wire Anemometer with Bandwidth Frequency Optimization

Analysis of the literature dealing with hot-wire anemometry points out the necessity to adjust the hot-wire anemometric measurement systems in a way that would ensure their optimal transmission bands depending on the configuration of the system itself and the conditions under which the measurements are performed. In case of the conventional hot-wire anemometers, the bandwidth optimization is adjusted only for specific probes, cable types, length, velocities, and specific laboratory conditions. The electrical anemometer parameters are selected to ensure high bandwidth for different types of probes. However, such adjustment does not provide the guarantee that this bandwidth will be optimal. Depending on different measurement conditions, bandwidth changes, and in some cases, it can be much lower than we expected and assumed in experiment. The development of hot-wire anemometric systems characterized by wide transmission bands, constant over possibly the widest range of fluctuations of factors influencing their dynamic response, has become an important problem. Simulation testing of the model of such a system was presented in [21]. It showed that it is possible to develop a hot-wire anemometric measurement system that would allow for implementation of algorithms enabling the automatic adjustment of the hot-wire anemometer, ensuring that the optimal transmission band would be obtained. For this purpose, a simplified model of such a measurement system was used. Figure 2 presents conception of one measurement channel of the anemometer.

Adjustment of dynamic properties of such a system are facilitated by including two potentiometers between the upper point and the ground point of the resistance bridge. In line with the assumptions, adjusting the P_1 potentiometer gives the possibility to obtain a stable operation of the system with a preliminarily optimized transmission band. Adjusting the P_2 potentiometer allows for a final optimization of the transmission band of the whole system to be obtained.

Based on this concept, an eight-channel measurement system enabling the measurement of flow velocity in a constant-temperature operational mode or the measurement of temperature in a constant-current mode was developed. The measurement system was presented in Figure 3 and described in Table 1. It consists of a communication module and eight measurement channels (two channels applied in Figure 3) that can work as a constant temperature hot-wire anemometer or a constant current thermometer. The measurement system includes digitally controlled potentiometers responsible for setting the overheat ratio of the wire sensor, amplification of measured signal, and the control of the frequency and amplitude of rectangular voltage wave generated in one of the branches of the hot-wire anemometer's bridge circuit. The processes of control and measurement data acquisition are realized with the use of a multifunction measurement card and dedicated software package developed in the LabVIEW environment.

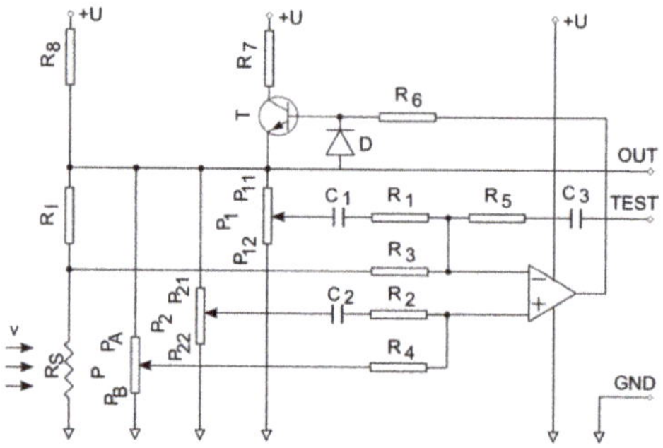

Figure 2. The simplified model of hot-wire anemometer featuring the optimization of bandwidth frequency.

Figure 3. Hot-wire anemometer with automatically adjusted dynamic properties.

Table 1. Hot-wire anemometer—specification.

Feature	Title 2
Communication: • Sensor resistance measurement • Overheat ratio setting • Bandwidth adjustment • CTA/CCT Channel control	Digital, I^{2}C
Signal Output	analog 1 . . . 5 V DC

The properties of the system were tested in a laboratory wind tunnel, dedicated to the analysis of the characteristics of hot-wire anemometric probes, which allows us to generate flows with a velocity ranging from 0.1 to 17 m/s. Single-wire hot-wire anemometric sensors including wires 3 and 5 μm in diameter were used in the study. The scheme of the measurement setup is presented in Figure 4.

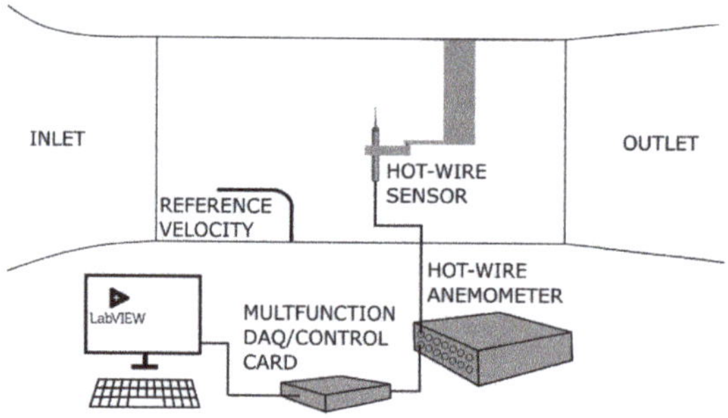

Figure 4. Measurement setup diagram.

Probes of this type are often used in flow experiments related to the study of turbulence. They are characterized by a large bandwidth up to 100 kHz depending on the diameter and material used. These probes have a unique properties such as resistance, which can cause changes in dynamic characteristics of the measurement channel, resulting in a non-optimal frequency measurement range.

During measurement experiments, the dynamic response of the system was generated by the rectangular voltage wave generator. The output signal was acquired under various measurement conditions. The frequency of the testing signal was adjusted such that it allowed the steady state of the system's response to the leading slope of the rectangular voltage wave to be obtained before the trailing slope occurrence. Such frequency was equal to 1 kHz. The output signal was registered as the voltage across the measuring wire U_{RS}, with the sampling frequency of 25 MHz.

4. Manual Adjustment of Dynamic Properties of the Measurement System

In the first stage of the study, the possibility to adjust the developed measurement system was verified. The system included a probe with a 3 µm-diameter measuring wire and was placed in a flow of which the velocity was 5 m/s. Subsequently, the setting of adjustment potentiometers P_1 and P_2 was changed. The adjustment started from starting values of resistance P_{11} (39.06 Ω) and P_{21} (9.82 kΩ), which ensured the stable yet slow response of the system to variable excitation—Figure 5a. During the initial stage of the experiment, the value of resistance P_{11} was increased until the moment, in which the response of the system was characterized by the occurrence of consecutive overshoots. One of the objectives of the regulation was to find such a value of P_{11} that oscillations in the dynamic response of the system began to appear, not leading to unstable operation of the system. Such oscillations should still be possible to reduce by adjusting the potentiometer P_2. Once such a boundary (Figure 5b) is found, it is possible to fine-tune the response of the measurement system.

In the case under investigation, such a response was recorded for P_{11} equal to 3.01 kΩ. Such a preoptimized measuring system was the basis for the final optimization using the P_2 potentiometer—Figure 5c. By decreasing the value of resistance P_{21}, one may obtain an optimal dynamic response of the measurement system characteristic of a fifteen-percent overshoot.

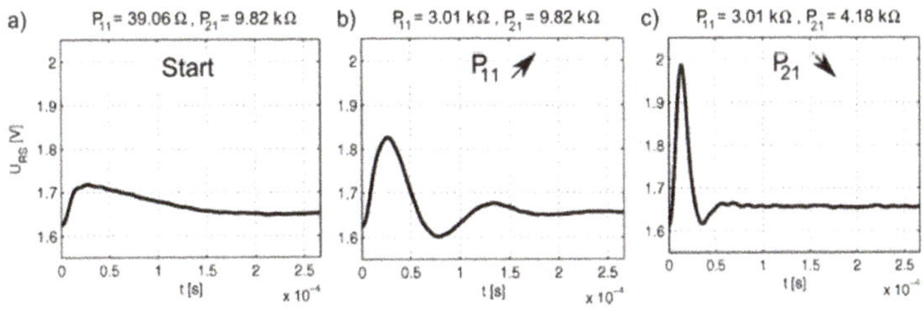

Figure 5. Successive stages of manual adjustment of dynamic properties of the hot-wire anemometer: (**a**) stable, slow response of the system; (**b**) oscillations in the dynamic response of the system; (**c**) optimal dynamic response of the measurement system.

During adjustment performed in the above-described way, it is crucial to determine the optimal value of resistance P_{11}, as the response of the measurement system with the characteristics ensuring the maximum bandwidth frequency can then be obtained by adjusting the setting of potentiometer P_2. Excessively increasing the value of resistance P_{11} or decreasing the value of resistance P_{21} results in the unstable operation of the hot-wire anemometer, shown by means of unfading oscillations of the voltage response of the system to testing excitation and may lead to damage of the measuring wire.

5. Automatic Adjustment Algorithm of Dynamic Properties of the Measurement System—Experimental Case

The performed experimental study confirmed the possibility of dynamic properties of analyzed measurement system to be adjusted. It allowed for the algorithm enabling the automatic adjustment of the measurement system to be applicated and successfully implemented. The aim of such automatic adjustment is to find optimal settings of potentiometers P_1 and P_2 to ensure that maximum bandwidth frequency of the measurement signal is obtained. The algorithm is divided into four stages, during the realization of which the character of the hot-wire anemometer's dynamic response changes from the initial stable but slow response to final optimal one, consistent with the one presented in Figure 1.

The detailed operation of the discussed adjustment algorithm can be explained with the use of the block diagram describing the individual steps of its implementation—Figure 6. In the initial phase, the P_1 potentiometer settings are increased with the assumed step Δp, until the dynamic response of the object with three maxima is obtained. If, in the response of the control system, the number of peaks is greater than three, it returns to the previous value of P_{11} and decreases the value of the step Δp. In the case where there are two maxima in the object response, the value of Δp is also decreased. The adjustment process is repeated until the obtained number of maxima in the dynamic response of the object is three. The fulfilment of this condition ends the control with the potentiometer P_1 and starts decreasing the value of the resistance P_{21} on the potentiometer P_2 with the set step Δp. For each of the successive values of the P_{21} resistance, the dynamic response of the system is recorded. The adjustment process lasts until the optimal dynamic response of the hot-wire measurement system is achieved with the assumed accuracy of the first overshoot. First, the number of maxima is checked, which in the ideal case should be one. Then, by appropriate selection of the control step value, the level of 15% of the first overshoot is reached. After achieving the assumed level of accuracy, the frequency response of the system is determined.

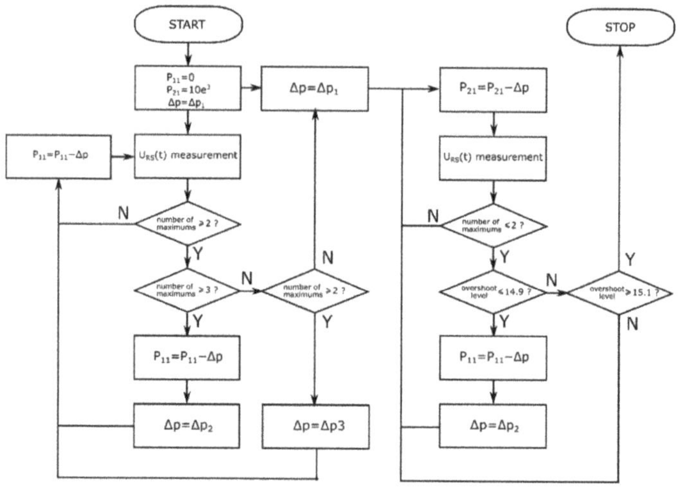

Figure 6. Adjustment algorithm—coarse adjustment.

This method of regulation ensures quick tuning of the system in terms of its dynamics. However, the manual configuration shows that it is possible to achieve a wider frequency response. This requires additional, fine system adjustment. The assumption of the final control is to obtain the maximum possible bandwidth for various configurations of the system and for various measurement conditions. The optimization algorithm (Figure 7) starts with the resistance values P_{11} and P_{21} obtained as a result of the preoptimization. In the first step, the resistance P_{11} is increased by the value Δp_1. Then, on the basis of the performed measurement of the dynamic response of the system, the overshoot value is checked. If it is lower than 14.9%, the value of Δp_2 is reduced by half, and the resistance P_{21} is increased by the new value of Δp_2. In the case when the obtained overshoot is greater than 15.1%, the resistance P_{21} is decreased by the value of Δp_2. Each time the resistance value changes, the dynamic response of the system is measured. The process of such regulation lasts until the response with overshoot in the range of 14.9–15.1% is obtained.

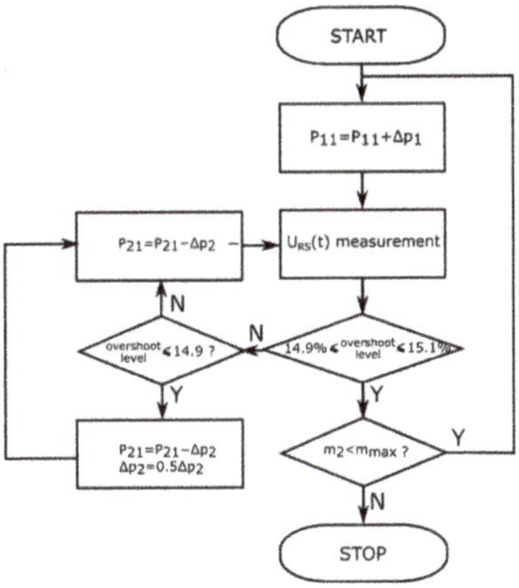

Figure 7. Adjustment algorithm—fine adjustment.

Obtaining the optimal frequency response is associated with obtaining an ideal dynamic response characteristic with one maximum. Appearance of the next maximum determines its overshoot, therefore the level of the next maximum in relation to the set value can be treated as a criterion of the final regulation. If this criterion is met, the value of resistance P_{11} is increased by Δp_1, and the variable Δp_2 is assigned an initial value. The optimization algorithm ends its operation when it is impossible to find for the next P_{11} value increased by the step Δp_1 the resistance value P_{21} ensuring the reduction of the second maximum level below the limit m_{max}.

6. Automatic Adjustment of Dynamic Properties of the Measurement System—Experimental Case

The presented algorithm was implemented in the measurement system and tested in various configurations (3.5 and 8 μm probes, various cables connecting the sensors with the anemometer, velocities, etc.). Figure 8 shows the next steps of the algorithm implementation. It contains a graph of changes in the resistance of the potentiometers during the automatic adjustment process. The example of the process of automatic adjustment shown here refers to study performed upon the measurement system connected to the sensor including the wire of 5 μm in diameter, placed in the air flow of velocity of 15 m/s.

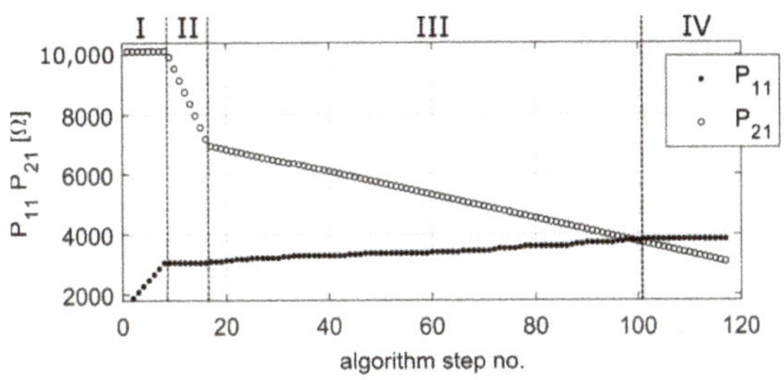

Figure 8. Successive stages of automatic adjustment of dynamic properties of the hot-wire anemometer.

In the first stage of the process of automatic adjustment, the value of resistance P_{11} is increased in eight iterations from 1.69 to 3.05 kΩ with the value of resistance P_{21} being simultaneously kept constant at the value of 10.13 kΩ. At the same time, the dynamic response of the measurement system is automatically analyzed. The second stage of the adjustment process starts once the system's response characterized by the occurrence of third maximum of assumed level is obtained (Figure 9a). The value of resistance P_{21} is gradually reduced until the recorded signal of the dynamic response is fifteen percent over-regulated (Figure 9b). In the example of the presented measurement, this takes place after the next eight iterations, following which the second stage of the process of preliminary optimization of the hot-wire anemometric measurement system is finished, and the resistances P_{11} and P_{21} are set to values 3.05 and 7.20 kΩ, respectively. The bandwidth frequency obtained in this way equals 20.15 kHz.

The third stage of adjustment realizes the algorithm of optimization of the hot-wire anemometer's dynamics, consisting of increasing the resistance P_{11} with the least possible step, the value of which is given by the resolution of settings of the adjustment potentiometers. For gradually increasing values of the resistance P_{11}, appropriate values of resistance P_{21} (decreasing) ensuring the model characteristic of the dynamic response of the system are sought. Individual consecutive settings allowing for the above-mentioned dynamic response to be obtained are recorded until the maximum value of bandwidth frequency is obtained. In the case of the example presented, such a frequency was equal to 44.92 kHz (Figure 9c).

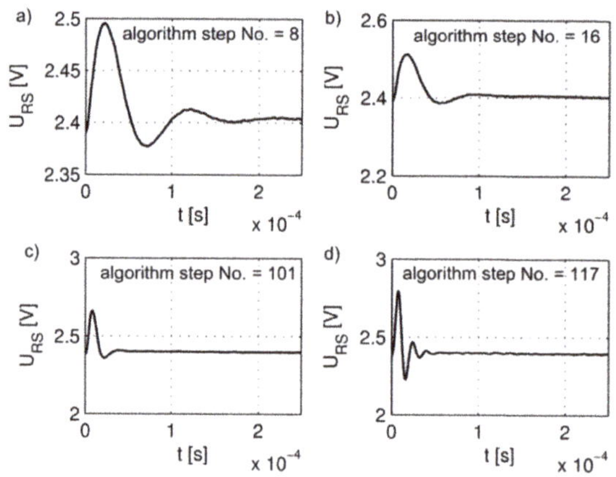

Figure 9. The dynamic response of the hot-wire anemometer in individual stages of the process of automatic adjustment: (**a**) first adjustment stage; (**b**) second adjustment stage; (**c**) third adjustment stage; (**d**) fourth adjustment stage.

Further adjustment causes additional over-regulation maxima to appear in the dynamic response of the hot-wire anemometric system, which as a consequence leads to an unstable operation of the system. In order to prevent the system from reaching this state, the number of consecutive maxima and amplitude of over-regulations are analyzed in the fourth stage of the process of automatic adjustment. If the parameters of the response of the measurement system, i.e., the number of recorded maxima and their amplitude level (Figure 9d), exceed beyond the preset boundary values, the optimization process is terminated and the values for which the optimal value of bandwidth frequency of the system was obtained in the third stage of the process are taken as final settings.

7. Application of the Hot-Wire Anemometer with Automatic Adjustment of Its Dynamic Properties

One of the main factors influencing the dynamic properties of hot-wire anemometric measurement systems is the mean value of flow velocity. It is related to the intensity of heat exchange between the measuring wire and flowing medium. Due to the optimization of hot-wire anemometric measurement systems to low velocities, the initially obtained response of the system, which is ideally adjusted for high velocities, becomes distorted. In the dynamic response of the system, additional overshoots begin to appear, which leads to a considerable confinement of the whole system's transmission band. Therefore, at the stage of planning of hot-wire anemometric measurement systems, their dynamic response is adjusted to high values of velocities. This allows the risk of damage to the measurement probe to be eliminated when performing measurement in high-velocity flows. It does not eliminate the problem of reduction of bandwidth frequency in measurements in which the mean value of velocity is lower than the one for which the system was optimized at the stage of planning. When performing the multi-point measurements, it is also necessary to use various measurement probes. As shown by the study [22], the type of measurement probe used for the measurement is of crucial importance with respect to dynamic characteristics of the measurement system, and the use of different types of measurement probes (i.e., different wires, properties of material of which the measuring wires are made) makes the development of a universal measurement system impossible.

The basic test of the developed system is its ability to maintain the optimal bandwidth in the case of a change in the mean velocity value and the use of different types of sensors. The measurement system, together with the algorithm for automatic adjustment of dynamic properties, was tested in a wide range of velocities, during which two types of measuring probes including the wires of 3 and 5 μm in diameter were used. Figure 10 presents the results of the performance of the above-described algorithms in real-life measurement experiments.

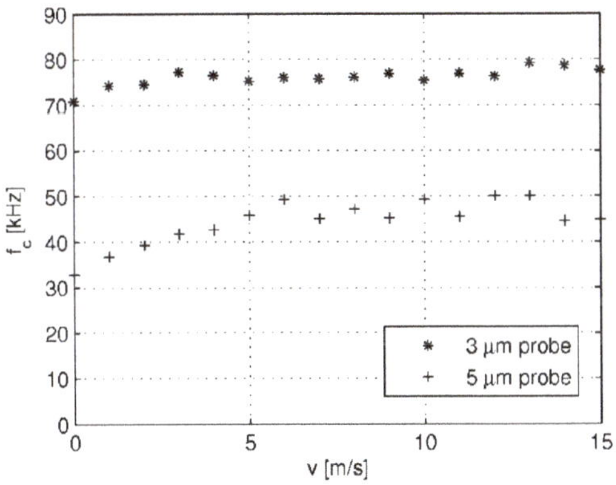

Figure 10. The bandwidth frequency in the function of flow velocity for measurement probes with 3 and 5 μm wires.

The presented results for bandwidth frequency of the measurement system in the function of flow velocity confirm the efficacy of the developed procedure of automatic adjustment. Bandwidth frequencies obtained for two types of commonly used measurement probes are characterized by their constancy in a wide range of flow velocities (above 3 m/s), reaching the level of 78 kHz and 46 kHz for 3 μm and 5 μm probes, respectively. In the case of lower velocities, the values of the bandwidth frequencies of the measurement system are insignificantly lower, reaching the minimal values in the case of optimization under the no-flow conditions. This effect may be explained by the fact that the characteristic of heat exchange between the sensor and its surroundings changes under these conditions. As the flow velocity decreases, the role of free convection in such a heat exchange increases, resulting in a change in thermal conditions around the sensor and thus hindering the possibility of compensation.

The use of the algorithm for the automatic optimization of the transmission band of the hot-wire anemometric measurement system thus allowed for optimal values of bandwidth frequency of the system to be obtained in a single channel, irrespective of installed measurement probe or adopted flow velocity.

8. Conclusions

In routinely performed hot-wire anemometric measurements, the measurement system's dynamics are influenced by factors related to properties of flowing medium in which such measurements are performed (the mean velocity value), the used type of measurement probe, or the length of used signal wires. Every change of measurement system configuration leads to perturbations in the ideal characteristics of the originally designed measurement system and changes its dynamic properties in an unspecified manner. Performing measurements using a measurement apparatus with unknown dynamic properties undermines the reliability of results obtained from measurements of high-frequency phenomena, including determination of the intensity of turbulence. One solution to this problem may exist in the presented measurement system, allowing for the effect of various factors interfering with dynamic characteristic of hot-wire anemometric measurement systems to be compensated for. As a result of performed investigations, the full ability of the system to automatically compensate for the effect imposed by various measurement-related factors on the bandwidth frequency of the system was confirmed.

Author Contributions: Conceptualization, P.L.; methodology, P.L. and P.J.; software, P.J.; validation, P.L. and P.J.; formal analysis, P.J.; investigation, P.L. and P.J.; data curation, P.J.; writing—original draft preparation, P.L. and P.J.; writing—review and editing, P.L. and P.J.; visualization, P.J.; supervision, P.L. All authors have read and agreed to the published version of the manuscript.

Funding: This research received no external funding.

Data Availability Statement: The data presented in this study are available on request from the corresponding author. The data are not publicly available due to company's policy.

Conflicts of Interest: The authors declare no conflict of interest.

References

1. Ligęza, P. Basic, Advanced, and Sophisticated Approaches to the Current and Forecast Challenges of Wind Energy. *Energies* **2021**, *14*, 8147. [CrossRef]
2. Sathyajith, M. *Wind Energy Fundamentals, Resource Analysis and Economics*; Springer: Berlin/Heidelberg, Germany, 2006; 236p.
3. Tian, W.; Ozbay, A.X.; Wang, D.; Hu, H. Experimental investigation on the wake interference among wind turbines sited in atmospheric boundary layer winds. *Wind. Energy* **2017**, *33*, 742–753. [CrossRef]
4. Kumer, V.; Reuder, J.; Eikill, R. Characterization of turbulence in wind turbine wakes under different stability conditions from static Doppler LiDAR measurements. *Remote Sens.* **2017**, *9*, 242. [CrossRef]
5. Schümann, H.; Pierella, F.; Sætran, L. Experimental Investigation of Wind Turbine Wakes in the Wind Tunnel Energy Procedia. *Energy Procedia* **2013**, *35*, 285–296. [CrossRef]
6. Wolff, S.; Brunner, S.; Fottner, L. The Use of Hot-Wire Anemometry to Investigate Unsteady Wake-Induced Boundary-Layer Development on a High-Lift LP Turbine Cascade. *J. Turbomach.* **2000**, *122*, 644–650. [CrossRef]
7. Al-Abadi, A.; Kim, Y.J.; Ertunç, Ö.; Delgado, A. Turbulence Impact on Wind Turbines: Experimental Investigations on a Wind Turbine Model. *J. Phys. Conf. Ser.* **2016**, *753*, 032046. [CrossRef]
8. Hansen, K.S.; Barthelmie, R.J.; Jensen, L.E.; Sommer, A. The impact of turbulence intensity and atmospheric stability on power deficits due to wind turbine wakes at horns rev wind farm. *Wind. Energy* **2012**, *15*, 183–196. [CrossRef]
9. Wittwer, A.R.; Dorado, R.; Alvarez, G.A.; Degrazia, G.A.; Loredo-Souza, A.M.; Bodmann, B. Flow in the Wake of Wind Turbines: Turbulence Spectral Analysis by Wind Tunnel Tests. *Am. J. Environ. Eng.* **2016**, *6*, 109–115.
10. Kanevce, G.; Oka, S. Correcting Hot-wire Readings for Influence of Fluid Temperature Variations. *DISA Inf.* **1973**, *15*, 21–24.
11. Ligęza, P. *Simultaneous Velocity and Temperature Measurements in Transient Gas Flows*; Strata Mechanics Research Institute Polish Academy of Science: Krakow, Poland, 2009.
12. Kidron, I. Measurement of the transfer function of hot-wire and hot-film turbulence transducers. *IEEE Trans. Instrum. Meas.* **1966**, *15*, 76–81. [CrossRef]
13. Jamróz, P. Relationship Between Dynamic Coefficients of Two Temperature Sensors Under Nonstationary Flow Conditions. *IEEE Sens. J.* **2011**, *11*, 335–340. [CrossRef]
14. Nabielec, J. Error in the method of sensor dynamics parameter identification. *Przegląd Elektrotechniczny* **2011**, *87*, 99–103.

15. Ligęza, P. Dynamic Error Correction Method in Tachometric Anemometers for Measurements of Wind Energy. *Energies* **2022**, *15*, 4132. [CrossRef]
16. Jamróz, J.P. Nabielec Adaptive Sensors for Dynamic Temperature Measurements. In *Smart Sensors for Industrial Applications*; Iniewski, K., Ed.; CRC Press: Boca Raton, FL, USA, 2013.
17. Freymuth, P. Frequency response and electronic testing for constant-temperature hot-wire anemometers. *J. Phys. E.Sci. Instrum.* **1977**, *10*, 705–710. [CrossRef]
18. Watmuff, J.H. Investigation of the constant-temperature hot-wire anemometer. *Exp. Therm. Fluid Sci.* **1995**, *11*, 117–134. [CrossRef]
19. Li, D.J. Dynamic response of constant temperature hot-wire system in turbulence velocity measurements. *Meas. Sci. Technol.* **2004**, *15*, 1835–1847. [CrossRef]
20. Li, D.J. The effect of electronic components on the cut-off frequency of the hot-wire system. *Meas. Sci. Technol.* **2005**, *16*, 766–774. [CrossRef]
21. Jamroz, P.; Ligeza, P.; Socha, K. Dynamic properties of hot-wire anemometric measurement circuits in the aspect of measurements in mine condition. *Arch. Min. Sci.* **2012**, *57*, 699–714.
22. Jamróz, P.; Ligęza, P.; Socha, K. Advanced metrological tool designed to natural environment parameters measurements. Hot-wire-anemometer with bandwidth optimization. *Trans. Strat. Mech. Res. Inst.* **2010**, *12*, 197–208. (In Polish)

energies

MDPI

Article

Experimental Study on the Optimal-Based Vibration Control of a Wind Turbine Tower Using a Small-Scale Electric Drive with MR Damper Support

Paweł Martynowicz

Department of Process Control, AGH University of Science and Technology, Mickiewicza 30 Ave., 30-059 Kraków, Poland; pmartyn@agh.edu.pl

Abstract: The paper presents an experimental implementation of an optimal-based vibration control for a scaled wind turbine tower-nacelle structure. A laboratory model of the approximate power scale of 340 W, equipped with a nonlinear tuned vibration absorber (TVA), is analysed. For control purposes, a combined operation of a small-scale electric servo drive and a magnetorheological (MR) damper is used in the TVA system. Nonlinearities of both the electric drive and the MR damper are intrinsic parts of the adopted nonlinear control concept. The aim of the research is the simple-hardware real-time implementation and the experimental investigation of the simultaneous actuator and damper control, including the analysis of the influence of optimal control law parameters and quality function weights on the vibration attenuation efficiency and actuator energy demand. As a reference, an optimal-based, modified ground-hook control with the single goal of the primary structure deflection minimisation is used along with the passive system with zero MR damper current and idling electric actuator, proving the advantages of the proposed method. The regarded solutions guarantee 57% maximum structure deflection reduction concerning the passive TVA configuration, using an MR damper of 32 N maximum force and an electric drive of 12.5 N nominal force and 0.76 W nominal power. An interesting alternative is the optimal control concept tuned with regard to the actuator power minimisation—it provides 30% maximum structure deflection attenuation (concerning the passive TVA configuration) while using a passive damper of 3.3 N maximum force and an actuator of 0.17 W nominal power only. It makes evident the advantage of the properly tuned optimal control algorithm over the modified ground-hook law—it requires 51% less actuator energy than the latter parametrised to exhibit the same vibration attenuation properties.

Keywords: real-time vibration control; optimal-based control; wind turbine; hybrid tuned vibration absorber; electric drive; magnetorheological damper

Citation: Martynowicz, P. Experimental Study on the Optimal-Based Vibration Control of a Wind Turbine Tower Using a Small-Scale Electric Drive with MR Damper Support. *Energies* **2022**, *15*, 9530. https://doi.org/10.3390/en15249530

Academic Editor: Mohamed Benbouzid

Received: 15 October 2022
Accepted: 12 December 2022
Published: 15 December 2022

1. Introduction

Mechanical vibrations are problematical phenomena concerning slender structures such as towers, masts, chimneys, bridges [1,2], skyscrapers [3–5], wind turbines [6–10], as well as plate structures [11,12], etc. Most of them are fitted with dedicated solutions for vibration minimisation and fatigue reduction, such as tuned vibration absorbers/tuned mass dampers (TVAs/TMDs), tuned liquid column dampers (TLCDs), bracing systems, etc. [13–18]. TVAs are more and more widely spread vibration reduction systems. A standard (passive) TVA is built as an additional mass connected with the protected structure by a spring and a viscous damper (in parallel), the parameters of which are tuned to the selected mode of the vibration [19]. Passive TVAs cope reasonably well with the vibration of a single frequency but cannot adapt to a broader spectrum [8]. During the structure exploitation lifespan, its frequency response may vary due to i.a. temperature fluctuations, icing, or external loading, apart from the defects that may arise. Thus, more advanced TVA systems have been investigated. Hybrid TVAs (H-TVAs), being the parallel connection of a passive TVA with active actuators [5], are the most dependable systems used in civil engineering [15,20–23]. The active force of the H-TVA increases the vibration attenuation efficiency and robustness of the TVA, while its energy, power, and force requirements are much lower than those of an active TVA (i.e., a TVA in which an active actuator replaces a viscous damper) of comparable performance. The devices used for the structural vibration control include active electro-hydraulic/-pneumatic/-magnetic actuators, semi-active magnetorheological (MR) or changeable-orifice dampers, or passive devices such as viscoelastic/hydraulic/friction/granular dampers, etc. [13–15,24].

Most of the (active or semi-active) actuator real-time control solutions are based on the bang-bang strategy [25], fuzzy logic, or two-stage algorithms that consist of the determination of an actuator's required force and its accurate tracking [2,26,27]. The latter concepts experience an inability to produce the required (by the first stage algorithm) force pattern due to the actuator nonlinearities/constraints (including the ever-present force and stroke constraints), the impossibility of generating active forces for semi-active actuators, etc. The stroke limitation of the real-world vibration reduction system, specifically the TVA, is frequently addressed by the use of end-stop bumpers or spring-damper buffer systems [28], which prevent the collision with the primary structure but compromise the control quality at the same time (the presence of the additional springs also alters the TVA tuning frequency within the buffer ranges). As a result, the force pattern determined to be optimal (at the first stage) is not the same as the actuator's output. Some sophisticated algorithms require real-time frequency determination, which may be problematic for polyperiodic or random vibrations.

The idea of a concurrent, parallel operation of the MR damper and the active actuator has only occasionally been investigated, to mention just a few references. Kim et al. [29] used a parallel, concentric connection of the three actuators: a passive air spring, an MR damper, and an electromagnetic coil actuator in a precision machine mount to isolate it from unwanted vibrations. The pneumatic forces constantly supported the heavy weight of an upper structure, the MR damper handled the transient response, while the electromagnetic actuator reduced the resonance response, which was switched mutually with regard to the velocity threshold (the control signal was applied either to the MR damper or the active device). Switching logic was implemented to resolve the problem of interference between the MR damper and active actuator control forces. A simple proportional controller was applied to the MR damper, while a proportional-derivative circuit was applied to the electromagnetic and pneumatic devices. Sophisticated hardware was required to treat the high sampling rate. The parallel combination of the MR damper and active air spring in a hybrid mount system designed for vibration isolation was investigated in [30]. The air spring was used to support the precision stage and to isolate the large loads by controlling the spring coefficient. Additionally, the MR damper force was produced to control the extensive vibrations. The isolation performance was investigated by the utilisation of the simple PID controller for the air servo valve and the MR damper current. The vibration was reduced at low frequencies, yet the response amplitude at frequencies above 30 Hz was slightly increased. In turn, the simultaneous operation of the MR damper arranged in series with the hydraulic cylinder in a bracing system of a shear frame model representing a monopile wind turbine structure was investigated in [31], adopting the linear control theory for the linearised system and actuator—the MR damper was used to emulate the behaviour of the idealised linear dash-pot. The hydraulic cylinder provided the correct displacement across the MR damper, while the bang-bang force tracking algorithm determined the MR damper control current. The controller activated high-frequency modes and generated drift in the actuator displacement, though. Thus, only a fraction of the measured damper force could be used as input to the designed integral force feedback in the real-time hybrid simulations.

With regard to the above considerations, the current study concerns a complex, hybrid MR TVA system (H-MR-TVA) utilising a parallel, concurrent operation of a small-scale electric servo drive [13,22,28,32–34] and an MR damper [2,4,35–41]. An MR damper exhibits a wide range of resistance forces compared with a viscous damper, as well as a millisecond response time and only signal-level energy requirements, although it cannot add energy to the system when necessary (being a semi-active device); it also suffers from a nonzero remanent force [27,41–45]. On the other hand, a small-scale electric actuator may be used to generate the active forces and cancel the MR damper remanent force while using a modest amount of energy. Simulations and experiments have shown that implementing both an MR damper and a small electric drive in the TVA system may lead to state-of-the-art vibration reduction efficiency. The experimental implementation of a parallel operation of the simultaneously controlled MR damper and electric drive in the TVA system may be considered the originality of this work.

To cope with the control limitations (discussed above), the author devised a concept [34,40,41] to embed the nonlinearities of the actuators (i.e., the MR damper and the electric servo in the current study), including their force constraints, into a control problem formulation, removing efficiency and robustness issues that arise when a determined optimal-based control is imprecisely mapped or beyond the permissible actuator range. This requires the use of nonlinear control methods, which include maximum-principle-based methods [9,12,46], Lyapunov function-based methods [1,3,32,39,44], linearisation methods with linear optimal control theory (LQR/LQG/$H2/H\infty$) [3,13,28,33,37,47,48], etc. Each method group has advantages and disadvantages, including the high computational load necessary for real-time operation and control authority degradation due to disturbances or unmodelled dynamics. The deployment of a nonlinear maximum-principle-based control method that incorporates actuators' nonlinearities while providing relatively simple real-time hardware implementation is the major contribution of this study.

The concept of the maximum-principle-based nonlinear optimal (or optimal-based) control was previously investigated by the author [40], considering, in particular, a scaled wind turbine tower-nacelle model [41,49–54]. Wind turbines experience varying external loads, such as wind variations, wind shear, Karman vortices, blade passing, changeable inflow conditions for the blades, sea waves, and ice load, etc. Additionally, internal factors, such as rotating machinery unbalance, contribute to the structural vibration and fatigue wear of towers and blades. In order to investigate the problem of wind turbine tower vibrations, a scaled tower-nacelle laboratory model that exhibits partial dynamic similarity (i.e., similarity of motions of tower tips) with a real-world 1.5 MW Vensys82 structure was developed and built. It was assumed that a nacelle, a hub, a shaft, a generator, blades, and a gearbox would all be represented by the rigid body fixed to the top of the rod, modelling a tower. A horizontal force produced by a modal shaker may be applied to the rod modelling a tower or to the rigid body, representing a nacelle assembly. The laboratory model enabled the analysis of two initial bending modes of the tower-nacelle system; however, only the first mode was investigated in the current study, and H-MR-TVA was tuned to its frequency. Current research results may be transferred to a real-world wind turbine thanks to previously determined time, length, and force scale factors [53,54]. The approximate power scale of the laboratory model was 340 W.

The paper is organised as follows. In the next section, a regarded system is described. Subsequently, the optimal vibration control problem is formulated and solved, covering simultaneous MR damper and actuator control. Then, the implementation procedure, the experimental setup, and the test conditions are discussed. This is followed by the key section covering real-time control results. The paper is summarised with several conclusions.

2. A Regarded System

A scaled wind turbine tower-nacelle model is regarded as a protected structure whose first bending mode modal parameters are: mass m_1, stiffness k_1, and damping c_1. An H-MR-TVA of absorber mass m_2 and spring stiffness k_2 is considered (Figure 1). The movement of both m_1 and m_2 was constrained to be linear displacement x_1 and x_2 (accordingly) along the common axis (horizontal in Figure 1) of an applied excitation force F_e, modelling the resultant load applied to the nacelle. An MR damper and a force actuator (F_a) were both built parallel to the spring. Although the force actuator was carefully selected to exhibit the lowest motion damping for the required output nominal force F_{nom}, its influence (excess damping) on the TVA operation was significant; thus, mass m_2 was increased with regard to the previous research [41] to 14.1% of the mass m_1 to obtain two local maxima of the primary system displacement (x_1) frequency response (see Section 7) [19], while stiffness k_2 was tuned accordingly. Both the MR damper and force actuator were used for control purposes. The values of the adopted system parameters are presented in Table 1.

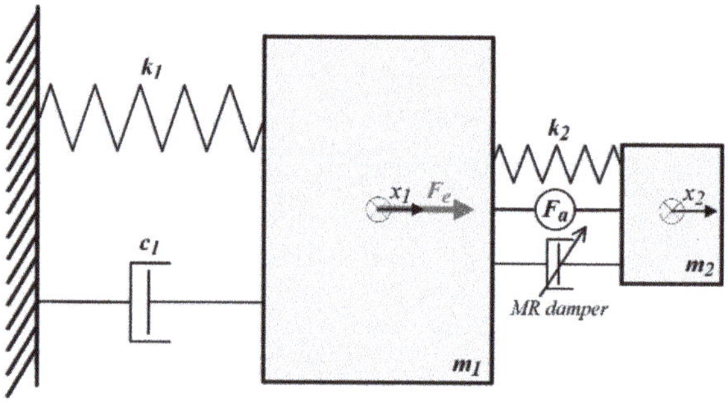

Figure 1. Two-body diagram of a regarded system with an H-MR-TVA.

Table 1. The adopted system parameters.

Parameter	Value
m_1	170.34 kg
k_1	82,554 N/m
c_1	49.53 Ns/m
m_2	24.01 kg (14.1% m_1)
k_2	8000 N/m
F_{nom}	12.5 N/6.25 N

Figure 2 presents the diagram of the theoretical passive TVA efficiency (vs. mass ratio, i.e., m_2/m_1), according to the authors of [19], represented by maximum nacelle displacement amplitudes $A(x_1)$ for the regarded primary structure parameters (Table 1); $A(\bullet)$ states for the amplitude. Two dots indicate mass ratios assumed in the current (14.1%) vs. previous research (7.7%) [41]. As it is commonly known, the TVA efficiency characteristic is nonlinear. The reduction in $A(x_1)$ due to the mass ratio increase to 14.1% was expected to be 23% only (concerning TVA of 7.7% mass ratio). However, a larger mass ratio contributes to lower TVA sensitivity to detuning [55].

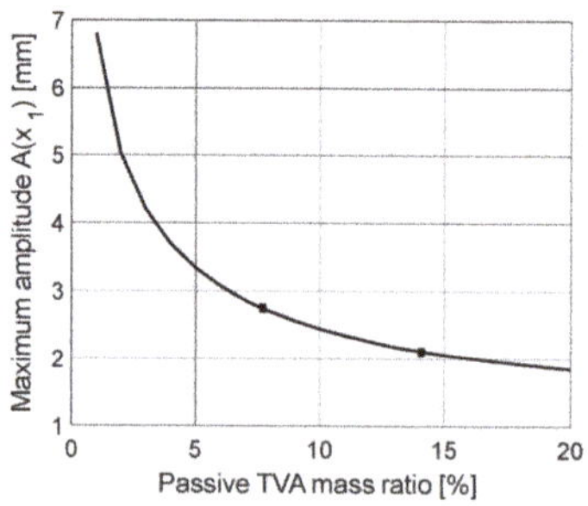

Figure 2. Maximum nacelle horizontal displacement amplitude $A(x_1)$ vs. passive TVA mass ratio (tuned according to [19]).

3. Control Problem Formulation and Solution

Consider the equation of motion of a vibrating structure with an H-MR-TVA:

$$\dot{z}(t) = f(z(t), u(t), t),\ t \in [t_0, t_1] \tag{1}$$

where $z(t)$ is a state vector:

$$z(t) = \begin{bmatrix} z_1(t) & z_2(t) & z_3(t) & z_4(t) \end{bmatrix}^T, \tag{2}$$

$u(t) = \begin{bmatrix} u_1(t) & u_2(t) \end{bmatrix}^T \in U\ (U = R^2)$ is a piecewise-continuous control vector, and a quality index to be minimised is:

$$G(z, u) = \int_{t_0}^{t_1} g(z(t), u(t), t)dt. \tag{3}$$

Following Section 2, assume: $z_1 = x_1, z_2 = \dot{x}_1, z_3 = x_2, z_4 = \dot{x}_2$, thus:

$$f(z(t), u(t), t) = \begin{bmatrix} z_2(t) \\ \frac{1}{m_1}(-(k_1 + k_2)z_1(t) - c_1 z_2(t) + k_2 z_3(t) + F_{mr}(z(t), u(t), t) + F_a(u(t)) + F_e(t)) \\ z_4(t) \\ \frac{1}{m_2}(k_2 z_1(t) - k_2 z_3(t) - F_{mr}(z(t), u(t), t) - F_a(u(t))) \end{bmatrix} \tag{4}$$

where:

$$\begin{aligned} F_{mr}(z(t), u(t), t) = (C_1 i_{mr}(u(t), t) + C_2)\tanh\{v[(z_4(t) - z_2(t)) + (z_3(t) - z_1(t))]\} + \\ +(C_3 i_{mr}(u(t), t) + C_4)[(z_4(t) - z_2(t)) + (z_3(t) - z_1(t))] \end{aligned} \tag{5}$$

is the MR damper force represented by the hyperbolic tangent model with the parameters: C_1, C_2, C_3, C_4, v [40]; $i_{mr}(u(t), t)$ is the MR damper control current, $F_a(t)$ is the actuator force, and $F_e(t)$ is the excitation force applied to the protected structure. To include the MR damper current restriction to $[0,\ i_{max}]$ range ($i_{max} > 0$), it was further assumed:

$$i_{mr}(u(t), t) = i_{max}\sin^2(u_1(t)). \tag{6}$$

To include the actuator output (static) nonlinearity, i.e., the nominal force limitation to $[-F_{nom},\ F_{nom}]$ range (see the adopted F_{nom} design values in Table 1), it was assumed:

$$F_a(t) = F_{nom}\sin(u_2(t)). \tag{7}$$

Remark

For an electric servo drive with a ball screw slide mechanism used as the force actuator in this study (see Section 5), the output was considered to be linear within limits of $[-F_{nom},\ F_{nom}]$ (contrary to the electro-hydraulic actuator considered in [34]); the corresponding motor driving torque range was $\left[-\frac{rF_{nom}}{\pi} - M_0,\ \frac{rF_{nom}}{\pi} + M_0\right]$, where M_0 is no-load driving torque of the slide unit, while r is the spindle radius. In addition to the actuator's static non-linearity (force limitation), its linear dynamics were not considered in this study.

The regarded quality function was:

$$\begin{aligned} g(z(t), u(t), t) = g_{11}{z_1}^2(t) + g_{12}{z_2}^2(t) + g_{13}(z_1(t) - z_3(t))^2 + g_{14}(z_2(t) - z_4(t))^2 + g_{21}{i_{mr}}^2(u(t), t) + \\ g_{221}{F_{mr}}^2(z(t), u(t), t) + g_{222}{F_a}^2(u(t)) + g_{23}{P_a}^2(u(t)) \end{aligned} \tag{8}$$

to account for the protected structure displacement z_1 and velocity z_2 minimisation, the TVA stroke $z_1 - z_3$ minimisation, the MR damper coil current i_{mr} and resistance force F_{mr} minimisation, and the actuator force F_a and power P_a minimisation, where:

$$P_a(t) = F_a(t)(z_2(t) - z_4(t)).$$

Assume the Hamiltonian in the form:

$$H(\xi(t), z(t), u(t), t) = -g(z(t), u(t), t) + \xi^T(t) f(z(t), u(t), t). \tag{9}$$

If $(z^*(t), u^*(t))$ is an optimal control process, there exists a co-state vector function ξ satisfying the equation:

$$\dot{\xi}(t) = -f_z^{*T}(z^*(t), u^*(t), t)\xi(t) + g_z^T(z^*(t), u^*(t), t),\ t \in [t_0, t_1] \tag{10}$$

with a terminal (transversality) condition:

$$\xi(t_1) = 0 \tag{11}$$

so that $u^*(t)$ maximises the Hamiltonian over the set U for almost all $t \in [t_0, t_1]$ (f_z and g_z are f and g derivatives with respect to z; f and g are continuously differentiable with respect to state and continuous with respect to time and control) [56]. For the analysed system, the co-state vector was:

$$\xi(t) = [\xi_1(t)\ \ \xi_2(t)\ \ \xi_3(t)\ \ \xi_4(t)]^T, \tag{12}$$

while:

$$f_z^{*T}(z^*(t), u^*(t), t) = \\ \begin{bmatrix} 0 & -\frac{1}{m_1}\left(k_1 + k_2 + \widetilde{F}_{mr}(z^*(t), u^*(t), t)\right) & 0 & \frac{1}{m_2}\left(k_2 + \widetilde{F}_{mr}(z^*(t), u^*(t), t)\right) \\ 1 & -\frac{1}{m_1}\left(c_1 + \widetilde{F}_{mr}(z^*(t), u^*(t), t)\right) & 0 & \frac{1}{m_2}\widetilde{F}_{mr}(z^*(t), u^*(t), t) \\ 0 & \frac{1}{m_1}\left(k_2 + \widetilde{F}_{mr}(z^*(t), u^*(t), t)\right) & 0 & -\frac{1}{m_2}\left(k_2 + \widetilde{F}_{mr}(z^*(t), u^*(t), t)\right) \\ 0 & \frac{1}{m_1}\widetilde{F}_{mr}(z^*(t), u^*(t), t) & 1 & -\frac{1}{m_2}\widetilde{F}_{mr}(z^*(t), u^*(t), t) \end{bmatrix} \tag{13}$$

with:

$$\widetilde{F}_{mr}(z^*(t), u^*(t), t) = \\ \nu(C_1 i_{mr}(u^*(t), t) + C_2)\left\{1 - \tanh^2\left[\nu\left(z_4^*(t) + z_3^*(t) - z_2^*(t) - z_1^*(t)\right)\right]\right\} + +(C_3 i_{mr}(u^*(t), t) + C_4) \tag{14}$$

thus:

$$\widetilde{F}_{mr}(z^*(t), u^*(t), t) = \\ \frac{\partial F_{mr}(z^*(t), u^*(t), t)}{\partial z_3^*(t)} = \frac{\partial F_{mr}(z^*(t), u^*(t), t)}{\partial z_4^*(t)} = -\frac{\partial F_{mr}(z^*(t), u^*(t), t)}{\partial z_1^*(t)} = -\frac{\partial F_{mr}(z^*(t), u^*(t), t)}{\partial z_2^*(t)} \tag{15}$$

and:

$$g_z^T(z^*(t), u^*(t), t) \\ = \begin{bmatrix} 2g_{11}z_1^*(t) + 2g_{13}\left(z_1^*(t) - z_3^*(t)\right) - 2g_{221}F'_{mr}(z^*(t), u^*(t), t) \\ 2g_{12}z_2^*(t) + 2g_{14}\left(z_2^*(t) - z_4^*(t)\right) - 2g_{221}F'_{mr}(z^*(t), u^*(t), t) + 2g_{23}F_a^2(t)\left(z_2^*(t) - z_4^*(t)\right) \\ -2g_{13}\left(z_1^*(t) - z_3^*(t)\right) + 2g_{221}F'_{mr}(z^*(t), u^*(t), t) \\ -2g_{14}\left(z_2^*(t) - z_4^*(t)\right) + 2g_{221}F'_{mr}(z^*(t), u^*(t), t) - 2g_{23}F_a^2(t)\left(z_2^*(t) - z_4^*(t)\right) \end{bmatrix} \tag{16}$$

where:

$$F'_{mr}(z^*(t), u^*(t), t) = F_{mr}(z^*(t), u^*(t), t)\widetilde{F}_{mr}(z^*(t), u^*(t), t).$$

Thus, Hamiltonian (9) takes a form:

$$H(\xi(t), z(t), u(t), t) = -g_{11}{z_1}^2(t) - g_{12}{z_2}^2(t) - g_{13}(z_1(t) - z_3(t))^2 - g_{14}(z_2(t) - z_4(t))^2 - \\ g_{21}{i_{mr}}^2(u(t), t) - g_{221}{F_{mr}}^2(z(t), u(t), t) - g_{222}{F_a}^2(u(t)) - g_{23}{F_a}^2(u(t))(z_2(t) - z_4(t))^2 + \\ \xi^T(t) f(z(t), u(t), t),$$

where:

$$\xi^T(t)f(z(t),u(t),t)$$
$$= \begin{bmatrix} \xi_1(t) & \xi_2(t) & \xi_3(t) & \xi_4(t) \end{bmatrix} \begin{bmatrix} z_4(t) \\ \frac{1}{m_1}(-(k_1+k_2)z_1(t) - c_1 z_2(t) + k_2 z_3(t) + F_{mr}(z(t),u(t),t) + F_a(u(t)) + F_e(t)) \\ z_4(t) \\ \frac{1}{m_2}(k_2 z_1(t) - k_2 z_3(t) - F_{mr}(z(t),u(t),t) - F_a(u(t))) \end{bmatrix}$$

The Hamiltonian maximisation conditions [56] are:

$$\frac{\partial H(\xi(t),z^*(t),u(t),t)}{\partial u_1(t)} = \left\{\left(\frac{1}{m_1}\xi_2(t) - \frac{1}{m_2}\xi_4(t) - 2g_{221}F_{mr}(z^*(t),u(t),t)\right)\frac{\partial F_{mr}(z^*(t),u(t),t)}{\partial i_{mr}(u(t),t)} - 2i_{\max}g_{21}\sin^2(u_1(t))\right\}\sin(2u_1(t))i_{\max} = 0 \tag{17}$$

$$\frac{\partial H(\xi(t),z^*(t),u(t),t)}{\partial u_2(t)} = \left\{\frac{1}{m_1}\xi_2(t) - \frac{1}{m_2}\xi_4(t) - 2F_{nom}\left[g_{222} + g_{23}(z_2(t) - z_4(t))^2\right]\sin(u_2(t))\right\}\cos(u_2(t))F_{nom} = 0 \tag{18}$$

with the appropriate sign change regimes, where:

$$\frac{\partial F_{mr}(z^*(t),u(t),t)}{\partial i_{mr}(u(t),t)} = \mathrm{C}_1\tanh[\nu(z_4^*(t) + z_3^*(t) - z_2^*(t) - z_1^*(t))] + \mathrm{C}_3(z_4^*(t) + z_3^*(t) - z_2^*(t) - z_1^*(t))$$

Let us fix an attention on $u_1(t)$ range of $[0,\ \pi]$; thus, Equation (17) results in ($g_{21} \neq 0$ is assumed):

$$\sin(2u_1(t)) = 0$$

or:

$$\sin^2(u_1(t)) = \frac{1}{2i_{\max_{21}}}\left(\frac{1}{m_1}\xi_2(t) - \frac{1}{m_2}\xi_4(t) - 2g_{221}F_{mr}(z^*(t),u(t),t)\right)\frac{\partial F_{mr}(z^*(t),u(t),t)}{\partial i_{mr}(u(t),t)}a \tag{19}$$

Analogically to [31]:

$$i_{mr}^*(u^*(t),t) = \begin{cases} 0,\ if,\ RHS(19) < 0 \\ \frac{1}{2g_{21}}\left(\frac{1}{m_1}\xi_2(t) - \frac{1}{m_2}\xi_4(t) - 2g_{221}F_{mr}(z^*(t),u(t),t)\right)\frac{\partial F_{mr}(z^*(t),u(t),t)}{\partial i_{mr}(u(t),t)}, & if\ RHS(19) \in [0\ 1) \\ i_{\max},\ if\ RHS(19) \in \geq 1 \end{cases} \tag{20}$$

where $RHS(19)$ is the right-hand side of Equation (19).

Condition (18) yields ($g_{222} \neq 0$ is assumed):

$$\frac{\partial H(\xi(t),z^*(t),u(t),t)}{\partial u_2(t)} = \left\{\frac{1}{2F_{nom}\left[g_{222} + g_{23}(z_2(t) - z_4(t))^2\right]}\left(\frac{1}{m_1}\xi_2(t) - \frac{1}{m_2}\xi_4(t)\right) - \sin(u_2(t))\right\}\cos(u_2(t)) = 0. \tag{21}$$

To analyse Hamiltonian derivative (21) sign change conditions, let us fix an attention on $u_2(t)$ range of $[-\pi, \pi]$, regarding the period of both $\sin(u_2(t))$ and $\cos(u_2(t))$. This analysis, with the help of Figure 3, yielded proposition (22)(23)(24), covering three disjoint and complementary cases:

(1) *if* $\left\{\frac{1}{2F_{nom}\left[g_{222} + g_{23}(z_2(t) - z_4(t))^2\right]}\left(\frac{1}{m_1}\xi_2(t) - \frac{1}{m_2}\xi_4(t)\right)\right\} \leq -1,$ *then* (21) is fulfilled and $\frac{\partial H(\xi(t),z^*(t),u(t),t)}{\partial u_2(t)}$ exhibits $+/-$ sign change (Hamiltonian maximisation) for: $u_2{}^*(t) = -\frac{\pi}{2}$ only (see Figure 3); *thus*:

$$F_a^*(t) = -F_{nom}. \tag{22}$$

(2) *if* $\left\{ \frac{1}{2F_{nom}\left[g_{222} + g_{23}(z_2(t) - z_4(t))^2\right]} \left(\frac{1}{m_1}\xi_2(t) - \frac{1}{m_2}\xi_4(t) \right) \right\} \geq 1,$ *then* (21) is fulfilled and $\frac{\partial H(\xi(t), z^*(t), u(t), t)}{\partial u_2(t)}$ exhibits $+/-$ sign change (Hamiltonian maximisation) for: $u_2{}^*(t) = \frac{\pi}{2}$ only (see Figure 3); *thus*:

$$F_a^*(t) = F_{nom}. \tag{23}$$

(3) *if* $\left\{ \frac{1}{2F_{nom}\left[g_{222} + g_{23}(z_2(t) - z_4(t))^2\right]} \left(\frac{1}{m_1}\xi_2(t) - \frac{1}{m_2}\xi_4(t) \right) \right\} \in (-1,\ 1),$ *then* (21) is fulfilled and $\frac{\partial H(\xi(t), z^*(t), u(t), t)}{\partial u_2(t)}$ exhibits $+/-$ sign change (Hamiltonian maximisation) for: $u_2{}^*(t) = \arcsin\left\{ \frac{1}{2F_{nom}\left[g_{222} + g_{23}(z_2^*(t) - z_4^*(t))^2\right]} \left(\frac{1}{m_1}\xi_2(t) - \frac{1}{m_2}\xi_4(t) \right) \right\}$ only (see Figure 3); thus:

$$F_a^*(t) = \frac{1}{2\left[g_{222} + g_{23}(z_2^*(t) - z_4^*(t))^2\right]} \left(\frac{1}{m_1}\xi_2(t) - \frac{1}{m_2}\xi_4(t) \right) \tag{24}$$

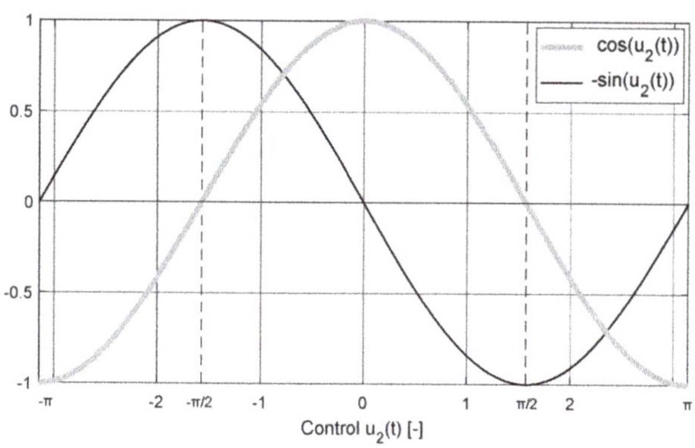

Figure 3. Hamiltonian derivative (21) sign analysis.

4. Control Implementation

Regarding the considerations presented in [34,40,41], the implementation of the above control proposition (20) and (22)–(24) may be as follows: the boundary value problem (1)(3)(9)÷(11) may be solved in every sampling step of the real-time control, while the horizon of the optimisation may be assumed as one integration step to cope with the large computational load. A numerical implementation of such a nonlinear optimal control for a structure equipped with either an MR-TVA (a standard TVA featuring an MR damper instead of a passive one) [40] or an H-TVA [34] was realised recently using MATLAB/Simulink level-2 s-function and *bvp4c* iterative scheme [57]; it was shown numerically (for the MR-TVA system also experimentally [41]) that the iteration procedure yielding high computational load may be omitted, adopting a short time horizon optimal problem task and zero initial conditions for co-state integrators. It was proved that the influence of the terminal condition (11) error was negligible for the considered control applications. In the present research, both the active electric actuator optimal-based control and the

MR damper optimal-based control were implemented simultaneously in the TVA system (Figure 1)—this solution will be designated hereinafter by *OPT*.

A modified two-level displacement ground-hook control (hereinafter designated by *GH*)—the optimal control direct implementation for the case when the protected structure deflection minimisation is the single objective—was additionally regarded during this study. The *GH* control law changes the actuator force between $-F_{nom}$ and $+F_{nom}$ with regard to x_1 sign changes and varies the MR damper control current between 0 and i_{max}th regard to x_1 and F_{mr} signs [34,40].

On the experimental ground, the *OPT* approach was compared with the *GH* technique for different levels of the nominal actuator force and maximum MR damper current, adopting the 1 ms sampling step.

5. The Experimental Setup

The regarded wind turbine tower-nacelle scaled laboratory model fulfilled a dynamic similarity condition of motions of tower tips with a real-world full-scale wind turbine Vensys82 structure [49,50,53,54].

The test rig (Figure 4) consisted of a vertically oriented Ti-Gr5 circular rod (no. 1, modelling the wind turbine tower) and a complex of steel plates (no. 2, modelling the nacelle unit, including the turbine) fastened to the top of the rod, with the H-MR-TVA system built-in. The Ti-Gr5 rod was fixed to a section steel foundation frame (no. 3). The H-MR-TVA (no. 4) consisted of an additional mass moving horizontally along linear bearing guides, joined to the nacelle-unit (no. 2) via springs (no. 5), Lord RD-1097-1 MR damper (no. 6) [45], and Festo EGSC-BS-KF-32-50-8P mini slide actuator of linear stroke (powered by Festo EMME-AS-40-S-LV-AS servo motor) (no. 7) [58] in parallel. The H-MR-TVA operated along the same direction as the force excitation applied. The structure was excited by the TMS2060E modal shaker (no. 8 [59]), whose force was transferred to the nacelle-unit (no. 2) with the use of the drive train system (no. 9) of the changeable leverage.

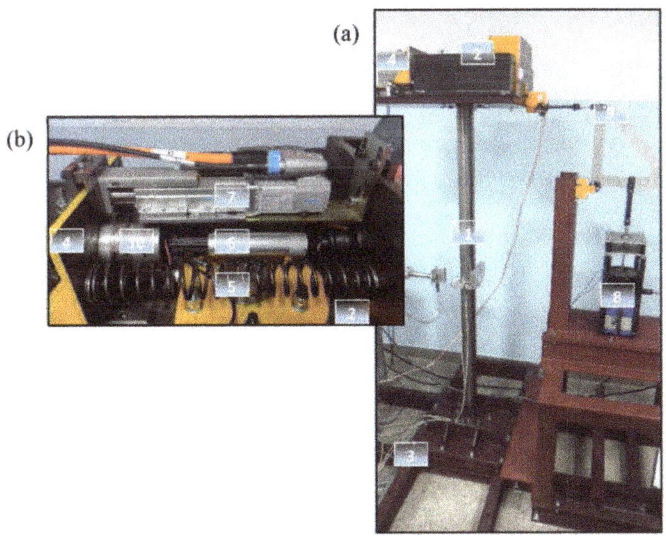

Figure 4. The laboratory test rig: (**a**) a general view, (**b**) the H-MR-TVA.

The data acquisition and control system used a laser sensor for the nacelle-unit absolute displacement (i.e., the tower tip deflection) x_1 measurement, a laser vibrometer for the H-MR-TVA mass absolute displacement x_2 measurement, the MR damper force sensor (no. 10), the shaker force sensor, the MR damper coil current Hall sensor, power supply and conditioning circuits for the sensors and actuators, and a measuring-control PC embedded with Inteco I/O board of the RT-DAC4 series [60] and MATLAB/Simulink/RT-CON applications. The RT-CON software ensured the fulfilment of real-time regimes using the 1 kHz sampling rate. The RT-DAC4 analogue input channels and RT-CON analogue input drivers mediated measurement data transfer to the Simulink control environment, where the demanded MR damper current/actuator force was calculated in each sample step. The RT-CON/RT-DAC4 analogue output drivers/channels were used for control output mediation. The MR damper current output was conditioned with the dedicated amplifier and PID controller to force the required electric current through the MR damper coil. The demanded actuator force was fed to the Festo CMMT-AS-C2-3A-EC-S1 servo controller connected with the EMME-AS-40-S-LV-AS motor. The shaker control output was generated using the TMS2060E force measurement signal compared with the demanded excitation pattern and conditioned with the PI controller implemented in the MATLAB/Simulink/RT-CON environment [36].

The system parameters, as in Table 1, along with the identified MR damper unit (Table 2) [41] and electric drive (21) parameters, were assumed for the experiments.

Table 2. The identified MR damper unit parameters [41].

Parameter	Value
C_1	44
C_2	1.0
C_3	225
C_4	7.0
ν	70

The electric drive ±15 N step response series was gathered (Figure 5) using the locked (with 0.9 A control current) MR damper, built-in parallel, and its force sensor (no. 10, Figure 4). The response was identified with a time-delayed transfer Function (21) of three poles and two zeros from MATLAB *System Identification App* (fit to estimation data 96%, data prefiltered, stability enforced):

$$G_a(s) = \frac{32.6s^2 + 1.55e4s + 9.66e4}{s^3 + 140s^2 + 1.71e4s + 1.06e5} e^{-0.006s} \tag{25}$$

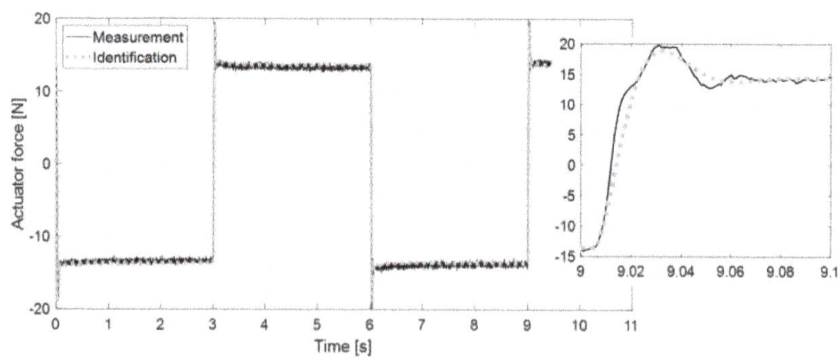

Figure 5. Electric drive force step response series: measurement vs. identification.

The actuator identification tests confirmed the considerable response delay of the used mini slide powered by a servo motor (the electro-hydraulic cylinder assumed in the previous numerical study [34] guaranteed a measurably faster response [61]), which was projected to be compensated by the MR damper output force (exhibiting a few-millisecond response time) while maintaining the simplicity of real-time hardware implementation—

the electric drive response delay influenced state and co-state variables values, which in turn amended the i^*_{mr} control current according to (20).

6. The Test Conditions

For the real-time vibration control of the wind turbine tower-nacelle model first bending mode, the approach described in Sections 3 and 4 was implemented using the damper control Formula (20) and the actuator control Formulas (22)–(24). Various *OPT* control *cases I–III* (see below) were regarded along with *GH* control, assuming different MR damper maximum currents and actuator nominal force values.

The test conditions parameters were as follows. The wind turbine tower-nacelle model was excited by a harmonic force of amplitude $A(F_e(t)) = 45.4$ N and a frequency range of [2.2, 4.5] Hz. The fixed sample step $t_s = 10^{-3}$ s was adopted. The nominal actuator output force $F_{nom} = 12.5$ N (thus servo motor nominal driving torque $M_{nom} = 17.2 \cdot 10^{-3}$ Nm) was assumed as a *baseline OPT/GH* configuration, along with $F_{nom} = 6.25$ N assumed as its more energy-efficient, *restricted force* alternative; the corresponding actuator control signal $F_{a\ ctr}$ maximum value of 15 N for the *OPT* system (13.5 N for the *GH* system), or 7.5 N for the *OPT* system (6.75 N for the *GH* system) (see time responses in Section 7), were assumed accordingly, based on the identification (25). The maximum value of $F_{a\ ctr}$ corresponding to the assumed F_{nom} value for the *GH* system was 10% lower than for the *OPT* system due to the 10% duty cycle of the *OPT* resetting function (zero initial conditions for co-state integrators are assumed). The MR damper maximum current $i_{max} = 1.0$ as was assumed for the *baseline* configuration ($F_{nom} = 12.5$ N), while $i_{max} = 0.5$ as was assumed for its *restricted force* counterpart ($F_{nom} = 6.25$ N)—see Table 3. The values of i_{max} elding F_{mr} ranges were tuned to the assumed F_{nom} ranges, as the MR damper, due to its few-millisecond response time (see, e.g., Figure 12b in [41] p. 12), was activated ahead of the electric drive, while the electric actuator may have cancelled the detrimental MR damper remanent force.

Table 3. The test cases: nominal actuator force F_{nom}/maximum MR damper current *i_max*.

Configuration \ Control Method	*OPT CASE I* $g_{221} = 0$ $g_{23} = 0$	*OPT CASE II* $g_{221} = 10^6$ $g_{23} = 0$	*OPT CASE III* $g_{221} = 0$ $g_{23} = 10^{10}$ or $g_{23} = 10^{11}$	*GH*
Baseline	12.5 N/1.0 A	12.5 N/1.0 A	12.5 N/1.0 A	12.5 N/1.0 A
Restricted force	6.25 N/0.5 A	6.25 N/0.5 A	6.25 N/0.5 A	6.25 N/0.5 A

The general weighting factors for the *OPT* control solution's quality index (8) were assumed as follows: $g_{11} = 10^{18}$, $g_{12} = 0$, $g_{13} = 10^{15}$, $g_{14} = 0$, $g_{21} = 4$, $g_{222} = 4 \cdot 10^{-12}$. A negligible but nonzero g_{222} value was selected to eliminate calculation problems for $z_2(t) = z_4(t)$ in (21), (22)(23)(24); the electric drive force magnitude was tuned through the F_{nom} assumption (*baseline* or *restricted force* configuration) in the current research.

The remaining weights were assumed to (*OPT cases I–III*, see Table 3):

I. minimise the primary structure deflection (nacelle-assembly displacement) x_1 amplitude as the primary objective while considering TVA stroke amplitude limitation (these were the two basic objectives for *cases II* and *III* as well);

II. minimise the MR damper force in addition to the *case I* objectives (to account for the MR damper's possible force limit and thermal constraints);

III. minimise the actuator power in addition to the *case I* objectives (to account for the actuator's energy constraints).

7. Real-Time Control Results

The efficiency of the adopted solutions was analysed using the frequency characteristics of the nacelle displacement (primary structure deflection) amplitude $A(x_1)$ (Figures 6–8), the TVA stroke amplitude $A(x_1 - x_2)$ (Figures 9–11), the maximum MR damper force (Figures 12–14), and the mean actuator power (Figures 15–17), along with the time patterns of x_1, $x_1 - x_2$, $F_{a\ ctr}$, F_a, and F_{mr} (Figures 18–21). If omitted, $i_{max} = 0$ d all weights as for control *case I* (Section 6) were assumed all over this section. Figures 6, 9, 12 and 15 present the frequency characteristics of $A(x_1)$, $A(x_1 - x_2)$, the maximum MR damper force, and the mean actuator power, respectively, obtained for the *OPT* and *GH* systems with $F_{nom} = 12.5$ N, relative to the passive system with constant MR damper control current values of 0.0 A, 0.1 A, 0.2 A, 0.5 A, and idling electric actuator (passive system omitted in Figure 15). Similarly, Figures 7, 10, 13 and 16 present the frequency patterns of $A(x_1)$, $A(x_1 - x_2)$, the maximum MR damper force, and the mean actuator power, respectively, obtained for the *OPT* solution with $F_{nom} = 12.5$ N, and different i_{max}/g_{221} values ($i_{max} = 0$, $i_{max} = 0.5$ A or $i_{max} = 1.0$; nonzero MR damper force weight $g_{221} = 10^6$ for $i_{max} = 1.0$ case only), the *OPT* solution with $F_{nom} = 6.25$ N, and $i_{max} = 0$ $i_{max} = 0.5$ A, and *GH* solution with $F_{nom} = 6.25$ N, and $i_{max} = 0$ or $i_{max} = 0.5$ A. Additionally, Figures 8, 11, 14 and 17 present the frequency characteristics of $A(x_1)$, $A(x_1 - x_2)$, the maximum MR damper force, and the mean actuator power, respectively, obtained for the *OPT* concept with $F_{nom} = 12.5$ N and different i_{max}/g_{23} values ($i_{max} = 0$, $i_{max} = 0.5$ A or $i_{max} = 1.0$; the actuator power weight $g_{23} = 10^{10}$ or $g_{23} = 10^{11}$).

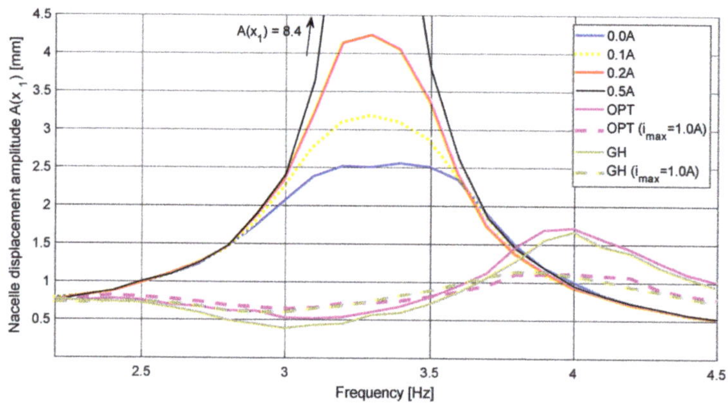

Figure 6. Nacelle horizontal displacement amplitude $A(x_1)$ frequency characteristics: passive system vs. *OPT* for $F_{nom} = 12.5$ N vs. *GH* for $F_{nom} = 12.5$ N.

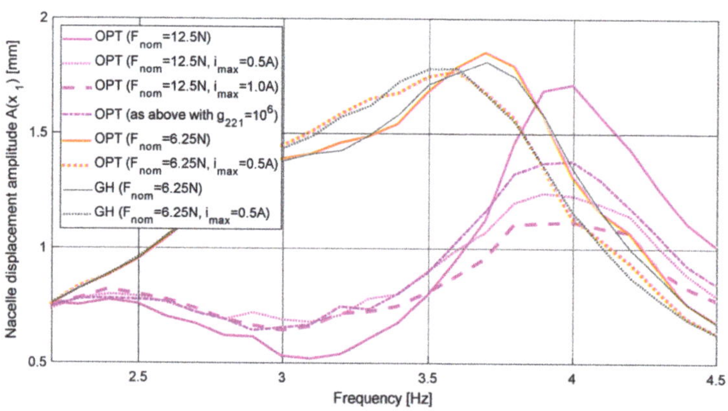

Figure 7. Nacelle horizontal displacement amplitude $A(x_1)$ frequency characteristics: *OPT* for different control parameter values vs. *GH* for $F_{nom} = 6.25$ N.

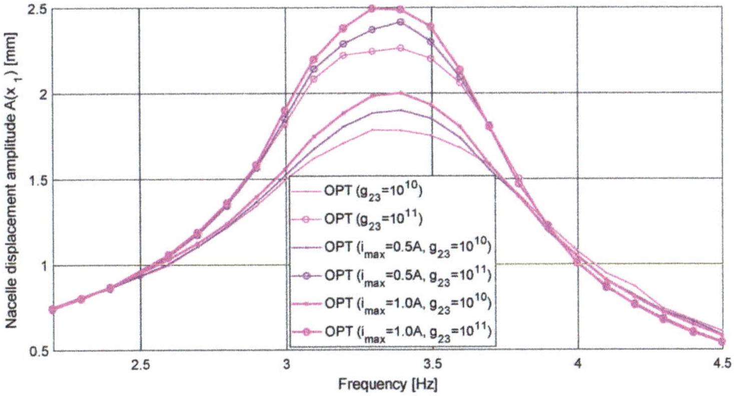

Figure 8. Nacelle horizontal displacement amplitude $A(x_1)$ frequency characteristics: *OPT* for different i_{max} d g_{23} values, and $F_{nom} = 12.5$ N.

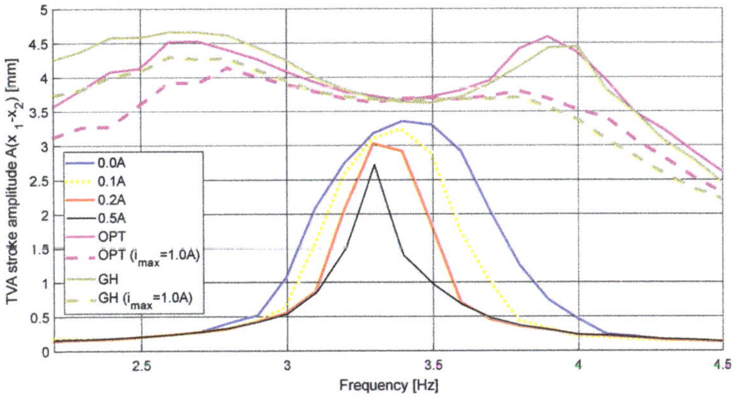

Figure 9. The TVA stroke amplitude $A(x_1 - x_2)$ frequency characteristics: passive system vs. *OPT* for $F_{nom} = 12.5$ N vs. *GH* for $F_{nom} = 12.5$ N.

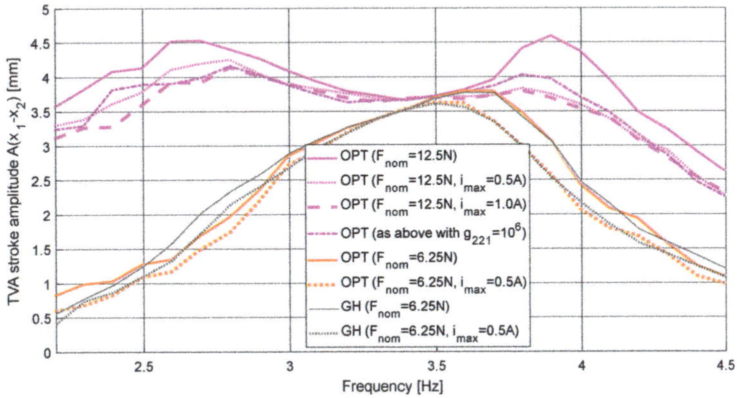

Figure 10. The TVA stroke amplitude $A(x_1 - x_2)$ frequency characteristics: *OPT* for different control parameter values vs. *GH* for $F_{nom} = 6.25$ N.

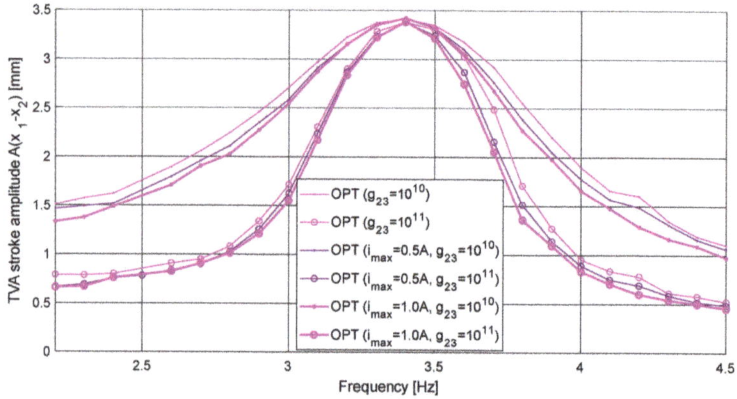

Figure 11. The TVA stroke amplitude $A(x_1 - x_2)$ frequency characteristics: *OPT* for different i_{max} an d g_{23} values, and $F_{nom} = 12.5$ N.

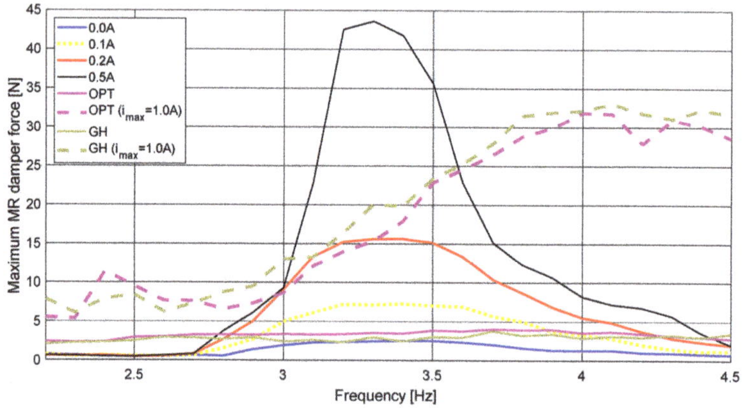

Figure 12. The maximum MR damper force frequency characteristics: passive system vs. *OPT* for $F_{nom} = 12.5$ N vs. *GH* for $F_{nom} = 12.5$ N.

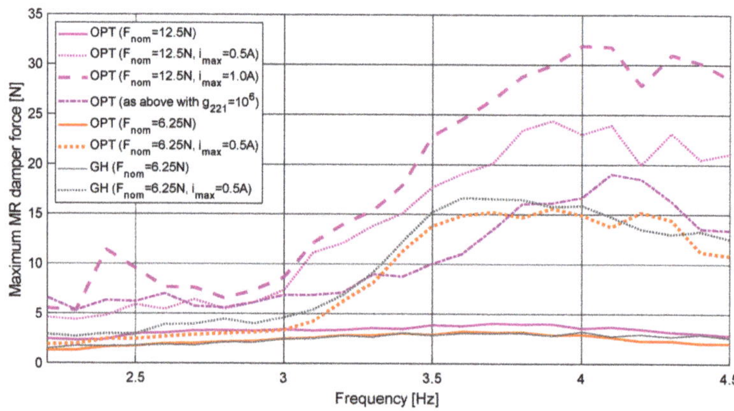

Figure 13. The maximum MR damper force frequency characteristics: *OPT* for different control parameter values vs. *GH* for $F_{nom} = 6.25$ N.

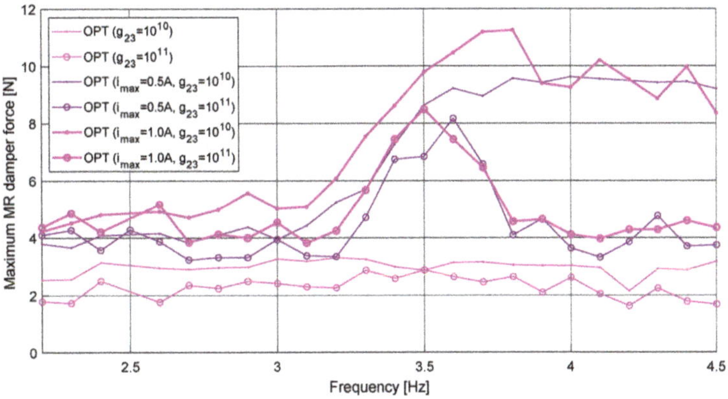

Figure 14. The maximum MR damper force frequency characteristics: *OPT* for different i_{max} and g_{23} values, and $F_{nom} = 12.5$ N.

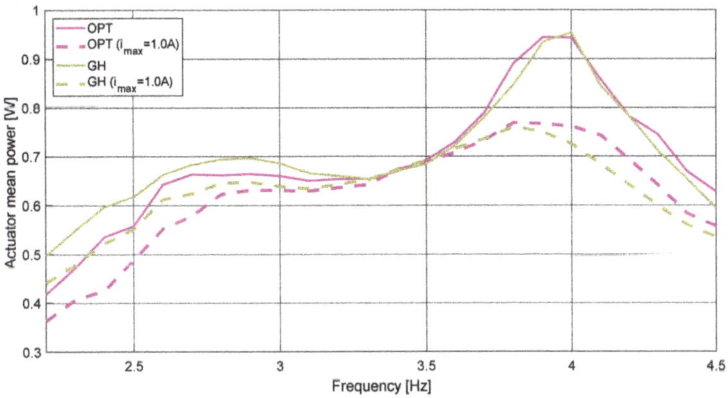

Figure 15. The mean actuator power frequency characteristics: *OPT* for $F_{nom} = 12.5$ N vs. *GH* for $F_{nom} = 12.5$ N.

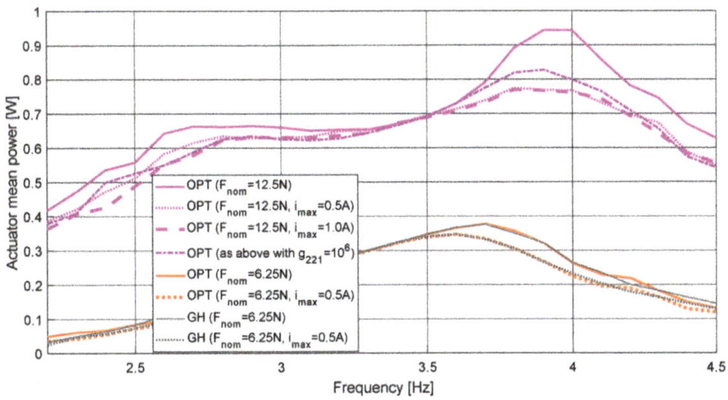

Figure 16. The mean actuator power frequency characteristics: *OPT* for different control parameter values vs. *GH* for $F_{nom} = 6.25$ N.

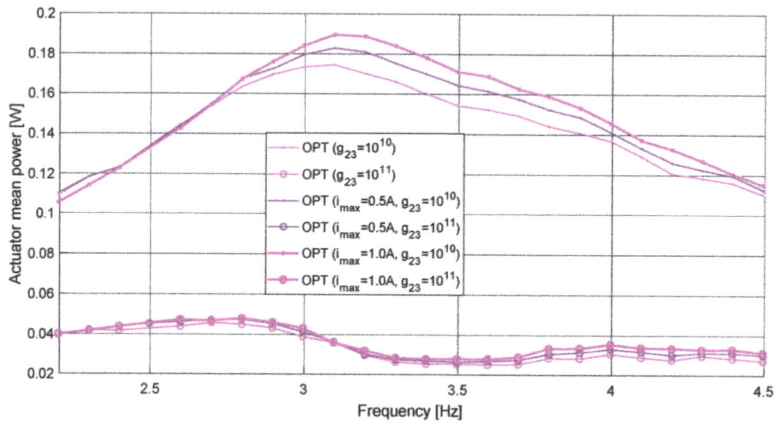

Figure 17. The mean actuator power frequency characteristics: *OPT* for different i_{max} and g_{23} values, and $F_{nom} = 12.5$ N.

Figures 18–21 present the comparison of the primary structure deflection (i.e., nacelle horizontal displacement) x_1, the TVA displacement $x_1 - x_2$, the actuator control signal $F_{a\ ctr}$ and force F_a, and the MR damper force F_{mr} time patterns obtained at the specific frequency points: 3.2 Hz (Figure 18), 4.0 Hz (Figure 19), 3.4 Hz (Figure 20), 3.6 Hz, and 3.7 Hz (Figure 21). Figures 18 and 19 present the comparison of the time responses obtained for the *OPT* system with $F_{nom} = 12.5$ N, $i_{max} = 0/0.5/1.0$ A, $g_{221} = 0$, the *OPT* system with $F_{nom} = 12.5$ N, $i_{max} = 1.0$ A, $g_{221} = 10^6$, and the *GH* system with $F_{nom} = 12.5$ N, $i_{max} = 0/1.0$ A. Fig. 20 presents the comparison of the time responses obtained for the *OPT* system with $F_{nom} = 6.25$ N, and $i_{max} = 0$ or 0.5 A. Figure 21 compares the time responses determined for the *OPT* system with $F_{nom} = 12.5$ N, $i_{max} = 0/1.0$ A, and $g_{23} = 0/10^{10}/10^{11}$.

As may be observed in Figures 6 and 7 for $F_{nom} = 12.5$ N (in Figure 7 for $F_{nom} = 6.25$ N), both *OPT* and *GH* solutions with an MR damper semi-active operation (i.e., with nonzero $i_{\max}$) exhibited significantly lower maximum $A(x_1)$ values within the regarded structure first bending mode frequency neighbourhood (especially in 4.0 Hz neighbourhood), than *OPT* and *GH* without the MR damper support (i.e., with $i_{max} = 0$, omitted in the legends). The latter, however, presented slightly lower $A(x_1)$ values in the 3.2 Hz neighbourhood. These phenomena are illustrated in Figures 18 and 19 for $F_{nom} = 12.5$ N and various i_{max}/g_{221} values. The additional force provided by the MR damper semi-active operation resulted in a slightly smaller TVA stroke ($x_1 - x_2$) amplitude (compared to Figures 9 and 10), which yielded somewhat lower TVA efficiency for nonzero i_{max} at 3.2 Hz (Figure 18). At a higher frequency of 4.0 Hz (Figure 19), however, the other factor was dominant in relation to the electric servo activation delay time regarding the shorter vibration period; thus, the MR damper few-millisecond response time was greatly beneficial, as previously expected (Section 5). Similar phenomena for $F_{nom} = 6.25$ N are depicted in Figure 20a,b (for 3.6 Hz) and Figure 20c,d (for 3.7 Hz). Figure 6 clearly indicates the TVA excess damping with regard to the optimally tuned TVA [19]—the maximum $A(x_1)$ for the zero-control passive system was greater than 2.5 mm. In contrast, Figure 2 suggests a value of 2.1 mm for a 14.1% mass ratio.

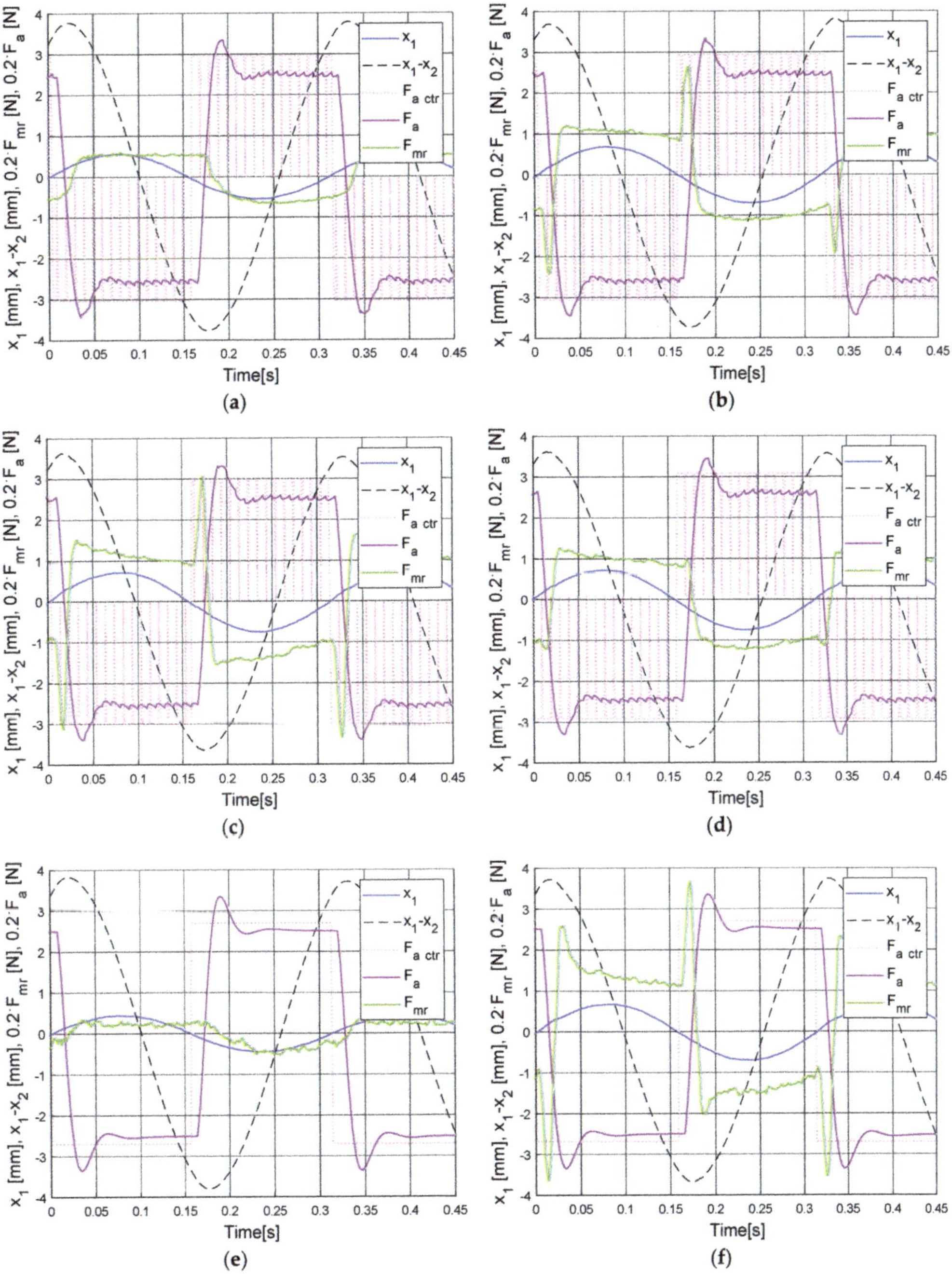

Figure 18. Time responses at 3.2 Hz: (**a**) *OPT*, $i_{max} = 0$ (**b**) *OPT*, $i_{max} = 0.5$ A (**c**) *OPT*, $i_{max} = 1.0$ A, (**d**) *OPT*, $i_{max} = 1.0$ A, $g_{221} = 10^6$, (**e**) *GH*, $i_{max} = 0$, (**f**) *GH*, $i_{max} = 1.0$ A ($F_{nom} = 12.5$ N).

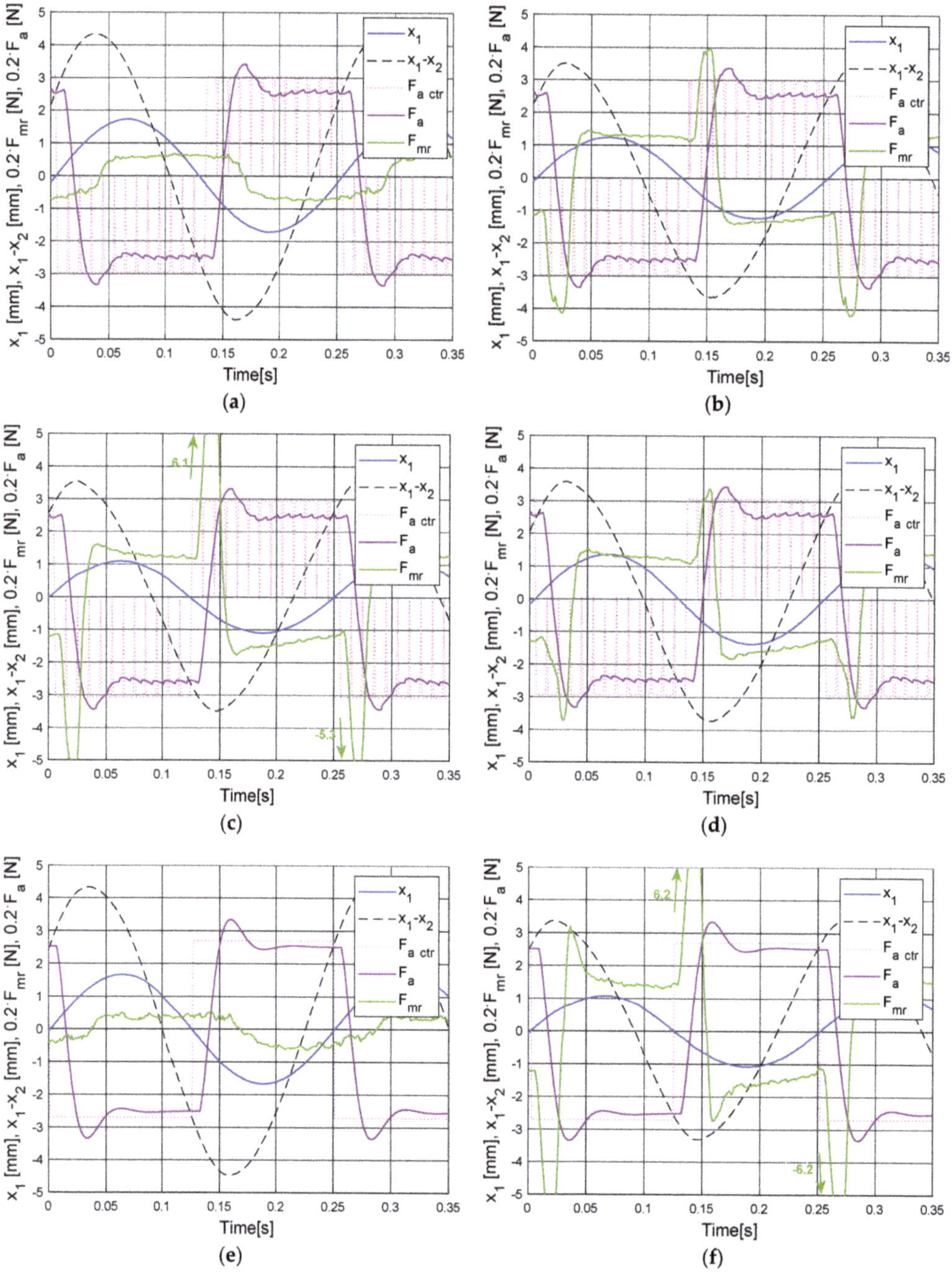

Figure 19. Time responses at 4.0 Hz: (**a**) *OPT*, $i_{max} = 0$ (**b**) *OPT*, $i_{max} = 0.5$ A (**c**) *OPT*, $i_{max} = 1.0$ A, (**d**) *OPT*, $i_{max} = 1.0$ A, $g_{221} = 10^6$, (**e**) *GH*, $i_{max} = 0$, (**f**) *GH*, $i_{max} = 1.0$ A ($F_{nom} = 12.5$ N).

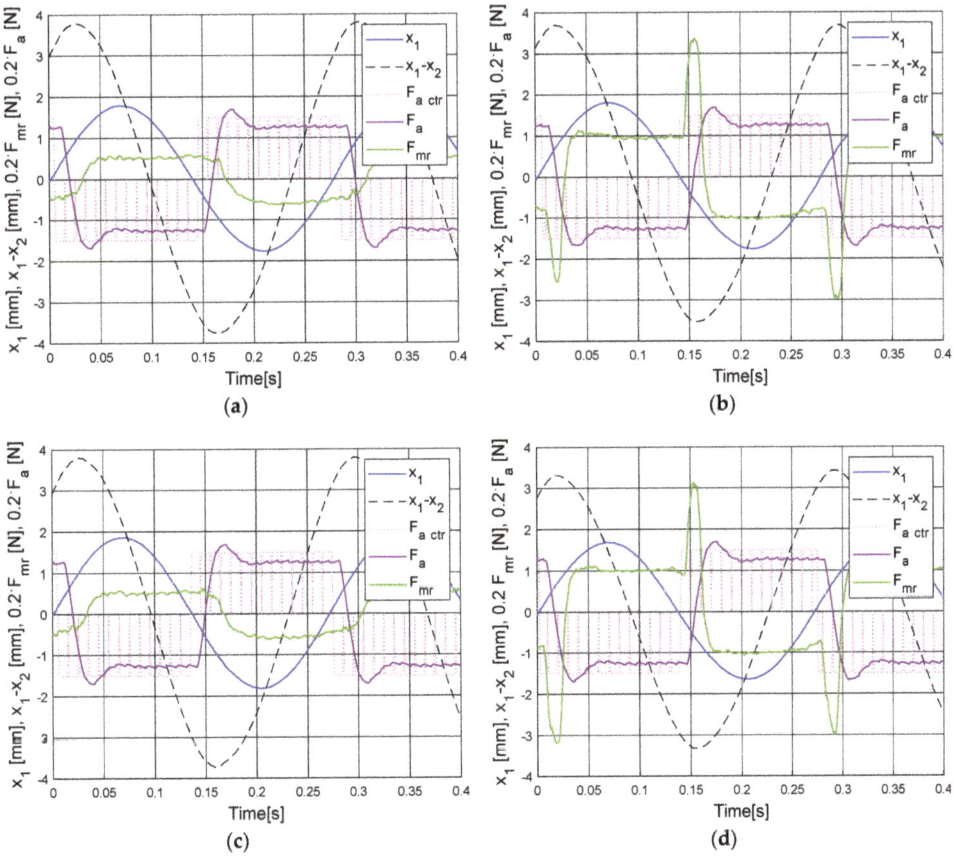

Figure 20. *OPT* time responses at 3.6 Hz: (**a**) $i_{max} = 0$, (**b**) $i_{max} = 0.5$ A vs. *OPT* time responses at 3.7Hz: (**c**) $i_{max} = 0$, (**d**) $i_{max} = 0.5$ A ($F_{nom} = 6.25$ N).

The combined electric drive and MR damper operation in the H-MR-TVA system (control *case I*, *baseline* configuration), i.e., with $F_{nom} = 12.5$ N, $i_{max} = 1.0$ A led to a 35% maximum $A(x_1)$ reduction in relation to the H-MR-TVA operation without MR damper semi-active support (i.e., with $i_{max} = 0$ a and 52% reduction with regard to the previous research with an MR-TVA system of a lower (7.7%) mass ratio, but less TVA excess damping [41]. This is regarded as a satisfactory result for such a real-time implementation, considering the nonlinearity of Figure 2 characteristics and the expected 23% $A(x_1)$ reduction only (Section 2). Both *GH* and *OPT* (control *case I*, *baseline* configuration) provided maximum $A(x_1)$ values close to 1.1 mm (i.e., 57% reduction concerning the 0.0 A passive configuration exhibiting the lowest $A(x_1)$ values, and 87% reduction regarding the 0.5 A passive configuration, see Figure 6). This result was obtained at ca. 4 mm TVA stroke amplitude (*OPT* configuration, Figure 9), ca. 32 N maximum MR damper force (Figure 12), and ca. 0.76 W maximum actuator power (Figure 15). The configurations without the MR damper semi-active support were characterised by more than 4.5 mm TVA stroke amplitude, only 4 N maximum MR damper force, and as much as 0.95 W maximum actuator power, while being significantly less efficient in primary structure deflection minimisation.

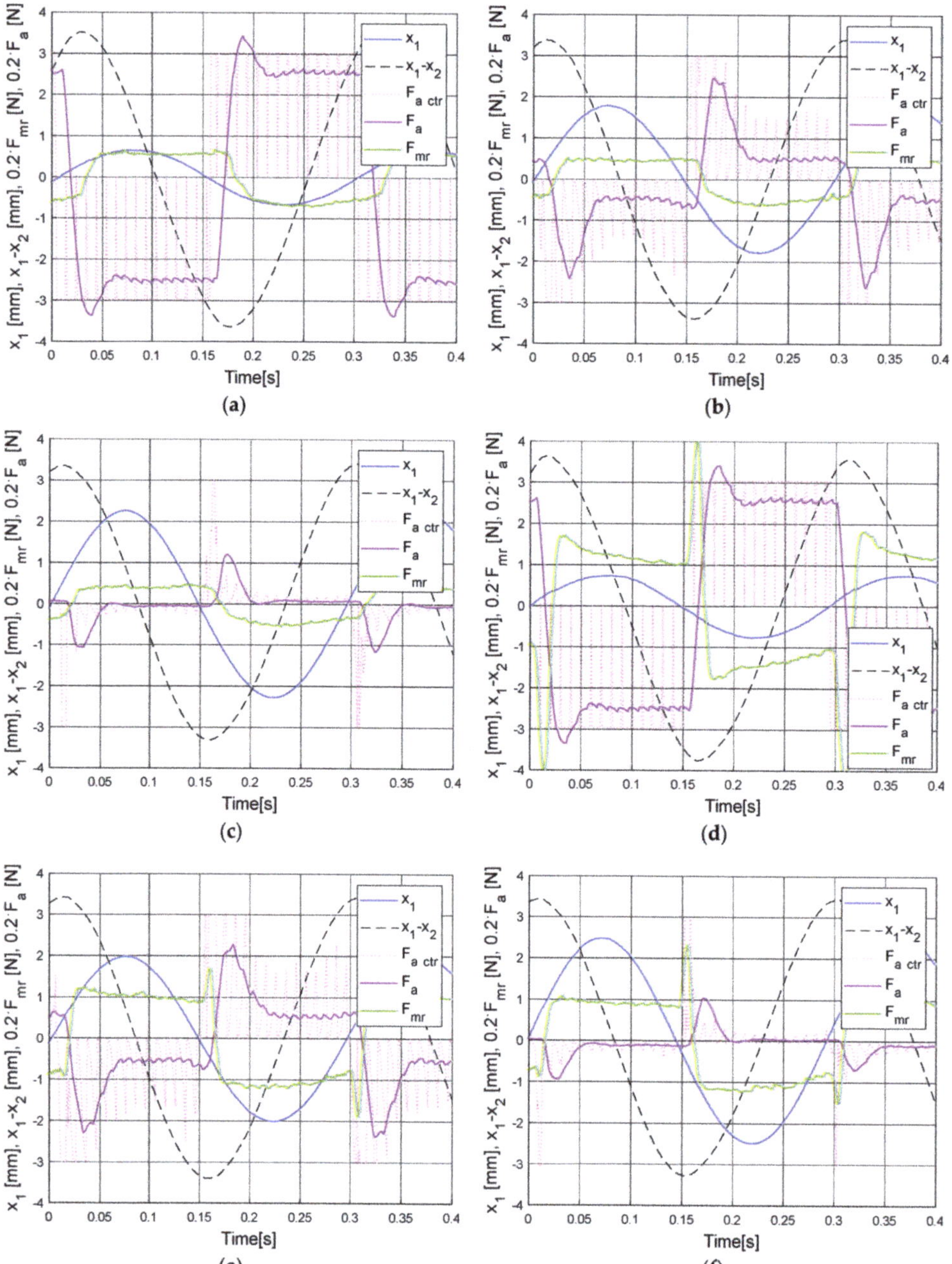

Figure 21. *OPT* time responses at 3.4 Hz: **(a)** $i_{max} = 0$ **(b)** $i_{max} = 0$, $g_{23} = 10^{10}$ **(c)** $i_{max} = 0$, $g_{23} = 10^{11}$ **(d)** $i_{max} = 1.0$ A **(e)** $i_{max} = 1.0$ A, $g_{23} = 10^{10}$, **(f)** $i_{max} = 1.0$ A, $g_{23} = 10^{11}$ ($F_{nom} = 12.5$ N).

As presented in Figures 6, 9, 12 and 15 for $F_{nom} = 12.5$ N (in Figures 7, 10, 13 and 16 for $F_{nom} = 6.25$ N), the results of *OPT* and *GH* operation were similar, with minor differences noticeable concerning $A(x_1)$, $A(x_1 - x_2)$, the maximum MR damper force, and the mean actuator power, especially for $F_{nom} = 12.5$ N. In Figure 18a,c vs. Figure 18e,f as well as in Figure 19a,c vs. Figure 19e,f, disparities in the time patterns may be spotted with $F_{a\ ctr}$ differences coming to the forefront, as explained in the final paragraph of Section 6.

The interesting alternative to the *baseline* configuration ($F_{nom} = 12.5$ N, $i_{max} = 1.0$ A with $g_{221} = 0$ (control *case I*) was a configuration with a lower $i_{max} = 0.5$ A value (control *case I*), and with $i_{max} = 1.0$ A and nonzero MR damper force weight $g_{221} = 10^6$ (control *case II*); $F_{nom} = 12.5$ N for both these configurations. In Figures 7, 10, 13 and 16, the respective frequency characteristics are included, while in Figure 18b,d as well as in Figure 19b,d, the time patterns are presented. It may be observed that these two solutions exhibited higher x_1 deflection amplitudes, similar $x_1 - x_2$ strokes, and lower MR damper forces (as intended, especially for the $i_{max} = 1.0$ A, $g_{221} = 10^6$ configuration) than the *baseline* configuration. The maximum actuator power used in the $i_{max} = 1.0$ A, $g_{221} = 10^6$ configuration was actually slightly higher, being also less efficient in $A(x_1)$ minimisation than the $i_{max} = 0.5$ A, $g_{221} = 0$ configuration. In Figure 18c vs. Figure 18d as well as in Figure 19c vs. Figure 19d, the differences in the F_{mr} patterns resulting from the different g_{221} (MR damper force) weights may be spotted. Due to the MR damper millisecond response time, its control signal i_{mr} patterns were omitted in the time characteristics Figures 18–21 (details may be found in [41] pp. 12–13).

In Figure 8, Figure 11, Figure 14, and Figure 17, the frequency characteristics determined for the *OPT* system with nonzero actuator power weights g_{23} (control *case III*) and different i_{max} values are presented. Figure 21 depicts time patterns obtained for zero and nonzero g_{23} values with the MR damper in passive mode ($i_{max} = 0$) and semi-active mode ($i_{max} = 1.0$ A) Due to lowering actuator power (Figure 17) (and so its mean force, see Figure 21), the maximum $A(x_1)$ values may be observed within the (3.3, 3.4)Hz neighbourhood (Figure 8), contrary to the configurations with $g_{23} = 0$ (Figures 6 and 7), for which $A(x_1)$ elevation may be observed at higher frequencies due to the more detrimental actuator response delay in the situation of its higher output force. However, for the (3.3, 3.4)Hz range, MR damper support was not beneficial (see Figures 8 and 21), as was discussed earlier (i.e., the MR damper semi-active operation resulted in a slightly smaller TVA stroke amplitude, while the actuator response delay was not that meaningful as for the higher frequencies). The influence of the actuator power weight on its operation is apparent in Figure 21—the actuator control signal $F_{a\ ctr}$ (and so its output force F_a) was clearly lowered for $g_{23} = 10^{10}$, and particularly for $g_{23} = 10^{11}$. This yielded significant actuator power reduction, i.e., below 0.19 W, regarding the control *case I* ($g_{23} = 0$), as shown in Figure 17 vs. Figures 15 and 16.

Three of the regarded control options deserve a special note: the *restricted force OPT* configuration, control *case I* ($F_{nom} = 6.25$ N, $i_{max} = 0.5$ A, $g_{23} = 0$) vs. the *restricted force GH* configuration ($F_{nom} = 6.25$ N, $i_{max} = 0.5$ A) vs. the *baseline* configuration, control *case III* ($F_{nom} = 12.5$ N, $i_{max} = 1.0$ A, $g_{23} = 10^{10}$). All three configurations yielded maximum x_1 deflection amplitudes of ca. 1.78 mm (Figures 7 and 8), which was more than a 30% reduction regarding the 0.0 A passive system (Figure 6), while the maximum TVA stroke amplitudes were close to those of the 0.0 A passive system. To obtain the primary structure deflection minimisation, both noted *restricted force OPT/GH* configurations required the MR damper of ca. 16 N maximum force and the actuator of 0.35 W nominal power, while the latter approach (*baseline* configuration, control *case III*) required the damper of 3.3 N maximum force and the actuator of 0.17 W nominal power, which is by far a preferable solution. Here comes forth the advantage of the properly tuned *OPT* solution (over the simple *GH* control with changeable F_{nom} and i_{max} values only) with its various optimisation fields (quality function components) to obtain a significantly better efficiency concerning the important aspect of the energy demand, associated with the g_{23} weight use, while the vibration attenuation efficiency was the same as for the *GH* configuration considered here.

Recapitulating, the control solutions that are particularly recommended are:

(1) *H-MR-TVA* system with *OPT* control *case I/GH baseline* configuration, characterised by the 57% maximum structure deflection reduction (regarding the passive configuration), ca. 0.76 W nominal actuator power, and ca. 32 N maximum MR damper force;
(2) *H-TVA* system with *OPT* control *case III baseline* configuration and $g_{23} = 10^{10}$, characterised by the 30% maximum structure deflection attenuation (regarding the passive configuration), 0.17 W nominal actuator power, and 3.3 N maximum (passive) damper force.

For higher actuator force solutions (as the former), the MR damper semi-active support with millisecond response time was greatly beneficial due to the substantial actuator response time delay related to the oscillation period at frequencies above 3.6 Hz (see Figures 5, 6 and 19). On the other hand, it may also be observed that the electric actuator cancelled the unwanted MR damper remanent force, as assumed previously—see Figures 17, 18, 19 and 20a,b,d.

Using a nonzero g_{221} value led to 40% MR damper force reduction ($F_{nom} = 12.5$ N, $i_{max} = 1.0$ see Figure 13); however, the cost was a 24% increased maximum structure deflection, and a 7.5% increased maximum actuator power, which rendered this solution aimless. The influence of the other weighting factors (g_{11}, g_{13}, g_{21}) was studied within the scope of the previous research [34,41]. The operation of the regarded system with MR damper control only (idling electric drive) was also studied before; in current research, it was regarded as aimless, as the already embedded electric actuator adds undesirable excess damping to the TVA system.

With the help of the dynamical similarity analysis that includes previously determined time and length scale factors ($s_T = 0.135$ and $s_L = 0.0176$, respectively) [53] in combination with force scale factor $s_F = 1.75 \cdot 10^{-3}$ [54], the results obtained in the current study may be used as the indicators of the demanded control forces (with regard to actual mass ratio) that have to be generated in the H-MR-TVA (H-TVA) system attached to the real-world Vensys82-class wind turbine structure. Full-scale implementation of the two exemplary control solutions (1) and (2) recommended above will require: (1) an actuator of 7.1 kN nominal force and 3.33 kW nominal power, plus an MR damper of 18.2 kN maximum force, (2) an actuator of 7.1 kN nominal force and 0.74 kW nominal power, plus a passive damper of 1.9 kN maximum force.

8. Conclusions

The purpose of this research was a real-time implementation and experimental study of the nonlinear optimal-based vibration control for a scaled wind turbine tower-nacelle model with the H-MR-TVA system, using the simultaneous operation of the small electric servo drive and the MR damper. With the use of commercially available hardware, the previously developed approach was successfully implemented and validated. The combined operation of the electric actuator and the MR damper in the H-MR-TVA *baseline* system led to a 35% maximum structure deflection reduction in relation to H-MR-TVA operation with MR damper in the passive mode (with zero current) and a 52% reduction compared to the previous research [41]. Both *GH* and *OPT* solutions guaranteed maximum deflection values close to 1.1 mm (the 57% reduction with regard to the best passive TVA configuration), using the MR damper of 32 N maximum force and actuator of 12.5 N nominal force and 0.76 W nominal power. This proved the effectiveness and validity of the adopted concept. No MR damper/actuator force tracking nor online vibration frequency determination were required, which decreased the control adequacy, especially during the multi-frequency and random vibration phases. The MR damper and actuator force constraints did not compromise the control quality. No offline calculation nor disturbance presumption were required for proper system operation.

An interesting alternative is the *OPT* system with the actuator power weight $g_{23} = 10^{10}$. It provided a 30% maximum structure deflection attenuation, regarding the passive configuration, while using a passive damper of 3.3 N maximum force and electric drive of

0.17 W nominal power, which was 22% of the *baseline* system nominal actuator power. This also makes evident the advantage of the properly tuned *OPT* solution (over the simple *GH* control) with its various optimisation fields, covered by the optimal control task quality index—it yielded significantly better energy efficiency marks compared to the *GH* approach exhibiting the same vibration attenuation properties. The *GH* control is devoted to the case when only the protected structure deflection has to be minimised, while the *OPT* concept coped well with different system constraints as, e.g., the TVA working space, the MR damper/actuator nonlinearities (including stroke and force limitations), the actuator power and energy demand, etc., which makes the proposed method a valuable real-time vibration control solution. However, the used electric drive response time is a substantial drawback of the regarded system; the actuator linear dynamics were not considered in this study, and this turned out to yield elevated vibration levels at higher frequencies (which may not be the case for other actuator types to be used, e.g., the electro-hydraulic one). Thus, the nearest research involves a (model) predictive optimal control with the actuator linear dynamics implemented to cope with this problem. Moreover, the possible increase in the sampling rate will benefit the *OPT* control.

Funding: This work was supported by AGH University of Science and Technology (research program no. 16.16.130.942).

Data Availability Statement: The data presented in this study are available on request.

Conflicts of Interest: The authors declare no conflict of interest.

References

1. Wang, X.; Gordaninejad, F. Lyapunov-Based Control of A Bridge Using Magneto-rheological Fluid Dampers. *J. Intell. Mater. Syst. Struct.* **2002**, *13*, 30–36. [CrossRef]
2. Weber, F.; Maślanka, M. Precise stiffness and damping emulation with MR dampers and its application to semi-active tuned mass dampers of Wolgograd Bridge. *Smart Mater. Struct.* **2014**, *23*, 015019.
3. Ali, S.F.; Ramaswamy, A. Hybrid Structural Control using magnetorheological dampers for base isolated structures. *Smart Mater. Struct.* **2009**, *18*, 055011. [CrossRef]
4. Esteki, K.; Bagchi, A.; Sedaghati, R. Semi-Active Tuned Mass Damper for Seismic Applications. In Proceedings of the Smart Materials, Structures & NDT in Aerospace, Montreal, Canada, 2–4 November 2011.
5. Kavyashree, B.G.; Patil, S.; Rao, V.S. Review on vibration control in tall buildings: From the perspective of devices and applications. *Int. J. Dynam. Control* **2020**, *9*, 1316–1331. [CrossRef]
6. Caterino, N. Semi-active control of a wind turbine via magnetorheological dampers. *J. Sound Vib.* **2015**, *345*, 1–17. [CrossRef]
7. Enevoldsen, I.; Mork, K.J. Effects of Vibration Mass Damper in a Wind Turbine Tower. *Mech. Struct. Mach.* **1996**, *24*, 155–187. [CrossRef]
8. Kirkegaard, P.H.; Nielsen, S.R.K.; Poulsen, B.L.; Andersen, J.R.; Pedersen, L.H.; Pedersen, B. Semiactive vibration control of a wind turbine tower using an MR damper. In *Structural Dynamics-EURODYN*; Grundmann, H., Schueller, G.I., Eds.; Swets & Zeitlinger: Lisse, The Netherlands, 2002.
9. Kucuk, I.; Yildirim, K.; Sadek, I.; Adali, S. Optimal control of a beam with Kelvin-Voigt damping subject to forced vibrations using a piezoelectric patch actuator. *J. Vib. Control* **2015**, *21*, 701–713. [CrossRef]
10. Xu, Z.D.; Zhu, J.T.; Wang, D.X. Analysis and Optimisation of Wind-Induced Vibration Control for High-Rise Chimney Structures. *Int. J. Acoust. Vib.* **2014**, *19*, 42–51. [CrossRef]
11. Bakhtiari-Nejad, F.; Meidan-Sharafi, M. Vibration Optimal Control of a Smart Plate with Input Voltage Constraint of Piezoelectric Actuators. *J. Vib. Control* **2004**, *10*, 1749–1774. [CrossRef]
12. Oates, W.S.; Smith, R.C. Nonlinear Optimal Control Techniques for Vibration Attenuation Using Magnetostrictive Actuators. *J. Intell. Mater. Syst. Struct.* **2008**, *19*, 193–209. [CrossRef]
13. Rotea, M.A.; Lackner, M.A.; Saheba, R. Active Structural Control of Offshore Wind turbines. In Proceedings of the 48th AIAA Aerospace Sciences Meeting Including the New Horizons Forum and Aerospace Exposition, Orlando, FL, USA, 4–7 January 2010.
14. Tsouroukdissian, A.; Carcangiu, C.E.; Pineda, A.I.; Martin, M. Wind Turbine Tower Load Reduction using Passive and Semiactive Dampers. In Proceedings of the European Wind Energy Association Annual Event, Brussels, Belgium, 14–17 March 2011.
15. Spencer, B.F., Jr.; Soong, T.T. New Applications and Development of Active, Semi-Active and Hybrid Control Techniques for Seismic and Non-Seismic Vibration in the USA. In Proceedings of the International Post-SMiRT Conference Seminar on Seismic Isolation, Passive Energy Dissipation and Active Control of Vibration of Structures, Cheju, Republic of Korea, 23–25 August 1999.
16. Zhang, Z.; Staino, A.; Basu, B.; Nielsen, S.R.K. Performance evaluation of full-scale tuned liquid dampers (TLDs) for vibration control of large wind turbines using real-time hybrid testing. *Eng. Struct.* **2016**, *126*, 417–431. [CrossRef]

17. Mensah, A.F.; Dueñas-Osorio, L. Improved reliability of wind turbine towers with tuned liquid column dampers (TLCDs). *Struct. Safety* **2014**, *47*, 78–86. [CrossRef]
18. Zhang, J.; Roschke, P.N. Active control of a tall structure excited by wind. *J. Wind. Eng. Ind. Aerodyn.* **1999**, *83*, 209–223. [CrossRef]
19. Den Hartog, J.P. *Mechanical Vibrations*; Dover Publications: Mineola, NY, USA, 1985.
20. Preumont, A.; Alaluf, D.; Bastaits, R. Hybrid Mass Damper: A Tutorial Example. In *Active and Passive Vibration Control of Structures*; Hagedorn, P., Spelsberg-Korspeter, G., Eds.; CISM International Centre for Mechanical Sciences; Springer: Vienna, Austria, 2014. [CrossRef]
21. Demetriou, D.; Nikitas, N. A Novel Hybrid Semi-Active Mass Damper Configuration for Structural Applications. *Appl. Sci.* **2016**, *6*, 397. [CrossRef]
22. Nakamura, Y.; Tanaka, K.; Nakayama, M.; Fujita, T. Hybrid mass dampers using two types of electric servomotors: AC servomotors and linear-induction servomotors. *Earthq. Eng. Struct. Dyn.* **2001**, *30*, 1719–1743. [CrossRef]
23. Nishimura, I.; Kobori, T.; Sakamoto, M.; Koshika, N.; Sasaki, K.; Ohrui, S. Active tuned mass damper. *Smart Mater. Struct.* **1992**, *1*, 306–311. [CrossRef]
24. Di Paolo, M.; Nuzzo, I.; Caterino, N.; Georgakis, C.T. A friction-based passive control technique to mitigate wind induced structural demand to wind turbines. *Eng. Struct.* **2021**, *232*, 111744. [CrossRef]
25. Shen, Y.J.; Wang, L.; Yang, S.P.; Gao, G.S. Nonlinear dynamical analysis and parameters optimization of four semi-active on-off dynamic vibration absorbers. *J. Vib. Control* **2013**, *19*, 143–160. [CrossRef]
26. Martynowicz, P. Vibration control of wind turbine tower-nacelle model with magnetorheological tuned vibration absorber. *J. Vib. Control* **2017**, *23*, 3468–3489. [CrossRef]
27. Martynowicz, P. Control of an MR Tuned Vibration Absorber for Wind Turbine Application Utilising the Refined Force Tracking Algorithm. *J. Low Freq. Noise Vib. Act. Control* **2017**, *36*, 339–353. [CrossRef]
28. Hu, Y.; He, E. Active structural control of a floating wind turbine with a stroke-limited hybrid mass damper. *J. Sound Vib.* **2017**, *410*, 447–472. [CrossRef]
29. Kim, H.T.; Kim, C.H.; Choi, S.B.; Moon, S.J.; Song, W.G. A novel triple actuating mechanism of an active air mount for vibration control of precision manufacturing machines: Experimental work. *Smart Mater. Struct.* **2014**, *23*, 077003. [CrossRef]
30. Han, C.; Choi, S.B.; Lee, Y.S.; Kim, H.T.; Kim, C.H. A new hybrid mount actuator consisting of air spring and magneto-rheological damper for vibration control of a heavy precision stage. *Sens. Actuator A Phys.* **2018**, *284*, 42–51. [CrossRef]
31. Brodersen, M.L.; Ou, G.; Høgsberg, J.; Dyke, S. Analysis of hybrid viscous damper by real time hybrid simulations. *Eng. Struct.* **2016**, *126*, 675–688. [CrossRef]
32. Hu, Y.; Chen, M.Z.Q.; Li, C. Active structural control for load mitigation of wind turbines via adaptive sliding-mode approach. *J. Frankl. Inst.* **2017**, *354*, 4311–4330. [CrossRef]
33. Fitzgerald, B.; Basu, B. Cable connected active tuned mass dampers for control of in-plane vibrations of wind turbine blades. *J. Sound Vib.* **2014**, *333*, 5980–6004. [CrossRef]
34. Martynowicz, P. Nonlinear Optimal-Based Vibration Control of a Wind Turbine Tower Using Hybrid vs. Magnetorheological Tuned Vibration Absorber. *Energies* **2021**, *14*, 5145. [CrossRef]
35. Koo, J.H.; Ahmadian, M. Qualitative Analysis of Magneto-Rheological Tuned Vibration Absorbers: Experimental Approach. *J. Intell. Mater. Syst. Struct.* **2007**, *18*, 1137–1142. [CrossRef]
36. Martynowicz, P. Study of vibration control using laboratory test rig of wind turbine's tower-nacelle system with MR damper based tuned vibration absorber. *Bull. Pol. Acad. Sci. Tech. Sci.* **2016**, *64*, 347–359.
37. Rosół, M.; Martynowicz, P. Implementation of LQG controller for wind turbine tower-nacelle model with MR tuned vibration absorber. *J. Theor. Appl. Mech.* **2016**, *54*, 1109–1123. [CrossRef]
38. Martynowicz, P.; Santos, M. Structural vibration control of NREL 5.0 MW FOWT using optimal-based MR tuned vibration absorber. In Proceedings of the 21st IFAC World Congress, Berlin, Germany, 11–17 July 2020.
39. Wang, L.; Liang, Z.; Cai, M.; Zhang, Y.; Yan, J. Adaptive Structural Control of Floating Wind Turbine with Application of MR Damper. *Energy Procedia* **2019**, *158*, 254–259. [CrossRef]
40. Martynowicz, P. Nonlinear optimal-based vibration control for systems with MR tuned vibration absorbers. *J. Low Freq. Noise Vib. Act. Control* **2019**, *38*, 1607–1628. [CrossRef]
41. Martynowicz, P. Real-time implementation of nonlinear optimal-based vibration control for a wind turbine model. *J. Low Freq. Noise Vib. Act. Control* **2019**, *38*, 1635–1650. [CrossRef]
42. Kciuk, S.; Martynowicz, P. Special application magnetorheological valve numerical and experimental analysis. *Control. Eng. Mater. Process. Diffus. Defect Data–Solid State Data Pt. B Solid State Phenom.* **2011**, *177*, 102–115. [CrossRef]
43. Laalej, H.; Lang, Z.Q.; Sapinski, B.; Martynowicz, P. MR damper based implementation of nonlinear damping for a pitch plane suspension system. *Smart Mater. Struct.* **2012**, *21*, 045006. [CrossRef]
44. Neelakantan, V.A.; Washington, G.N. Vibration Control of Structural Systems using MR dampers and a 'Modified' Sliding Mode Control Technique. *J. Intell. Mater. Syst. Struct.* **2008**, *19*. [CrossRef]
45. Lord RheoneticTM. *MR Controllable Friction Damper RD-1097-01 Product Bulletin*; Lord: Cary, NC, USA, 2002.
46. Bryson, A.E.; Ho, Y.C. *Applied Optimal Control*; Taylor & Francis: London, UK, 1975.
47. Itik, M. Optimal control of nonlinear systems with input constraints using linear time varying approximations. *Nonlinear Anal. Model. Control.* **2016**, *21*, 400–412. [CrossRef]

48. Shukla, P.; Ghodki, D.; Manjarekar, N.S.; Singru, P.M. A Study of H infinity and H2 synthesis for Active Vibration Control. *IFAC-PapersOnLine* **2016**, *49*, 623–628. [CrossRef]
49. Martynowicz, P.; Szydło, Z. Wind turbine's tower-nacelle model with magnetorheological tuned vibration absorber: The laboratory test rig. In Proceedings of the 14th International Carpathian Control Conference (ICCC), Rytro, Poland, 26–29 May 2013.
50. Martynowicz, P. Development of Laboratory Model of Wind Turbine′s Tower-Nacelle System with Magnetorheological Tuned Vibration Absorber. *Solid State Phenom.* **2014**, *208*, 40–51. [CrossRef]
51. Martynowicz, P.; Rosół, M. Wind Turbine Tower-Nacelle System with MR Tuned Vibration Absorber: Modelling, Test Rig, and Identification. In Proceedings of the ICCC 2019: Proceedings of the 20th International Carpathian Control Conference, Kraków-Wieliczka, Poland, 26–29 May 2019.
52. Rosół, M.; Martynowicz, P. Identification of the Wind Turbine Model with MR Damper Based Tuned Vibration Absorber. In Proceedings of the ICCC 2019: Proceedings of the 20th International Carpathian Control Conference, Kraków-Wieliczka, Poland, 26–29 May 2019.
53. Snamina, J.; Martynowicz, P.; Łatas, W. Dynamic similarity of wind turbine's tower-nacelle system and its scaled model. *Solid State Phenom.* **2014**, *208*, 29–39. [CrossRef]
54. Snamina, J.; Martynowicz, P. Prediction of characteristics of wind turbine's tower-nacelle system from investigation of its scaled model. In Proceedings of the 6WCSCM: Sixth World Conference on Structural Control and Monitoring–proceedings of the 6th edition of the World conference of the International Association for Structural Control and Monitoring (IACSM), Barcelona, Spain, 15–17 July 2014.
55. Maślanka, M. Optimised semi-active tuned mass damper with acceleration and relative motion feedbacks. *Mech. Syst. Signal Process.* **2019**, *130*, 707–731. [CrossRef]
56. Ioffe, A.D.; Tihomirov, V.M. *Theory of Extremal Problems. Studies in Mathematics and its Applications*; North-Holland Publishing Company: Amsterdam, The Netherlands, 1979.
57. Pinto, S.G.; Rodriguez, S.P.; Torcal, J.I.M. On the numerical solution of stiff IVPs by Lobatto IIIA Runge-Kutta methods. *J. Comput. Appl. Math.* **1997**, *82*, 129–148. [CrossRef]
58. Festo Ltd. Festo Products. 2022. Available online: https://www.festo.com/gb/en/c/products-id_pim1 (accessed on 12 October 2022).
59. *TMS 60 Lbf Modal Shaker*; The Modal Shop Inc.: Cincinnati, OH, USA, 2010.
60. *InTeCo RT-DAC4/PCI Multi I/O Board. User's Manual*; InTeCo Ltd.: Krakow, Poland, 2002.
61. Sahu, G.N.; Singh, S.; Singh, A.; Law, M. Static and Dynamic Characterization and Control of a High-Performance Electro-Hydraulic Actuator. *Actuators* **2020**, *9*, 46. [CrossRef]

energies

Article

Wind Microturbine with Adjustable Blade Pitch Angle

Stanisław Chudzik

Faculty of Electrical Engineering, Czestochowa University of Technology, 42-200 Czestochowa, Poland; stanislaw.chudzik@pcz.pl

Abstract: The article presents the results of research on the operation of a wind microturbine model with an adjustable blade pitch angle. The physical basics of wind turbine operation and the methods of its optimal control are discussed. The results of the measurements carried out for the selected blade geometry with the possibility of adjusting the pitch angle are presented. The tests were carried out for a resistive load with a linear characteristic and for a load with a non-linear characteristic of a Li-Po battery. The results of the operation of a simple MPPT control algorithm are presented. The practical methods of controlling larger wind turbines are not optimal for small and very small turbines. The conducted research focused on determining the possibility of using blades with an adjustable angle setting, depending on the rotational speed in wind microturbines. The use of a simple mechanism for changing the pitch angle of the blades depending on the rotational speed of the turbine can increase the efficiency of the microturbine in a wider range of wind speeds.

Keywords: small wind turbine control; maximum power point tracking; pitch angle wind turbine

Citation: Chudzik, S. Wind Microturbine with Adjustable Blade Pitch Angle. *Energies* **2023**, *16*, 945. https://doi.org/10.3390/en16020945

Academic Editor: Paweł Ligęza

Received: 16 December 2022
Revised: 12 January 2023
Accepted: 13 January 2023
Published: 14 January 2023

1. Introduction

The power of designed and built wind farms in the world is constantly increasing and reaches the value of MW. This pursuit is economically justified as larger turbines achieve higher efficiency of generated power. Modern commercial wind farms mostly have a horizontal axis of rotation and a three-blade turbine. The most frequently used generators are asynchronous machines and synchronous machines with permanent magnets (PMSG) together with power electronic converters and control systems. Modern wind turbines with a horizontal axis of rotation are usually adapted to work at wind speeds not exceeding 25 m/s. These turbines achieve their rated power at wind speeds between 10 and 15 m/s [1]. PMSG generators made of neodymium magnets are most often used in low-power wind turbines with high rotational speed, while in larger power plants, induction generators with a mechanical gear are used [2]. In modern wind turbines, several methods of adjusting the rotor speed and the generator power related to it are used in several ways, depending on the instantaneous wind speed [2]. To reduce mechanical power on the shaft in large turbines, pitch control is used, while in small and medium-sized ones, mainly passive stall control is used, breaking the laminar air stream [3,4].

The European Parliament issued a directive 2010/31/UE on the energy performance of buildings, introducing the nearly zero-energy building concept (zero-energy building). According to this directive, new buildings must meet the requirement that a significant amount of energy comes from renewable sources, including sources integrated with the building [5]. As a result, there is a growing interest in micro wind turbines that would supplement the electricity demand of small households [6–8]. The manufacturers' offer includes wind microturbines with a nominal power from 100 W to several kW. Miniature wind turbines can also often be seen on small yachts, recreational plots, and street-lighting lamps where there is no access to the power grid—Figure 1. Typically, such microturbines have three or more blades with as simple a profile as possible, are easy to manufacture, and directly mounted on the PMSG generator axis. Most often, the rated voltage of the generator is selected so that it charges the battery using an ordinary 6D rectifier.

(a) (b)

Figure 1. Hybrid power supply from renewable energy sources: (**a**) street-lighting lamp; (**b**) on the yacht.

The belief that the cost of manufacturing a controller for the optimal operation of a microturbine is high in relation to the possible improvement of the efficiency of the microturbine means that there is no such controller on the market. In addition, control methods for larger wind turbines are not optimal for small turbines [9–11]. Therefore, the author designed and built a small test stand that allows for preliminary research related to the selection of the optimal wind turbine geometry, and the development and testing of the operation of the control algorithm for the operation of a simple converter charging the battery so as to optimally use the power of the wind turbine [12].

The presented research focuses mainly on determining the possibility of using wind turbines with an adjustable blade pitch angle depending on the rotational speed in micro-turbines. In larger wind turbines, changing the pitch angle is only used to limit their power at high wind speeds [10,11]. Currently, such solutions in micro wind farms are not used for economic reasons. Wind microturbines are very often located in places where wind conditions are unfavorable: low height from the ground, and terrain obstacles limiting the speed of the wind stream and causing its turbulence [1,9]—Figure 1a. The use of a simple blade pitch angle adjustment mechanism depending on the angular velocity of the turbine can increase its efficiency in a wider range of wind speeds.

The test stand with small dimensions enables quick and cheap development of initial prototypes of turbine blades due to the use of 3D printing technology.

2. Materials and Methods

2.1. Working Principle of Wind Turbine

The operation of the wind turbine depends on the setting of the blade angle ϕ (Figure 2): it is the angle between the blade chord and the apparent wind V, being the sum of the vectors of the blade linear velocity $V_\omega = \omega \cdot r$ and the wind speed V. The value of the angle of attack α depends on the ratio of the linear blade speed V_ω to the wind speed V and the blade position angle. Two forces act on the turbine blade: the lifting force F_L, perpendicular to the direction of the resultant velocity V, and the thrust F_D, corresponding to its direction. The rotation of the turbine causes the component F_T of the resultant force F_L in the direction of the blade rotation. The value of F_T depends on the aerodynamic profile of the blade and is a non-linear function of the angle of attack. In the range of small values of the angle of attack, this force grows approximately linearly, but from a certain value of this angle it rapidly decreases; a stall state occurs [1,2,13]. Maintaining a constant value of the angle of attack, ensuring maximum aerodynamic efficiency, requires maintaining an approximately constant ratio of the angular velocity of the rotor to the wind speed, i.e., the change in rotor speed in proportion to the wind speed.

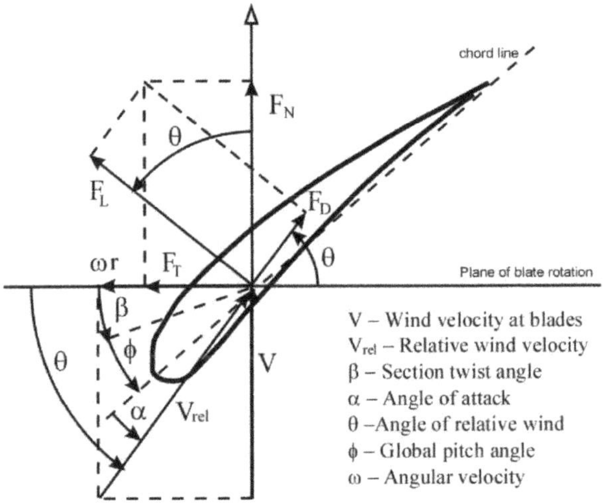

Figure 2. Aerodynamic forces on the blade profile.

The lift generated by the airfoil section is a function of the angle of attack to the incoming air stream. The angle of the air stream inflow depends on the rotational speed and the wind speed within a certain radius. The required twist angle of the blade depends on the ratio of the airfoil speed over a certain radius and the desired angle of attack of the airfoil. The part of the blade closer to the hub is tilted more against the wind due to the high ratio of wind speed to blade radial speed. The blade tip, on the other hand, will be almost perpendicular to the wind direction. In the case of blades with a constant angle of their setting, the optimal twist angle of the blade can be determined. However, when the angle of the blade is to be adjusted, the twist angle of the blade should also change. It would be too costly to build a wind turbine with variable-adjustable geometry and currently such solutions are not used in practice. Therefore, blade angle adjustment is only used in large turbines to limit their power. In the case of airplanes or helicopters where the adjustable pitch propeller is used as a drive, the blades of such propellers have a small or zero twist angle—this provides a wider speed range for effective operation of the propulsion [1,2]. For the purposes of the research on the impact of the adjustment of the blade angle on the operating efficiency range of a wind turbine, a propeller with a straight profile and a zero twist angle of the blade was made.

The aerodynamic properties of the wind turbine are determined by the power factor $Cp(\lambda, \beta)$, which depends on the tip speed ratio λ and the blade pitch angle β. The coefficient λ is defined as the ratio of the linear speed of the turbine blade tip to the wind speed,

$$\lambda = \frac{\omega_r \cdot R}{V_w}, \tag{1}$$

where ω_r is the angular velocity of the turbine and R is the radius of the turbine.

The power factor corresponding to the aerodynamic efficiency of a wind turbine is given by the expression,

$$C_P = \frac{P_m}{P_w}, \tag{2}$$

where P_m is the mechanical power of the turbine and P_w is the wind power:

$$P_w = 0,5\rho\pi R^2 V^3, \tag{3}$$

where ρ is the air density.

Figure 3 shows an example of the $Cp(\lambda, \beta)$ characteristic of a wind turbine. The maximum power generated by the turbine at a given wind speed V_w is achieved for the maximum value of the efficiency coefficient corresponding to a certain optimal value λ_{opt} of the tip speed ratio [1,2].

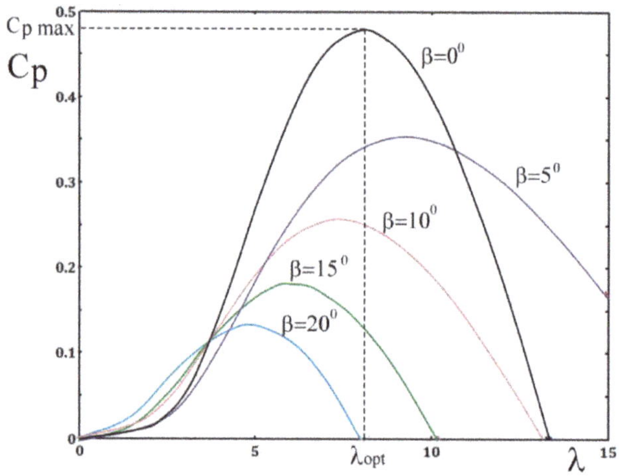

Figure 3. An example of the $Cp(\lambda, \beta)$ characteristics.

The mechanical power of the turbine and the output of the electrical system P_{out} is described by Equation (4), where M_t is the frictional moment, ω is the turbine angular velocity, J is the moment of inertia of the rotating mass, η is the overall electrical efficiency of the system from the generator input to the inverter output.

$$P_m = \frac{1}{\eta} P_{out} + \omega \cdot M_t + \omega \cdot J \frac{d\omega}{dt}, \quad (4)$$

The boundary layer is a thin layer of air and appears on the surface of objects in viscous flow. The boundary layer effect depends on the flow pattern which is related to the Reynolds number (5). Reynolds number is a function of fluid velocity,

$$Re = \frac{\rho \cdot V \cdot C}{\mu}, \quad (5)$$

where ρ is the fluid density, V is the mean velocity of the fluid, C is a chord length, and μ is the dynamic viscosity of the fluid.

Airfoil efficiency decreases as the Reynolds number decreases. Reducing the Reynolds number means that viscous effects become more dominant, increasing viscous drag. In addition, the thickness of the boundary layer will increase as the Reynolds number decreases, leading to increased shape drag or less lift. Small wind turbines usually operate at Reynolds number less than 5×10^5. In this Reynolds number range, laminar flow gets separated at the upper surface of the airfoil and is reattached to the surface as turbulent causing laminar separation bubble, which increases the drag of the airfoil. Large-scale wind turbines have become the development trend of wind power. At present, the radius of wind turbine rotors ranges to one hundred meters, or even more, which extends Reynolds number of the airfoil profile from the order of 10^6 to 10^7.

2.2. The Most Commonly Used Methods of Optimal Control of Wind Turbine Operation

The algorithms used to achieve maximum power point tracking (MPPT) by wind turbines can be divided into three main control methods: tip speed ratio (TSR) control, power signal feedback (PSF) control and, search method maximum power (hill-climb search—HCS) [14–16]. The TSR control method regulates the rotational speed of the wind turbine in order to maintain the optimal value of the tip speed ratio, at which the achieved turbine power is the highest [17,18]. This method requires measuring or estimating both the wind speed and the rotational speed of the turbine, and also requires knowledge of the characteristics of the optimal tip speed ratio for the designed turbine—Figure 4.

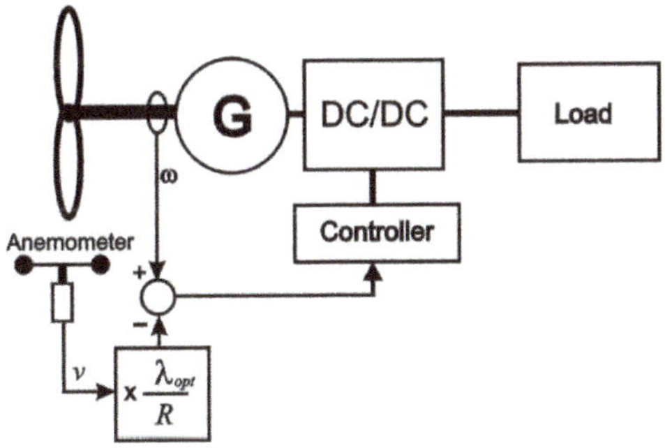

Figure 4. Tip speed ratio control.

In the PSF control method, it is required to know the maximum power characteristics of a wind turbine and to follow this curve with a control and measurement system [19,20]. The maximum power curves should be obtained by simulations or experiments with a disconnected wind turbine. In this method, the reference power is determined from the recorded maximum power curve or from the wind turbine power equation where the input speed is either wind speed or rotor speed—Figure 5.

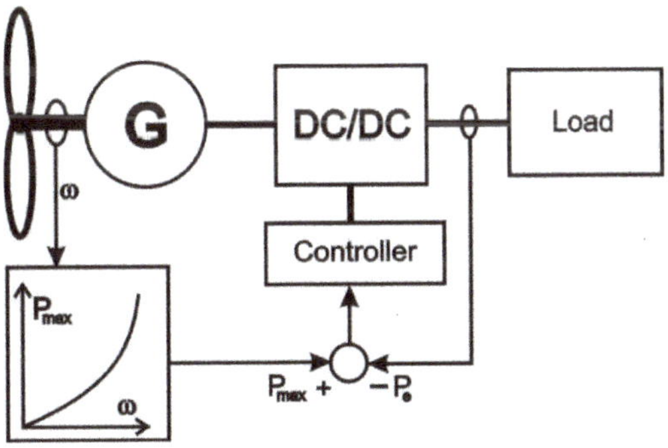

Figure 5. Power signal feedback control.

The HCS control algorithm is constantly looking for the peak power of the wind turbine—Figure 6. The tracking algorithm, depending on the position of the operating point of the turbine and the relationship between changes in power and rotational speed of the turbine, calculates the optimal signal to bring the wind turbine to the point of maximum power [21–25]. Unfortunately, HCS control can only work well when the moment of inertia of the wind turbine is very small so that the change in turbine speed occurs almost "instantaneously" to the change in wind speed. For wind turbines with higher inertia, the instantaneous power output of the power plant is related to the mechanical power of the turbine and changes in kinetic energy stored in the rotating elements, which often makes the HCS method ineffective. HCS control does not reach the maximum power points with rapid increases in wind speed and causes the so-called "stall" when wind speed decreases, which severely limits the usefulness of this method for wind turbines. Boundary layer flow characteristics are critical for determining inputs to control systems. The occurrence of the laminar-to-turbulent transition process, especially at high angles of attack and rapid changes of angles of attack, can significantly change the aerodynamic behavior of the blade and cause the phenomenon of stalling [26,27].

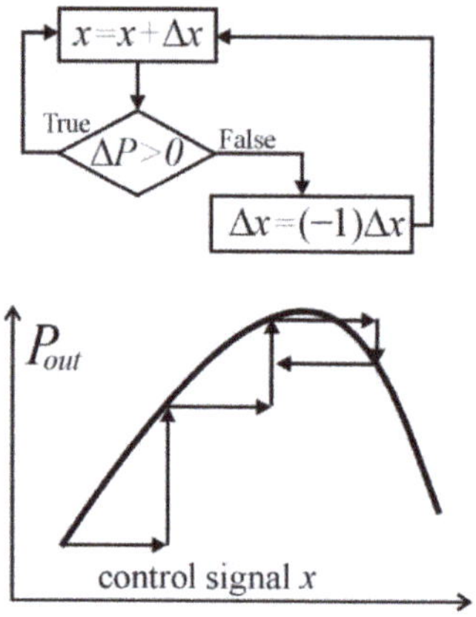

Figure 6. Hill-climb search algorithm.

It is therefore highly desirable to develop a maximum power output method for micro wind turbines that does not require measuring wind speed and turbine rotor speed, is independent of system characteristics, and is applicable to small wind turbines.

2.3. Experimental Stand

A small test stand was built for the purpose of testing the wind turbine model.

Figure 7 shows a schematic diagram of the stand. The test stand includes a virtual measurement and control device implemented on a PC class computer and a physical model of a miniature wind tunnel. The virtual measurement and control device was created in the LabView programming environment. Figure 8 shows the view of the front panel. The device can set the output value of the fan power, the value of the PWM duty cycle for the DC/DC converter, and the pitch value of the wind turbine blade. In addition, the instrument displays and records the following measured values: wind speed v, turbine speed n, generator output voltage U_G, output voltage of the DC/DC converter U_O, and load current I_O. The power delivered to the load P_O is also determined.

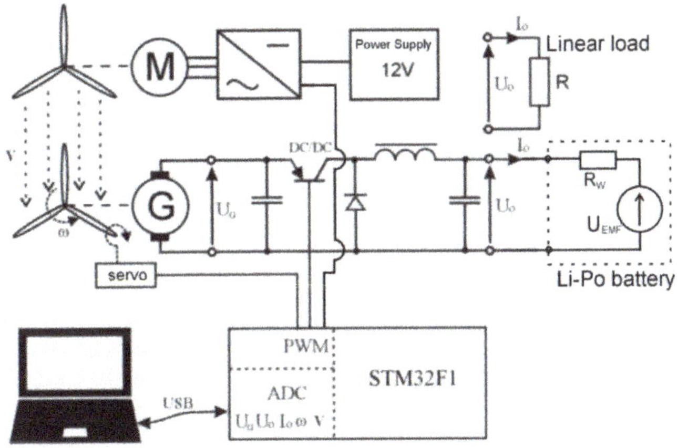

Figure 7. Schematic diagram of the stand for testing the wind microturbine model.

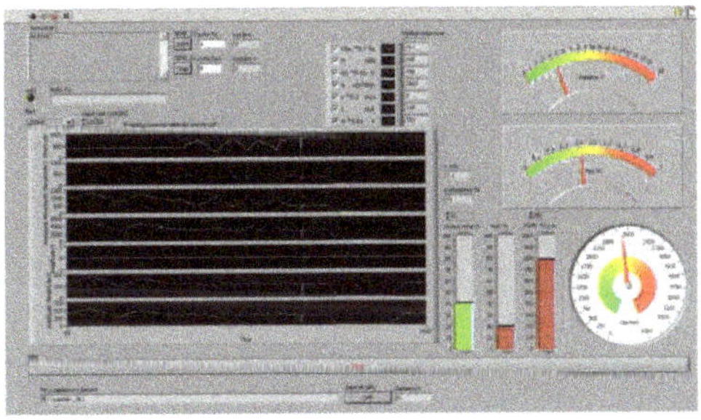

Figure 8. Virtual measuring and control device—front panel.

The model of the wind tunnel made, shown in Figure 9a,b, consists of a tunnel pipe (1) fixed in a frame (2) which also acts as a honeycomb (air stream straightener). Attached to the frame is a brushless BLDC fan motor (3) with a 203 mm (8-inch) diameter propeller and 101 mm (4-inch) pitch, and a DC brush motor (4) with permanent magnet excitation acting as a DC generator. The generator is powered by a propeller that functions as a wind turbine. At the entrance to the tunnel there is an anemometer sensor (5) measuring the speed of the wind. The BLDC motor of the fan is powered by a three-phase inverter (6). The research model has a controller (7) based on a 32-bit STM32F1 microcontroller which controls the operation of the fan, DC/DC converter, and the servomechanism (12) of the turbine propeller pitch (11). The controller also measures wind speed v, turbine rotational speed ω, generator output voltage U_G, converter output voltage U_O, and load current I_O (8). The controller, via the USB serial interface (9), communicates with the master computer on which the virtual measurement and control device program is running. The model is powered by a power supply (10) with a voltage of 12 V. Due to the small size of the model, some of its elements could be produced using a 3D printer.

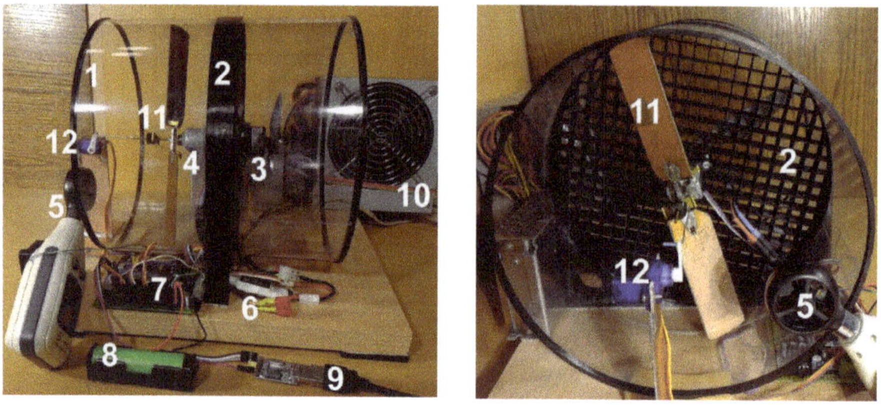

(**a**) (**b**)

Figure 9. (**a**) View of the research model; (**b**) View of the wind tunnel.

3. Results and Discussion

The virtual measurement and control device has the ability to automatically record measured values for ten set fan power values and one hundred set values of the PWM wave fill factor for the transistor in the DC/DC converter (buck converter). Because of this solution, a matrix of 1000 values is obtained for each measured value without the tedious involvement of the person conducting the measurements and data recording. Moreover, the measuring and control device has the possibility of automatic adjustment of the blade setting angle in accordance with the control function described by the expression dependent on the measured instantaneous rotational speed of the turbine $\beta = f(n)$. Ultimately, the practical solution of the microturbine provides for a possibly simple and cheap mechanical solution using the centrifugal force of the rotating mass, which would change the pitch of the blades. The optimal control of the power plant is based on achieving the maximum power for a given wind speed value. Power control, in this system solution of the power plant, is carried out by changing the duty cycle of the PWM signal that controls the operation of the switch in a simple DC/DC converter, which has a direct impact on the value of the converter output voltage. In the conducted research, measurements were carried out for a load with a non-linear current–voltage characteristic, which is a lithium-polymer battery cell. The electromotive force EMF of the cell used, depending on the state of charge, was in the range (3.3 ÷ 4.3) V and its internal resistance was determined as $R_w \approx 0.25\ \Omega$.

3.1. Wind Turbine with a Classic Aerofoil Profile Blade

In order to compare the results of the research on the influence of the adjustment of the pitch angle of the blades on the efficiency of the wind turbine operation, a propeller with a blade with a classic aerodynamic profile and a constant pitch angle was used in the experiment. Examples of power generated by a wind turbine for a resistive load as a function of the duty cycle of the PWM signal (control quantity) for different wind speeds are shown in Figure 10.

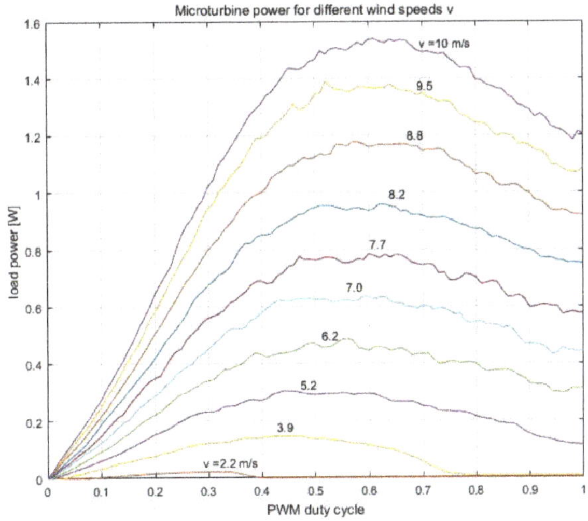

Figure 10. The power generated by a wind turbine for a linear load as a function of the control value—the duty cycle of the PWM signal, for different wind speeds.

It is worth noting that for a linear load such as a resistor, the optimal operation of the power plant for higher wind values is achieved for a narrow range of the PWM duty cycle (0.50 ÷ 0.65). The limitation of the power of the power plant in this case is mainly due to the relatively high resistance value of the generator winding. For lower wind speeds, the power limitation results mainly from the low value of mechanical power achieved by the wind turbine (propeller), which is transferred to the generator shaft. Examples of power generated by a wind turbine loaded with a Li-Po cell as a function of the duty cycle of the PWM signal (control quantity) for various wind speeds are shown in Figure 11. It is worth noting that for a non-linear load such as a Li-Po cell, the optimal control of the power plant operation requires a wider range of PWM duty cycle values within the range (0.4 ÷ 0.85). It is also necessary to take into account changes in the EMF voltage of the cell depending on the degree of its charge.

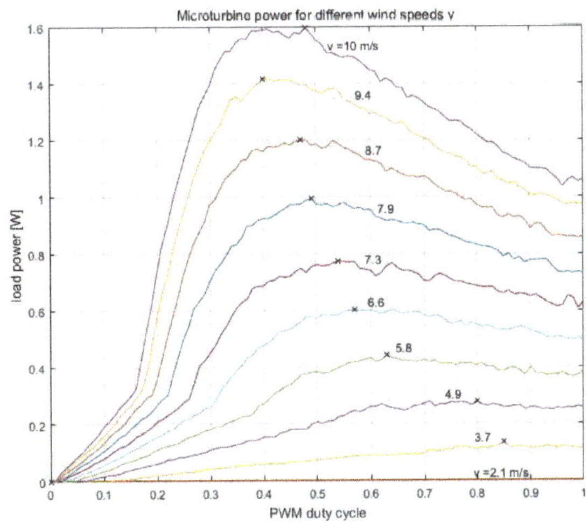

Figure 11. The power generated by a wind turbine for a non-linear load, as a function of the control value—duty cycle of the PWM signal, for different wind speeds.

Figure 12 presents the obtained characteristics of the maximum power of the power plant depending on the wind speed. It can be seen that at higher wind speeds, the slope of the characteristic no longer increases. The cause may be a significant increase in the thrust force F_r, resulting from the increase in the rotational speed of the turbine—the frontal resistance of the airfoil. In the conducted experiment, the turbine achieves a high tip speed ratio λ >10: the turbine blades are positioned almost perpendicular to the wind direction. In addition, taking into account the static friction of bearings and brushes and the cogging torque resulting from changes in the reluctance of the generator's magnetic circuit, the start-up of such a turbine is very difficult and takes place only at a relatively high wind speed of 4 m/s, compared to the wind speed of 1.5 m/s at which it stops.

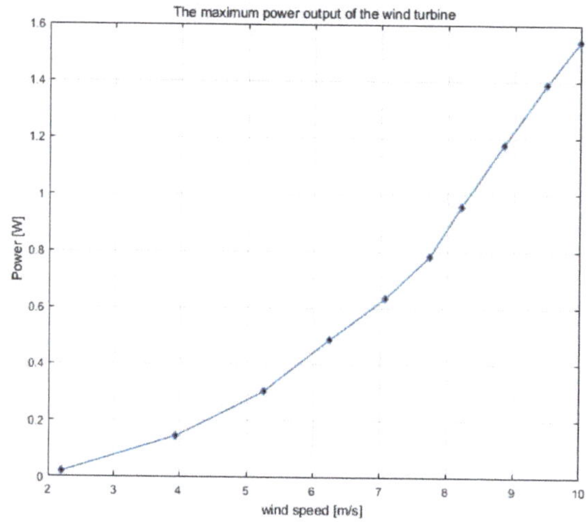

Figure 12. The maximum power of a wind turbine.

Preliminary results of the measurements indicate that the direct connection of the battery to the generator, often used in microturbines, whose charging characteristics are strongly non-linear and varies depending on the battery charge level, may be the cause of poorer efficiency of micro power plants. The solution to this problem may be the use of a possibly simple DC/DC converter, which could linearize the load characteristics of the wind microturbine.

Tests were carried out of a micro wind power plant model, controlled by a simple HCS algorithm presented in Figure 6. Figure 13 shows selected results of the experiment in the form of graphs: microturbine wind power, PWM duty factor, and turbine rotation speed. The results were obtained at a constant wind speed v = 5 m/s.

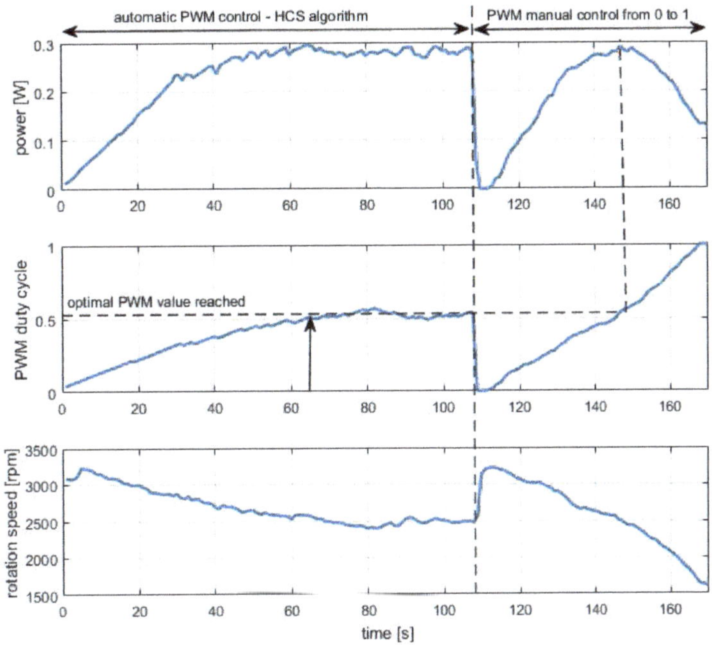

Figure 13. Graphs: microturbine power, duty factor PWM, and turbine rotation speed of a microturbine model controlled by HCS algorithm.

The results of the experiment indicate the possibility of using a simple HCS algorithm to control the operation of a wind microturbine only for a stable value of wind speed. In the case of rapid and greater decreases in wind speed, the presented simple HCS algorithm no longer works optimally, even leading to the wind turbine stopping and the need to restart it. Some improvement in operation can be achieved in this case by using the measured turbine rotational speed in the control algorithm.

3.2. *Wind Turbine with Adjustable Pitch*

For the purposes of the research on the impact of the adjustment of the blade angle on the operating efficiency range of a wind turbine, a propeller with a straight profile and a zero twist angle of the blade was used in the experiment—Figure 9. Figure 14 shows the power characteristics obtained in the tests carried out on a wind turbine with a constant pitch $\beta = 15^{\circ}$ as a function of the value of the PWM signal duty factor (control variable) for various wind speeds.

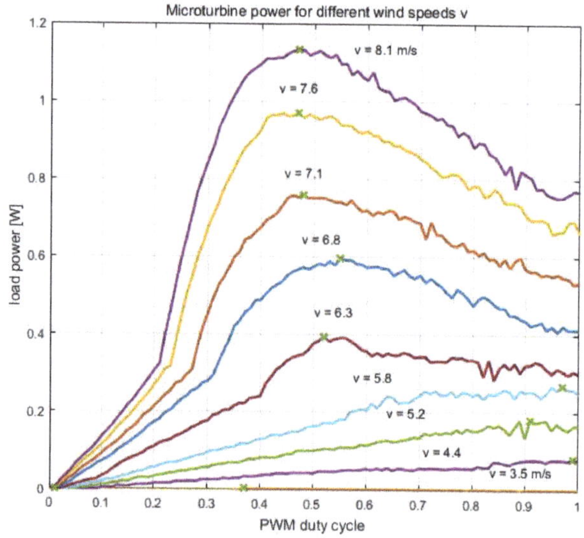

Figure 14. The power generated by a wind turbine with a constant pitch, as a function of the control value—duty cycle of the PWM signal, for different wind speeds.

On the other hand, Figure 15 shows the power characteristics obtained in the tests carried out on a wind turbine with adjustable pitch depending on the rotation speed n [rpm] of the turbine $\beta(n) = 30^{\circ} - n/150$, as a function of the duty cycle of the PWM signal for various wind speeds. It is worth noting that for a non-linear load such as a Li-Po cell, optimal operation of the power plant requires a wider range of PWM control variable values in the range (0.45 ÷ 0.95). It is also necessary to take into account changes in the EMF voltage of the cell depending on the degree of its charge.

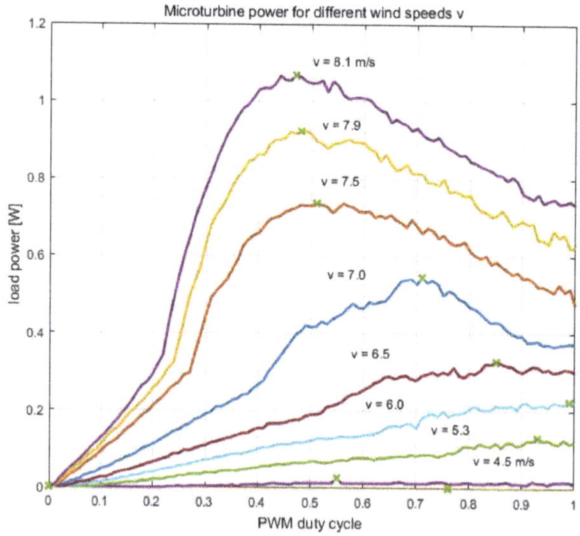

Figure 15. The power generated by a wind turbine with adjustable pitch, as a function of the duty cycle, for different wind speeds.

Figure 16 compares the obtained two characteristics of the maximum turbine power achieved as a function of wind speed for the case of fixed (β = *const*) and adjustable angle of the pitch ($\beta = f(n)$) of turbine blades. It can be seen that for the case of the adjustable angle of the pitch, the efficiency of the power plant is higher. In the experiment, the turbine achieves a high value of tip speed ratio $\lambda \approx 10$; the blades of the turbine are positioned almost perpendicular to the direction of the wind. In addition, taking into account the static friction of bearings and brushes and the cogging torque resulting from changes in the reluctance of the generator's magnetic circuit, the start-up of such a turbine is very difficult and takes place only at a relatively high wind speed of 4 m/s, compared to the wind speed of 1.5 m/s at which it stops. Adjusting the pitch angle of the blades can greatly facilitate the start-up of a wind turbine and allow it to operate at lower wind speeds. In addition, through the appropriate selection of the dependence $\beta = f(n)$, it is also possible to limit the maximum rotational speed of the turbine at which $\beta(n_{max}) = 0$.

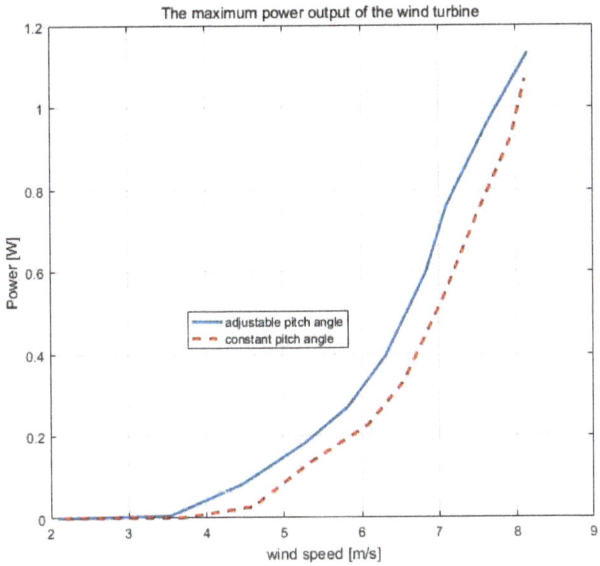

Figure 16. The maximum power of a wind microturbine for the case of constant and adjustable pitch.

The results of the measurements indicate that the direct connection of the battery to the generator, often used in wind microturbines, whose charging characteristics is strongly non-linear and changes depending on the battery charge level, may be the cause of poorer performance of the microturbine. The solution to this problem may be the use of a possibly simple converter that could linearize the load characteristics of the microturbine.

Tests were carried out of a wind microturbine model, controlled by an HCS algorithm, presented in Figure 6. Figure 17 shows selected results of the experiment in the form of graphs: power of the wind farm, value of the PWM duty factor, and the rotation speed of the turbine. The results were obtained at constant wind speed v = 6.5 m/s.

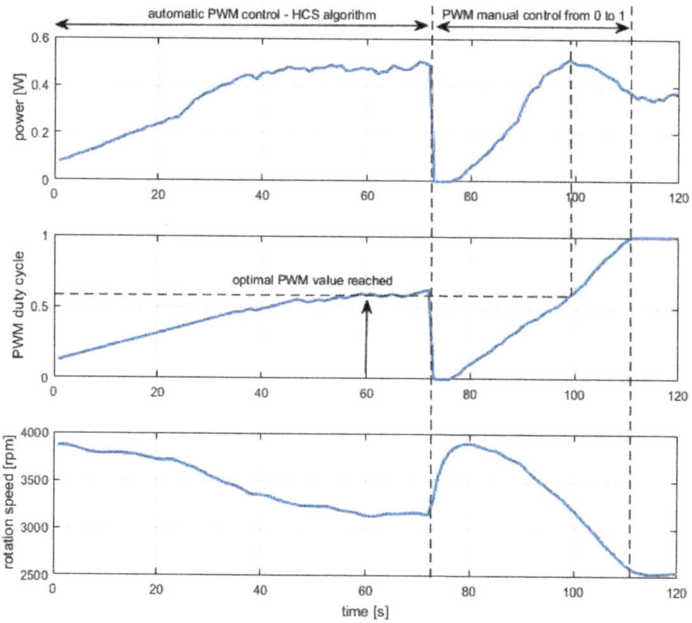

Figure 17. Graphs: microturbine power, duty factor PWM, and turbine rotation speed of a microturbine model controlled by HCS algorithm for adjustable pitch angle of the turbine blade.

The obtained results of the experiment show the possibility of using a simple HCS algorithm to control the operation of a wind microturbine only for a stable value of wind speed. In the case of rapid and greater decreases in wind speed, the presented simple HCS algorithm no longer works optimally, leading to a significant decrease in the turbine's rotational speed and its re-acceleration. A slight improvement in operation can be achieved in this case by using a simple mechanism for adjusting pitch angle of the blade $\beta = f(n)$, which prevents sudden stalling and facilitates the acceleration of the turbine. Some improvement can also be obtained by introducing additional actions to the control algorithm resulting from additional measurement of the instantaneous rotational speed of the turbine.

4. Conclusions

The obtained measurement results showed that the direct connection of the battery to the generator may cause poorer efficiency of the wind microturbine. The solution to this problem may be the use of a simple converter, which can linearize the load characteristics of the wind microturbine. The possibility of changing the pitch of the wind turbine depending on its rotation speed gives some improvement in the efficiency of the wind microturbine. However, in practical implementation, the additional costs of manufacturing the blade pitch adjustment mechanism and its reliability in difficult environmental conditions related to icing should be taken into account. The use of a simple mechanism for adjusting the pitch of the turbine using the centrifugal force of the rotating mass may be beneficial for two-blade microturbines with a high tip speed ratio. Resigning from the third blade in the wind turbine in favor of the blade angle adjustment mechanism, it is possible to achieve a higher rotational speed of the turbine, and thus to use a cheaper generator with a smaller number of pole pairs. The DC brush motor with permanent magnets used in the model as a generator proved its usefulness. The additional resistance to movement caused by the brushes did not cause significant limitations at lower wind speeds. Research has shown that DC motors used in radiator fans from scrap cars can be used in micro wind turbines for recycling. The use of a simple DC/DC converter in this case would allow optimal adjustment of the load characteristic to achieve the maximum power of the wind turbine in a wide range of wind speeds. The conducted research showed that the built model of the wind microturbine allows us to observe and measure the physical phenomena related to the aerodynamics of the wind turbine and the operation of the electricity generator. Test results that use only simplified mathematical models of the wind turbine often do not take into account the phenomenon of stalling, which is significant when the maximum load power of the wind turbine for a given wind speed is exceeded.

The research model in the presented scale allows you to quickly and cheaply make and test initial prototypes of the designed wind turbine blades using popular 3D printers. In addition, the controller made enables the implementation, in the program memory of the used STM32F1 microcontroller, the developed algorithms for optimal control of the wind microturbine. The computing power of the 32-bit microcontroller is sufficient even for more complex algorithms. The specificity of the operation, construction, and limitations of micro wind turbines is in some aspects completely different than in the case of larger wind turbines, for which numerous scientific studies have already been prepared.

Due to the growing demand for autonomous and portable hybrid micro power plants, combining small photovoltaic panels and wind micro turbines, it is advisable to conduct research on improving the energy efficiency of these devices [6–8].

Further research work will focus on the use of the PMSG permanent magnet synchronous generator, the start-up of a wind microturbine with a high tip speed ratio, and algorithms for optimal MPPT control of the wind microturbine.

Funding: This research received no external funding.

Data Availability Statement: Data sharing is not applicable to this article.

Conflicts of Interest: The author declares no conflict of interest.

References

1. Wu, Q.; Sun, Y. (Eds.) *Modeling and Modern Control of Wind Power*; John Wiley & Sons: Chichester, UK, 2018.
2. Bianchi, F.; De Battista, H.; Mantz, R. *Wind Turbine Control Systems. Principles, Modelling and Gain Scheduling Design*; Springer: London, UK, 2007.
3. Kadri, A.; Marzougui, H.; Bacha, F. MPPT control methods in wind energy conversion system using DFIG. In Proceedings of the 4th International Conference on Control Engineering & Information Technology (CEIT), Hammamet, Tunisia, 16–18 December 2016; pp. 1–6.
4. Zammit, D.; Spiteri Staines, C.; Micallef, A.; Apap, M. MPPT with Current Control for a PMSG Small Wind Turbine in a Grid-Connected DC Microgrid. In *Wind Energy Exploitation in Urban Environment*; Battisti, L., Ricci, M., Eds.; Springer International Publishing: Berlin/Heidelberg, Germany, 2018; pp. 205–219.
5. Carbon Trust. *Small-Scale Wind Energy: Policy Insights and Practical Guidance*; Carbon Trust: London, UK, 2008.
6. Juma, M.I.; Mwinyiwiwa, B.M.M.; Msigwa, C.J.; Mushi, A.T. Design of a Hybrid Energy System with Energy Storage for Standalone DC Microgrid Application. *Energies* **2021**, *14*, 5994. [CrossRef]
7. Mohamed, S.A.; Tolba, M.A.; Eisa, A.A.; El-Rifaie, A.M. Comprehensive Modeling and Control of Grid-Connected Hybrid Energy Sources Using MPPT Controller. *Energies* **2021**, *14*, 5142. [CrossRef]
8. Al-Quraan, A.; Al-Qaisi, M. Modelling, Design and Control of a Standalone Hybrid PV-Wind Micro-Grid System. *Energies* **2021**, *14*, 4849. [CrossRef]
9. Rolak, M.; Kot, R.; Malinowski, M.; Goryca, Z.; Szuster, J. AC/DC converter with maximum power point tracking algorithm for complex solution of small wind turbine. *Prz. Elektrotech.* **2011**, *87*, 91–96. (In Polish)
10. Shankareppagol, L.; Hampannavar, S.; Doadamani, S. Performance Analysis of P&O and INC MPPT for WECS. In Proceedings of the 3rd International Conference for Convergence in Technology (I2CT), Pune, India, 6–8 April 2018; pp. 1–8.
11. Baran, J.; Jąderko, A. Układ sterowania turbiny wiatrowej o regulowanej prędkości obrotowej i stałym kącie ustawienia łopat z liniowym obserwatorem momentu aerodynamicznego. *Prz. Elektrotech.* **2017**, *93*, 59–62. (In Polish) [CrossRef]
12. Chudzik, S. Stanowisko pomiarowe do testowania modeli mikroelektrowni wiatrowych. *Prz. Elektrotech.* **2021**, *97*, 154–157. (In Polish) [CrossRef]
13. Schubel, P.; Crossley, R. Wind turbine blade design. *Energies* **2012**, *5*, 3425–3449. [CrossRef]
14. Pande, J.; Nasikkar, P.; Kotecha, K.; Varadarajan, V. A Review of Maximum Power Point Tracking Algorithms for Wind Energy Conversion Systems. *J. Mar. Sci. Eng.* **2021**, *9*, 1187. [CrossRef]

15. Majout, B.; El Alami, H.; Salime, H.; Zine Laabidine, N.; El Mourabit, Y.; Motahhir, S.; Bouderbala, M.; Karim, M.; Bossoufi, B. A Review on Popular Control Applications in Wind Energy Conversion System Based on Permanent Magnet Generator PMSG. *Energies* **2022**, *15*, 6238. [CrossRef]
16. Abdullah, M.A.; Yatim, A.H.M.; Tan, C.W.; Saidur, C.W. A review of maximum power point tracking algorithms for wind energy systems. Renew. Sustain. *Energy Rev.* **2012**, *16*, 3220–3227.
17. Calabrese, D.; Tricarico, G.; Brescia, E.; Cascella, G.L.; Monopoli, V.G.; Leuzzi, R. Variable Structure Control of a Small Ducted Wind Turbine in the Whole Wind Speed Range Using a Luenberger Observer. *Energies* **2020**, *13*, 4647. [CrossRef]
18. Bekiroglu, E.; Yazar, M.D. MPPT Control of Grid Connected DFIG at Variable Wind Speed. *Energies* **2022**, *15*, 3146. [CrossRef]
19. Kim, J.-S.; Chung, I.-Y.; Moon, S.-I. Tuning of the PI controller parameters of a PMSG wind turbine to improve control performance under various wind speeds. *Energies* **2015**, *8*, 1406–1425. [CrossRef]
20. Bubalo, M.; Bašić, M.; Vukadinović, D.; Grgić, I. Experimental Investigation of a Standalone Wind Energy System with a Battery-Assisted Quasi-Z-Source Inverter. *Energies* **2021**, *14*, 1665. [CrossRef]
21. Yaakoubi, A.E.; Amhaimar, L.; Attari, K.; Harrak, M.; Halaoui, M.; Asselman, A. Non-linear and intelligent maximum power point tracking strategies for small size wind turbines: Performance analysis and comparison. *Energy Rep.* **2019**, *5*, 545–554. [CrossRef]
22. Aissaoui, H.E.; Ougli, A.E.; Tidhaf, B. Neural Networks and Fuzzy Logic Based Maximum Power Point Tracking Control for Wind Energy Conversion System. *Adv. Sci. Technol. Eng. Syst. J.* **2021**, *6*, 586–592. [CrossRef]
23. Syahputra, R.; Soesanti, I. Performance Improvement for Small-Scale Wind Turbine System Based on Maximum Power Point Tracking Control. *Energies* **2019**, *12*, 3938. [CrossRef]
24. Priyadarshi, N.; Ramachandaramurthy, V.K.; Padmanaban, S.; Azam, F. An Ant Colony Optimized MPPT for Standalone Hybrid PV-Wind Power System with Single Cuk Converter. *Energies* **2019**, *12*, 167. [CrossRef]
25. Daili, Y.; Gaubert, J.P.; Rahmani, L. Implementation of a new maximum power point tracking control strategy for small wind energy conversion systems without mechanical sensors. *Energy Convers. Manag.* **2015**, *97*, 298–306. [CrossRef]
26. Nandi, T.N.; Brasseur, J.; Vijayakumar, G. Prediction and Analysis of the Nonsteady Transitional Boundary Layer Dynamics for flow over an Oscillating Wind Turbine Airfoil using the γ-Reθ Transition Model. In Proceedings of the 34th Wind Energy Symposium, San Diego, CA, USA, 4–8 January 2016. [CrossRef]
27. Liu, J.; Xiao, Z.; Fu, S. Unsteady Transition Studies over a Pitching Airfoil Using a k-ω-γ Transition Model. *AIAA J.* **2018**, *56*, 3776–3781. [CrossRef]

energies

MDPI

Article

Investigation of the Mechanical Behavior of a New Generation Wind Turbine Blade Technology

Cihan Çiftci [1,2], Ayşe Erdoğan [3] and Mustafa Serdar Genç [3,4,*]

1 Department of Civil Engineering, Abdullah Gul University, Kayseri 38080, Turkey
2 Techno-CC R&D Innovation Ltd. Co., Erciyes Teknopark, Kayseri 38039, Turkey
3 Wind Engineering and Aerodynamic Research Center, Department of Energy Systems Engineering, Erciyes University, Kayseri 38039, Turkey
4 Energy Conversion Research and Application Center, Erciyes University, Kayseri 38039, Turkey
* Correspondence: musgenc@erciyes.edu.tr

Abstract: Wind turbine blades are one of the largest parts of wind power systems. It is a handicap that these large parts of numerous wind turbines will become scrap in the near future. To prevent this handicap, newly produced blades should be recyclable. In this study, a turbine blade, known as the new generation of turbine blade, was manufactured with reinforced carbon beams and recycled, low-density polyethylene materials. The manufacturing addressed in this study reveals two novelties: (1) it produces a heterogeneous turbine blade; and (2) it produces a recyclable blade. In addition, this study also covers mechanical tests using a digital image correlation (DIC) system and modeling investigations of the new generation blade. For the mechanical tests, displacement and strain data of both new generation and conventional commercial blades were measured by the DIC method. Instead of dealing with the modeling difficulty of the new generation blade's heterogeneity we modeled the blade structural system as a whole using the moment–curvature method as part of the finite element method. Then, the behavior of both the new generation and commercial blades at varying wind speeds and different angles of attack were compared. Consequently, the data reveal that the new generation blades performed sufficiently well compared with commercial blades regarding their stiffness.

Keywords: wind turbine blade; new generation blade technology; recyclable composite structures

Citation: Çiftci, C.; Erdoğan, A.; Genç, M.S. Investigation of the Mechanical Behavior of a New Generation Wind Turbine Blade Technology. *Energies* **2023**, *16*, 1961. https://doi.org/10.3390/en16041961

Academic Editor: Paweł Ligęza

Received: 28 January 2023
Revised: 10 February 2023
Accepted: 13 February 2023
Published: 16 February 2023

1. Introduction

Increasing environmental concern about the harmful consequences of global warming and carbon emissions has recently created new demands for renewable and sustainable energy sources such as wind, solar, biomass and geothermal energy [1]. Among these renewable energy sources, wind has become the focus of attention in recent years [2–7]. With the developing technology, wind power plants, which are a renewable energy alternative, are rapidly progressing to become one of the main energy sources in the world. For example, when the installations of wind power plants in Turkey are analyzed over periods of years, it can be seen that the interest in wind power plants in Turkey has increased over time. While the power of wind turbines in Turkey was 9253 MW according to the data of May 2021, it has increased to 10,976 MW according to the data for December 2022 [8,9]. WindEurope predicts that the installed power of the European Wind Power Plant, which is currently 220 GW in total, will reach 318 GW in 2025. It is also predicted that Turkey will reach a total installed power of 14 GW in the same period, ranking sixth in Europe [10]. In the report of the International Renewable Energy Agency (IRENA), titled "*Global Energy Transformation: A Roadmap to 2050*", it is stated that the electricity produced from wind energy is expected to constitute 36% of total electricity production in 2050. Accordingly, the share of wind energy in renewable energy sources is estimated to be 42% [10].

Wind turbines, which are still being installed around the world, typically have a lifespan of about 20 to 30 years. Although some components of wind power plants, such as towers and generators, that have reached the end of their lives can be recycled with some processes, it is very difficult to recycle thermoset-based composite materials used in the turbine blades [11]. Because of these difficulties, these turbine blades are usually buried on land, while only a few can be used in the construction industry. The fact that these large parts of a large number of wind turbines installed throughout the world in the last thirty years cannot be recycled and will turn into scrap also creates a handicap in terms of environmental and climate change. For example, it is predicted that there will be 43 million tons of wind turbine blade waste worldwide by 2050 [12]. According to this prediction, 40%, 25% and 16% of the world's total turbine blade waste will belong to China, Europe, and the United States, respectively. Considering that each kilowatt of wind energy needs about 10 kg of wind turbine blade material, it is expected that shortly, humanity will waste about 200,000 tons of blades [12]. It is also estimated that the amount of blade material that will need to be recycled annually between 2029 and 2033 may reach 400,000 tons [13]. Finally, the amount of blade waste is expected to increase to 800,000 tons per year after 2050 [13].

The commonly used intervention for the recycling of wind turbine blades consists of transporting the blades to landfills as waste, as shown in Figure 1. In addition, the production of various products, such as fibers, pellets, construction materials and panels from composite fiber wastes, contributes to the study of recycling. These forms of applications, also referred to as cyclical and zero-waste solutions, demonstrate attempts at reducing the carbon footprint of turbine blades [14].

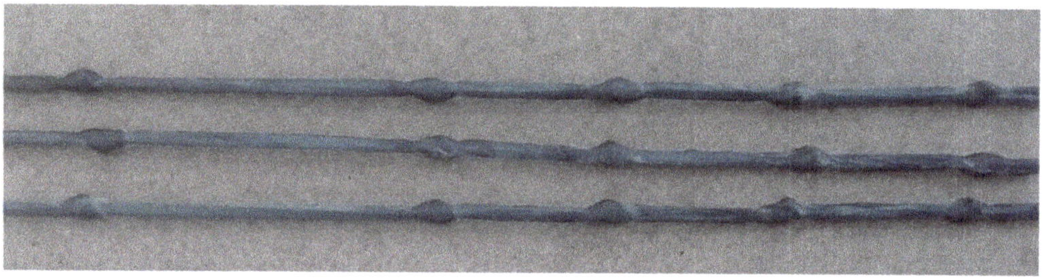

Figure 1. Knotted carbon-based bars used in the production of the new generation blade.

Energy production of wind turbines is largely dependent on the aerodynamics of the blades, and energy output can be increased by improving the blade's aerodynamic performance [15–21]. Another way to increase the amount of energy is to increase the blade dimensions, though this depends on the blade strength [22–25]. Considering the amount of energy produced by wind power plants and the dimensions of the turbine blades, the blades of medium- and large-scale turbines is produced with a different manufacturing concept compared with the blades of small-scale turbines, since they are subjected to higher bending moments and shear forces [26–28]. In the manufacturing of the blades of these medium- and large-scale blades, each layer of the blades consists of glass and carbon fiber reinforced epoxy materials. In these layered composites, the matrix material is epoxy, while the reinforcing materials are glass and carbon. Although the layers in these composite structures are heterogeneous within themselves, considering that turbine blades are formed by the repetition of these composite layers, it can be stated that the blades are produced in a homogeneous structure [29–33]. Small-scale turbine blades are produced by embedding chopped glass or carbon reinforcements into the matrix material homogeneously using plastic injection production technology in order to make production practical and fast.

It is vital that the composite structures used in the production of turbine blades can provide the desired strength values when exposed to wind force. According to the literature, the mechanical properties of the blade structures are commonly obtained by flapwise and edgewise tests. Considering that the usage of these test setups is expensive, the digital image correlation (DIC) technique offers a convenient and economical alternative, especially for a classical flapwise test. The digital image correlation (DIC) technique was invented in the early 1980s. It is still a common non-contact and non-destructive experimental measurement technique for measuring the displacement of structures under various loads. It can be seen in the literature that the DIC technique has been extensively applied to tests of structures where the sample size is small, and the experimental setup is well established [34]. For example, Winstroth et al. [35] successfully measured the displacements of turbine blades using the DIC technique, even in a large-scale wind turbine of 3.2 MW. Wu et al. [36] used a new and economical optical technique based on three-dimensional digital image correlation (3D-DIC) for the health monitoring of wind turbine blades. In their study, the measurements of a 5-kW wind turbine with a diameter of 4 m were obtained using 3D-DIC. First, they painted rotor blades with random black and white dot patterns and placed two digital cameras in front of the wind turbine to measure the rotor blade deformations. With this test setup, the displacement and strain values of the blades were obtained dynamically. Mastrodicasa et al. [37] also used the DIC system to measure the deformation and three-dimensional stresses on a 50-m blade subjected to periodic loading. They also improved the calibration of the DIC system considering the size (~4 m $\times$ 3 m) of the focal areas for the cameras. The results obtained in their study show that the DIC technique can provide reliable information about the changes in strains and deformations of the structures that are subjected to even fatigue loads. Khadka et al. [38] tried to develop a new non-contact technique using a drone to take dynamic measurements from the DIC system. In their experimental study, the integration of DIC and drone was used to obtain strain data on wind turbine blades at remote locations.

Composites show themselves in structures produced by the combination of two or more different materials. In this context, when we look at the history of the use of composite materials, it is known that they are used in almost different areas and many different products. Reinforced concrete beam or column structures used in the construction industry are also examples of composite structures. In these reinforced concrete structures, this concrete material needs to be reinforced due to the low performance of the concrete material under tensile load, although it is sufficiently resistant against the compressive load. In order to meet this need, the concrete is reinforced by using steel bars in its material. With this concept, massive weights are easily carried by reinforced concrete beams and columns in buildings. Reinforced concrete design specifications place a limitation on the distances between steel bars used in the concrete material to prevent delamination or splitting-type failures within the reinforced concrete beams or columns. Considering that turbine blades exhibit a kind of cantilever beam feature under wind load and also that the strength of layered composite structures should be increased despite the delamination failures, it is predicted in this study that the turbine blades can be produced in an analogous way to reinforced concrete beams.

A heterogeneous turbine blade, which in this study we call a new generation turbine blade, is produced in way that is inspired by the reinforced concrete beams that are frequently used in the construction industry and is unlike traditional turbine blades with a homogeneous structure. In this production, the first to be reported in the relevant literature, continuous carbon-based bars were used instead of reinforcing steel bars in reinforced concrete beams, and recycled low-density polyethylene was also used instead of the concrete material. The most important reason for producing the new generation blades in a heterogeneous structure inspired by reinforced concrete beams is to prevent delamination-type failures, which can often be seen in traditional turbine blades. In other words, it is predicted that delamination failures will be minimized in the new generation blades, whose reinforcing material is heterogeneously embedded in the matrix material, so that the total contact surface area between the materials is increased. Another novelty of this study is based on the originality of the production technique of the new generation blades, as the manufacturing technique used in the production of the new generation blade differs from the production methodology of the traditional turbine blades. Unlike traditional blades, almost no resin or epoxy material was used in the production of this new generation blade. This new generation blade was produced by bonding the matrix material (recycled polyethylene) to its reinforcement by exposing it to heat and physical melting it. Another advantage of using this technique is to supply a recyclable turbine blade since the matrix material of the blade can be easily stripped from the reinforcing material through heat application. In addition, in this study, we aimed to produce a new generation blade that is more mechanically durable than a conventional one. The modelling of a classical heterogeneous system will require information regarding several characteristics, obtained from coupon tests of the different materials in composite structures and also from interfacial bonding tests between these different materials. Since the heterogeneity of the new generation blades makes their modelling difficult, all the obtained mechanical data must be defined during modelling. Instead of obtaining all these data, this study facilitates the modelling of the blade structural system as a whole. While the strength and displacement of the wind turbine blades are obtained by using the DIC system, static loads were applied to the blades. While applying these loads, the deformations were measured with the DIC system and moment–curvature curves were calculated. In ADINA finite element method software, and by using the moment–curvature data, the behavior of the new generation and commercial blades was modelled as a whole based on pressure distributions obtained from CRADLE Computational Fluid Dynamics (CFD) software at different wind speeds. Thereby, there was no need to define the mechanical properties of materials in the finite element model of the blades. This is another novelty of our study.

In this study, first, the new generation blade was produced in the same profile and size as the purchased commercial blade, with the manufacturing and measurement methods explained in Section 2. Then, experimental studies were conducted on the strength of both new generation and commercial blades using the DIC technique. An analysis of the mechanics and flow are described in Section 2. Then, commercial and new generation blades are modelled and their behaviors under different wind speeds and angles of attack are compared in Section 3.

2. Materials and Methods

The new generation blade mentioned in this study differs from traditional turbine blades both in terms of production technique and the difference of materials used in production. In order to measure and evaluate how these differences contribute to the new generation turbine blades, a commercially available traditional turbine blade was purchased first. Then, the new generation blade was produced with the same size and profile as the commercial one. The details on the production of the new generation blade are given as follows.

2.1. *Manufacturing of the New Generation Blade*

In order to minimize delamination type fractures, which are frequently encountered in traditional turbines, new generation blades are inspired by the structure of composite reinforced concrete beams, which are widely used in the construction industry. In this production, continuous carbon-based bars are used instead of reinforcing steel bars in reinforced concrete beams, and recycled low-density polyethylene is also used instead of the concrete material. In addition, as is known, deformations are formed on the steel bars of the reinforced concrete structures in order to increase the interfacial bonding strength between the steel bars and concrete material. In other words, due to the usage of the deformed steel bars, we can ensure that the steel reinforcements do not slide into the concrete material. This important because the reliability of the design calculations of the reinforced concrete beams depends on the fact that the concrete material in the beams works as a whole with the adjacent steel reinforcement without slipping. With this context, in order to produce the new generation blade of this study so that it operates under a similar principle, sliding of the carbon reinforcement within the blade must be minimized.

For this reason, knotted structures were arbitrarily formed on each carbon-based bar to be used for the production of the new generation blade. The visuals of these knotted structures are given in Figure 1. Additionally, only this much information will be given in terms of the creation of these knotted structures, which were developed within the company established by one of the authors, since the know-how on this subject is evaluated to be within the scope of trade secrets. As a matter of fact, it is vital to minimize any reduction in the strength capacity of carbon rods during the creation of knotted structures, and the know-how in this regard belongs to the company.

Figure 2 is plotted to represent the locations of the knotted bars within the airfoil of the new generation blade. According to Figure 2, the new generation blade was produced in two separate parts using two different molds. In order to realize the production of these separate parts, two molds with different shapes were first obtained (see Figure 2). Both blade pieces were produced by heating in a furnace after the placement of the knotted carbon bars and recycled polyethylene material in the molds. These pieces were then glued together using a trace amount of epoxy material as shown in Figure 2. It can be emphasized that the amount of epoxy in the new generation blade is negligible when compared with the amount of resin or epoxy used in traditional turbine blades. Additionally, it is thought that this epoxy, which is used a negligible amount, does not have a negative effect on the recycling of new generation blades. Because, as is known from the recycling industry, the materials are never recycled in pure form. Moreover, the new generation blade in this study has already been produced using an impure recycled form of low-density polyethylene (F2–12).

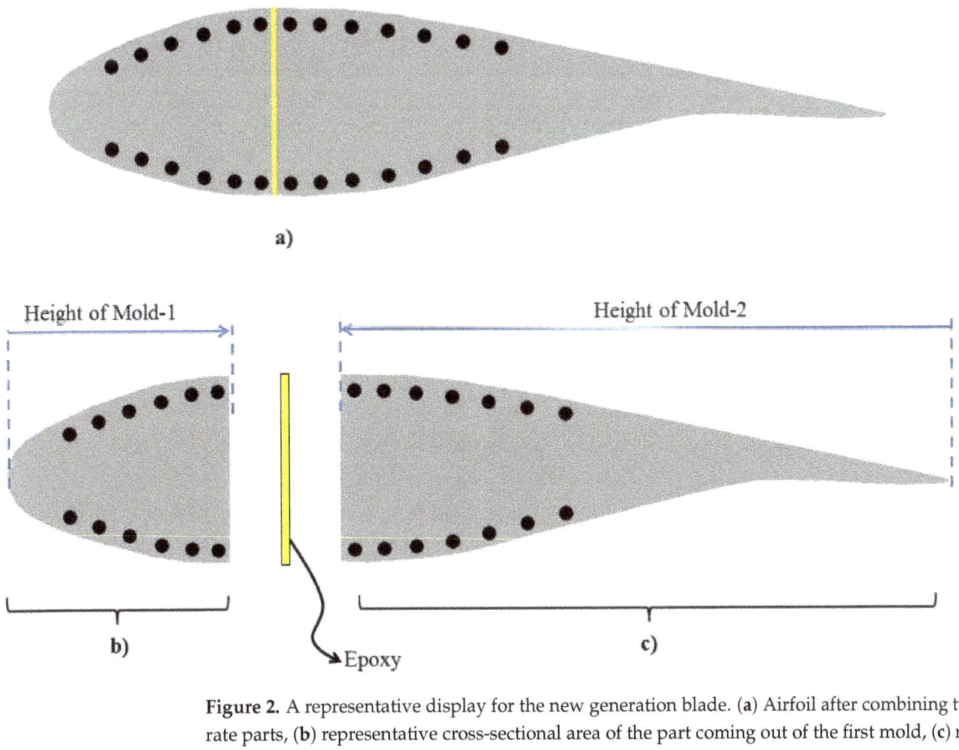

Figure 2. A representative display for the new generation blade. (**a**) Airfoil after combining two separate parts, (**b**) representative cross-sectional area of the part coming out of the first mold, (**c**) representative cross-sectional area of the part coming out of the second mold.

2.2. Digital Image Correlation Method

In this study, a digital image correlation (DIC) system was used to obtain the strength and displacements of both new generation blade and purchased commercial blades under static load. The DIC system captures repetitive images and then calculates the displacements of samples by tracking the deformation of the white dots applied to the surface of the samples using a cross-correlation method [39,40]. The DIC system used in this study includes 2 × 1 megapixel (MP) high speed cameras, 2 spot lamps, Zeiss precision F-Mount 35 mm lens, and ISTRA 4D software with a calibration target. The maximum frame rate of the camera is 1000 frames per second with a resolution of 1280 × 800 pixels. In order to activate the system, first the calibration process is performed. Calibration was performed before the experiment using a square work area of 500 mm × 500 mm. In the calibration process, first, the calibration board was displayed with 2 compact cameras and 2 spot lamps, and the speckled structure was defined in the system by transferring it to the software.

Both of the blades were made ready by being painted as a white dot (pattern) for the application of the DIC system as in Figure 3. The purpose of this process is to obtain deformation and displacement measurements by referring to the white dots of the cameras in the DIC system.

(**a**) (**b**)

Figure 3. Making the necessary preparations for the experimental analysis of the blades in the DIC system, (**a**) painting work required for the DIC system, (**b**) blade surface after painting.

According to Figure 4a, the images of both the purchased commercial blade and the manufactured new generation blade—before being subjected to any load—were taken by using the commercial software of the DIC system. These first images will be used as the reference image for the blades, and according to this reference image, the deformation values of the blades under static load can be calculated. The color changes in the images in Figure 4b indicate the strain values formed on the surface of the blades under loading.

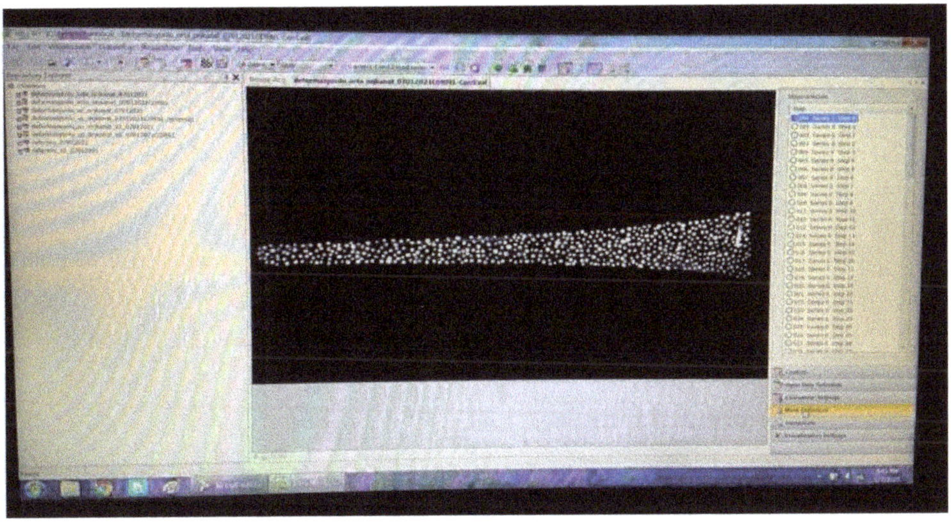

(**a**)

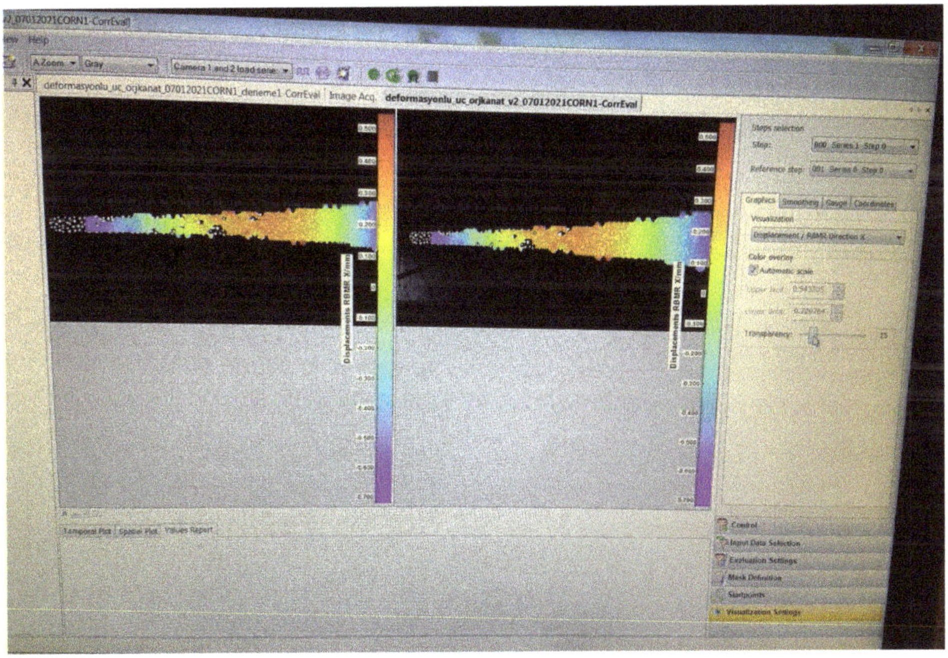

(**b**)

Figure 4. Application of the DIC method on a selected blade (**a**) reference image (**b**) strain distributions on the blades.

While the strength and displacement of the wind turbine blades are obtained by using the DIC system, the static loads applied on the blades are given in Figure 5. In order to compare the 2-kW small-scale ISTRA BREZEE brand commercial wind turbine blade with the new generation blade which has the same size and profile, the loads shown in Figure 5 are separately applied at the root, middle and tip regions of both blades. Information about the locations of the applied loads is given in Figure 5. While applying these loads, the deformations were measured with the DIC system.

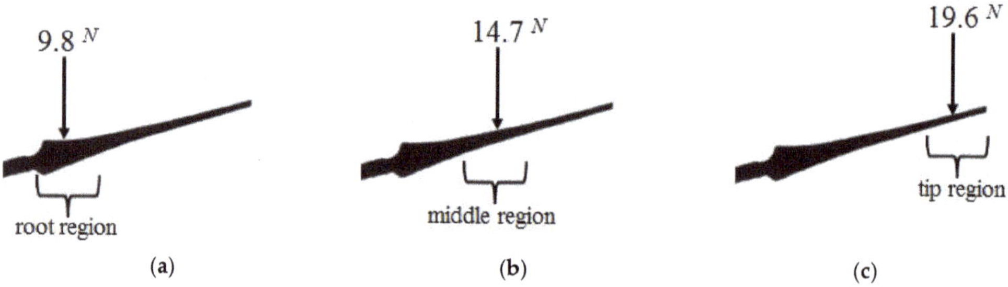

Figure 5. Representative images of the loads of (**a**) 9.8 N, (**b**) 14.7 N, (**c**) 19.6 N applied on the blades when using the DIC system.

2.3. Finite Element Modeling for the Blades

It is important for the energy sector that the new generation blades can be modeled as in the traditional blades. In addition, the heterogeneity of the new generation blades makes their modeling difficult. The modeling of a classical heterogeneous system will require information about several characteristics, which can be obtained from the coupon tests of the different materials in composite structures and also from interfacial bonding tests between these different materials. Then all these obtained mechanical data must be defined during modeling. Instead of obtaining all these data, this study deals with facilitating the modeling of the blade structural system as a whole. In this modeling, ADINA R.D. Inc. commercial engineering simulation software program was used. As mentioned in the tutorial examples of this program, it is possible to combine the mechanical properties of the entire structural system into just one batch of moment–bending curve data by considering the beam element structure as a whole in the finite element method. In other words, the moment–curvature curve shows how the output data of a simple beam structure changes according to the input data when bending, and it briefly illustrates the relationship between input and output. In this respect, by using the moment–curvature data, the behavior of beam structures with a variable cross-sectional area when bending can be modeled as a whole. For this modeling, it is sufficient to model the blades as a cantilever beam element without even defining the cross-sectional areas. The blades are modeled in two dimensions and the displacements in the flapwise testing direction could be calculated by the finite element method. These calculated displacement values will then be compared with the experimental values obtained from the DIC flapwise tests. By the way, to maintain control over the mesh size in the finite element modelling, mesh sizes were increased until consistent displacement output data were obtained. In addition, there is no need to define the mechanical properties of different materials in the finite element model of the blades. The only data needed here is to know the moment–curvature data of both commercial and new generation blades. Experimental data using the DIC system were used to learn these moment–curvature data. In light of these experimental data, the moment–curvature data of the root, middle and tip regions were obtained by using the strain values at the blade surfaces. In other words, the moment–curvature calculation of each region was roughly calculated by using the strain data of the root, middle and tip regions of the turbine blades obtained from the DIC experiments in the formula in Equation (1) [41]. Finally, the bending behavior of composite blades, whether homogeneous or heterogeneous, can be easily modeled using the learned moment–curvature curve.

$$\text{Curvature} = \frac{\text{Strain}}{(\text{blade thickness}) \times (0.5)} \tag{1}$$

2.4. Flow Analyses

In wind turbines, flow energy is converted into mechanical energy, and mechanical energy is converted into electrical energy in the generator. During this transformation, the effect of air as a fluid on the blades and the aerodynamic forces resulting from this action are basically based on pressure distributions. Flow analysis is required to obtain these pressure distributions, and these analyses can be done with CFD software [17,42,43]. In this study, MSC CRADLE, one of the MSC One software products, was used to obtain pressure distributions. In the numerical flow analysis, the flow area was firstly created according to the blade length. The same mesh structure was applied to the blade and the created flow area as seen in Figure 6 in each analysis. The hexagonal network structure formed in the blade and flow area was determined as a total of 922,673 octant (number of elements). In Figure 7, it is seen that while the mesh structure is denser in the blade, it is evenly distributed over the flow area in order to measure the changes in the blade more precisely. The angles of attack (α) were selected as 5° and 10° in the analyses. Pressure distributions were obtained by giving 5 m/s, 10 m/s, 15 m/s and 20 m/s fluid velocities at each angle of attack. The pressure distribution data obtained from the CRADLE program were applied to the blade beam models in order to determine the behavior of the blades under different wind speeds and different angles of attack and also to simulate the structure-fluid interaction.

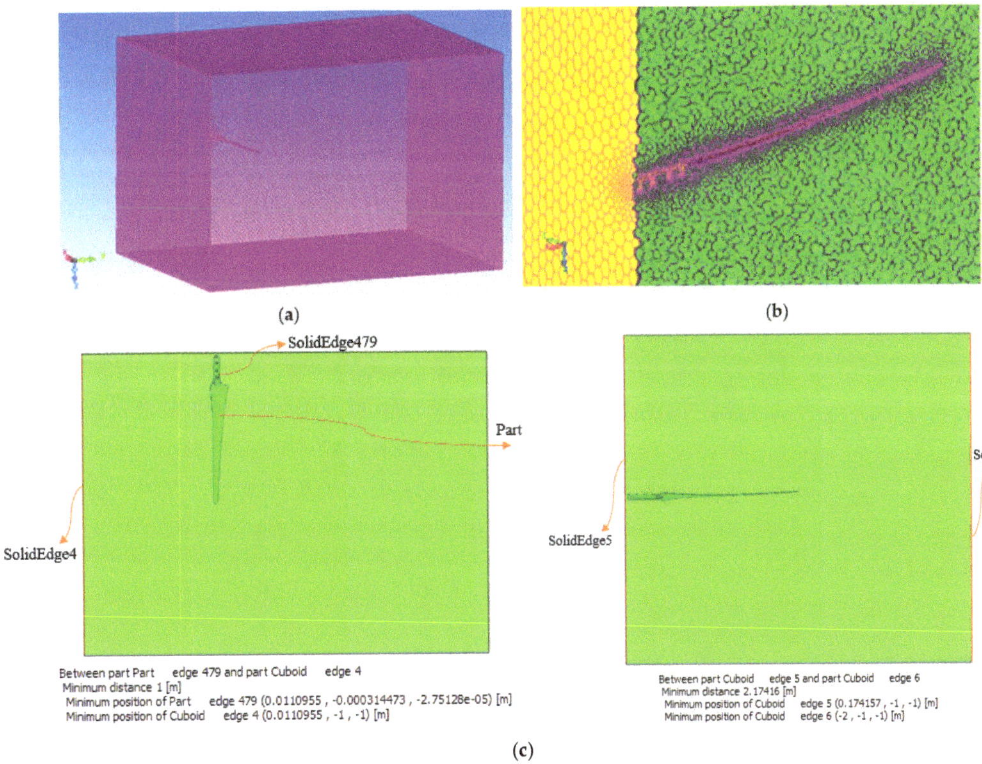

Figure 6. The mesh structure (**a**) the flow domain (**b**) the meshed blade and (**c**) length scale of the meshed blades.

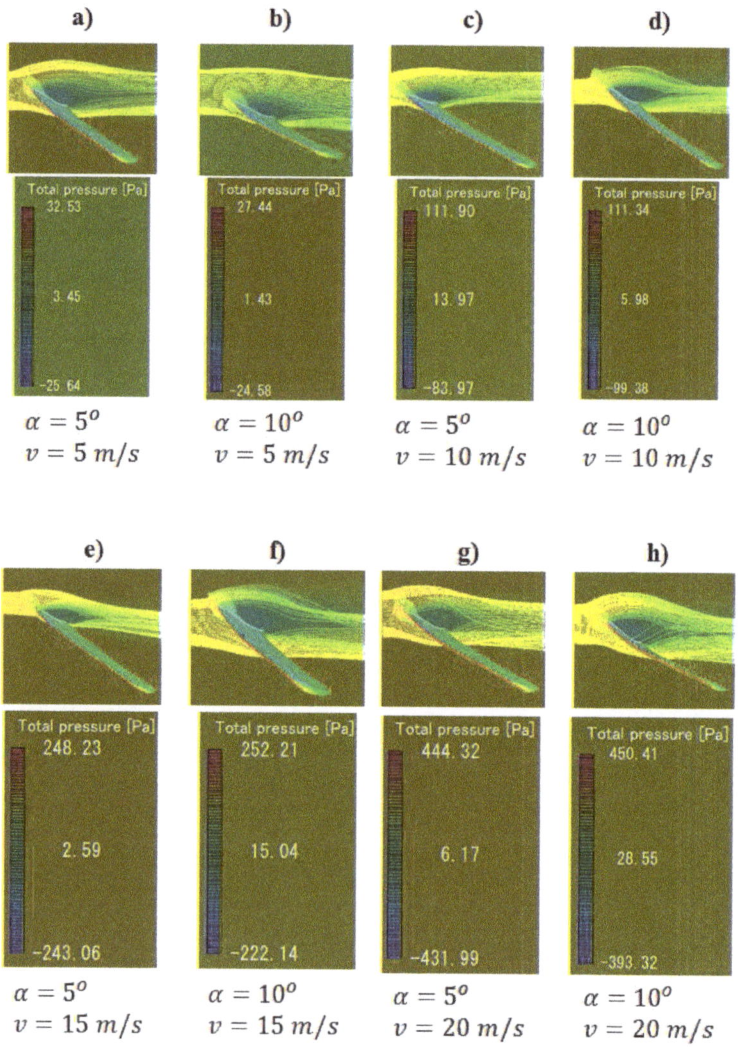

Figure 7. Pressure distribution on varying wind speeds and angles of attack: (**a**) $\alpha = 5°$ v = 5 m/s (**b**) $\alpha = 10°$ v = 5 m/s (**c**) $\alpha = 5°$ v = 10 m/s (**d**) $\alpha = 10°$ v = 10 m/s (**e**) $\alpha = 5°$ v = 15 m/s (**f**) $\alpha = 10°$ v = 15 m/s (**g**) $\alpha = 5°$ v = 20 m/s (**h**) $\alpha = 10°$ v = 20 m/s.

3. Results and Discussion

The results obtained in this study include (1) the data results of the experimental tests of the blades with the DIC system, (2) the results of the flow analyses, and (3) the behavior of the blades modeled in the finite element method.

3.1. Experimental Results Obtained from the DIC System

The strengths of both commercial and new generation blades under static load (see Figure 5) were compared using the DIC system. Table 1 has been created to better understand this comparison. In Table 1, the displacement and surface strain values of both the commercial and new generation blades are shown corresponding to each test case. When these displacement and strain values are compared, it is revealed that the new generation blade has higher stiffness than a commercial one. For example, after applying a single load to the tip region of both blades, the displacement values along the longitudinal axis of commercial and new generation blades vary between 5 mm–60 mm and 0.8 mm–30.3 mm, respectively. Since this example shows that the commercial blade makes about two-times more displacement than the new generation blade under the same load, it turns out that the stiffness value (stiffness = force/displacement) of the commercial blade is approximately equal to half of the stiffness of the other blade. In addition, if the strain values formed on the surfaces of the blades are to be compared with each other, the average strain values occurring in the root region of commercial and new generation blades are shown as approximately 0.0070 and 0.0029, respectively (see Table 1). The average strain values in the middle region of commercial and new generation blades are also shown as approximately 0.0147 and 0.0059, respectively. Finally, the average strain values in the tip region of commercial and new generation blades are approximately 0.0247 and 0.0129, respectively.

Table 1 also includes the displacement data for commercial and new generation blades after applying a single load to the middle and root regions. When the single load is applied to the middle region of the blades, the maximum displacements of the commercial and new generation blades are calculated as approximately 14 mm and 11.57 mm, respectively. This shows that the new generation blade again exhibits higher endurance performance. However, this is not the case when a single load is applied to the root zone. In other words, the maximum displacements of commercial and new generation blades were calculated as approximately 0.41 mm and 0.55 mm when the single load was applied to the root regions of the blades. In addition, considering that the displacement values calculated at such low values will be within the margin of error of the DIC system, it would be difficult to claim anything through comparison of the new generation and commercial blades with these data.

When the data in Table 1 are examined, the surface strain distributions that occur when a single load is applied to the middle regions of both blades show that the new generation blade is 77% stiffer than the commercial blade. In addition, the surface strain distributions formed when a single load is applied to the root regions of both blades show that the new generation blade is 43% more rigid than the commercial blade. However, as stated above, the new generation blade displaces more when a single load is applied to the root region of both blades. In this context, it can be thought that there is a mismatch between the displacement data from the DIC system and the stain data. However, it is thought that the main reason for this incompatibility is not related to the DIC system or commercial software, but that a possible fixation problem may have occurred during the testing of the new generation blade under the single load at its root region. In other words, since the strain data are independent of the perfection of the fixation of the blades, it is thought that the displacement data may have been greatly affected by a problem with the fixation of the blade root.

According to Table 1, it can be seen that there is more variation in the different color distributions on the blade surface for the commercial blade. It is thought that the excess in this change is due to the heterogeneity of the internal structure of the new generation blade. In order to further reduce the standard deviation of the strain values on the surface of the new generation blade and also to standardize its production, it is thought that an optimization and design study should be carried out on the number of knotted carbon reinforcements and the distance between these reinforcements.

Table 1. Displacements and surface strain values of both blades corresponding to the DIC tests.

<table>
<tr><th rowspan="2">Blade Type</th><th rowspan="2">Application of DIC Tests</th><th rowspan="2">Displacements of the Blades Corresponding to the DIC Tests</th><th colspan="3">Blade Surface Strain Values Corresponding to the DIC Tests</th></tr>
<tr><th>Blade Root Region</th><th>Blade Middle Region</th><th>Blade Tip Region</th></tr>
<tr><td>Commercial Blade</td><td>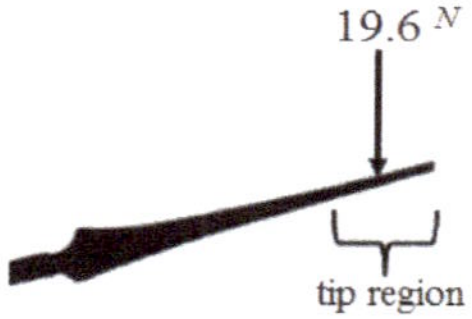
</td><td>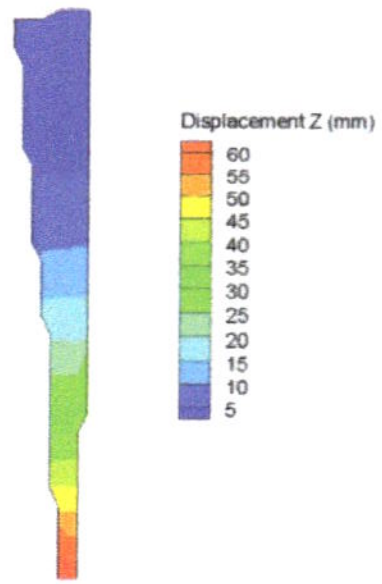
</td><td>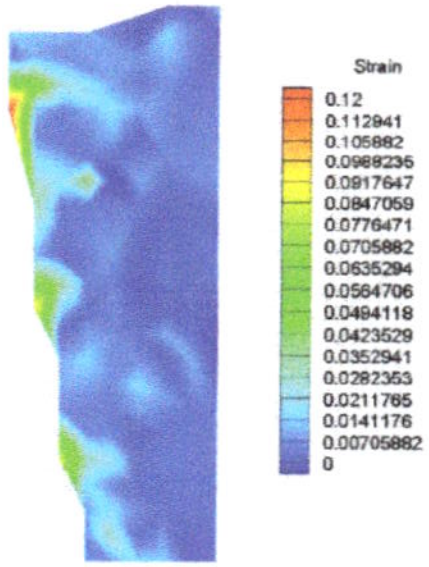
</td><td>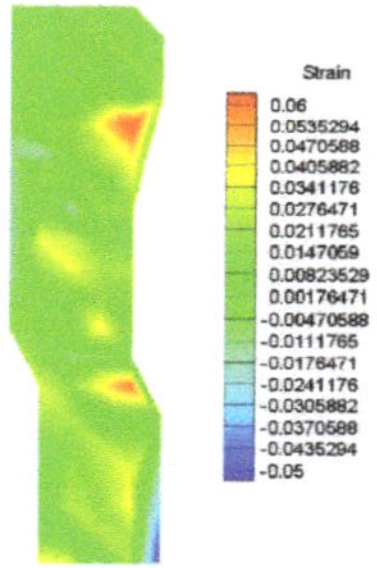
</td><td>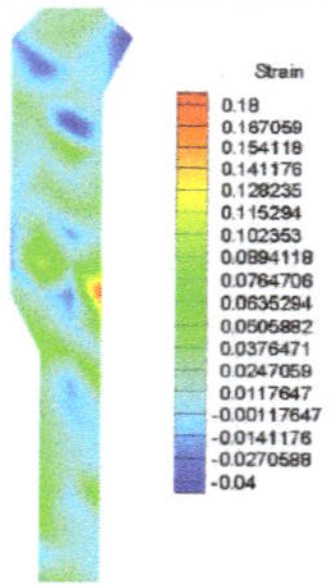
</td></tr>
<tr><td>New Generation Blade</td><td>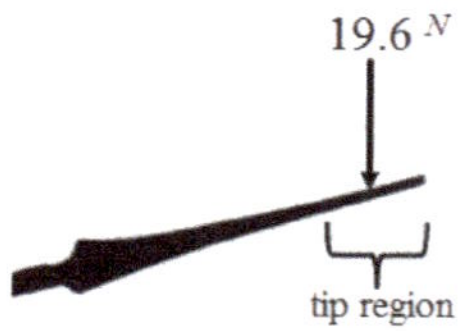
</td><td>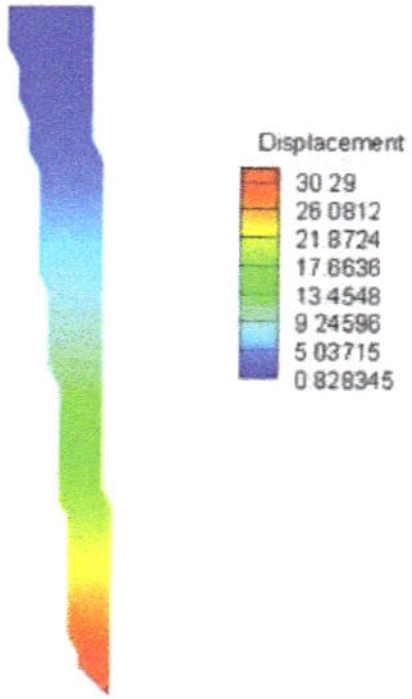
</td><td>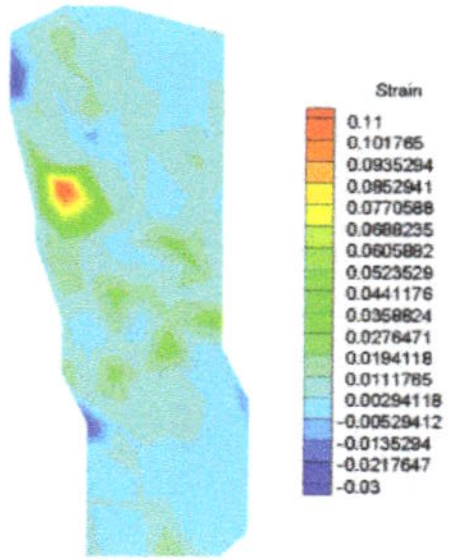
</td><td>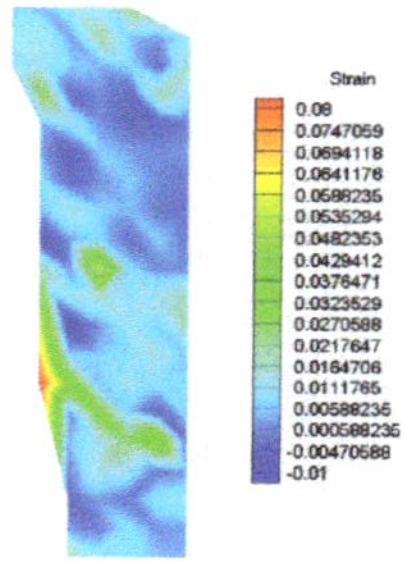
</td><td>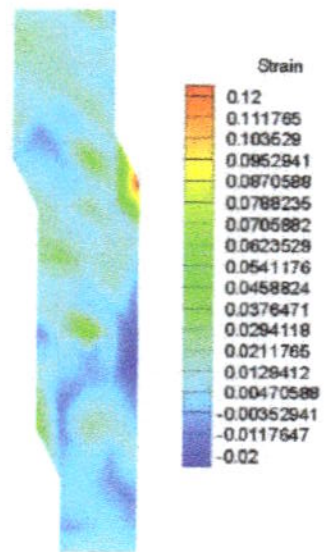
</td></tr>
</table>

Table 1. *Cont.*

Blade Type	Application of DIC Tests	Displacements of the Blades Corresponding to the DIC Tests	Blade Surface Strain Values Corresponding to the DIC Tests		
			Blade Root Region	Blade Middle Region	Blade Tip Region
Commercial Blade	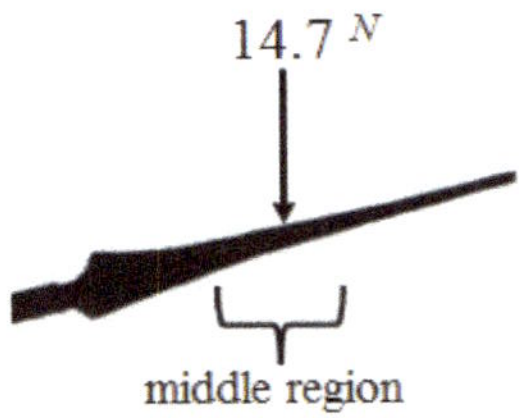	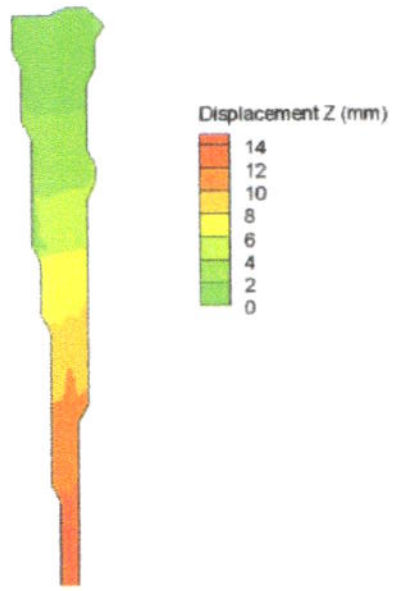	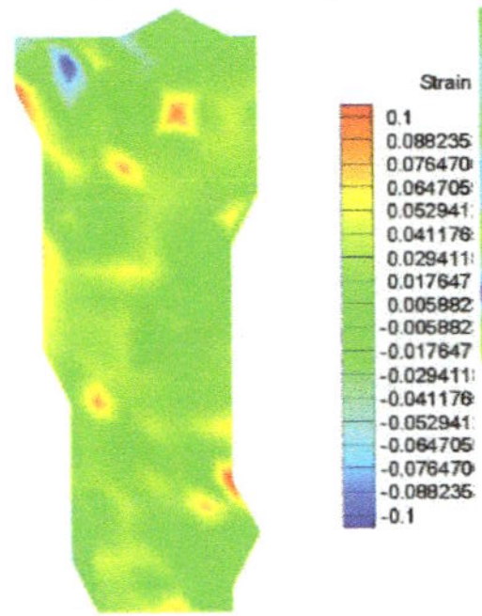	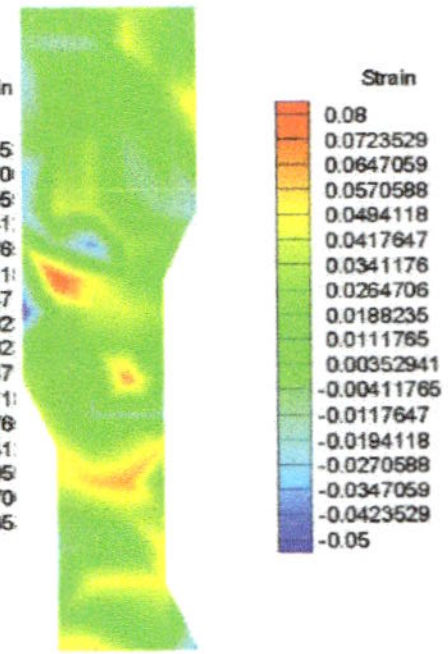	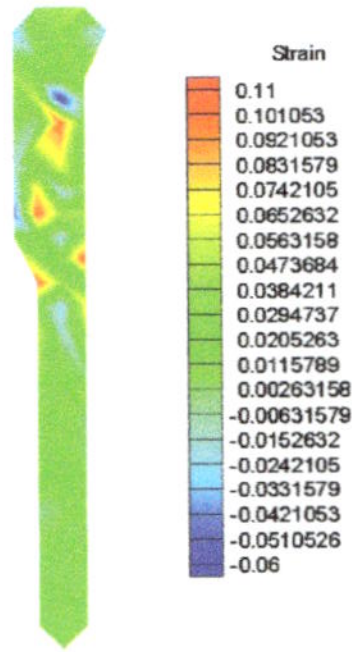
New Generation Blade	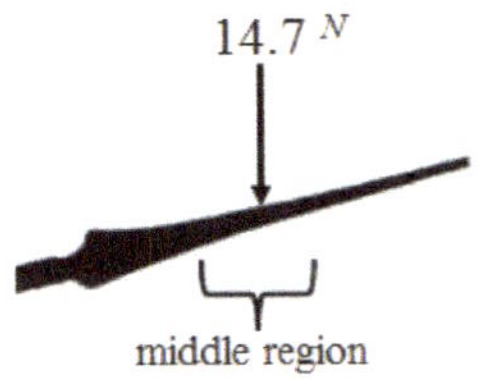	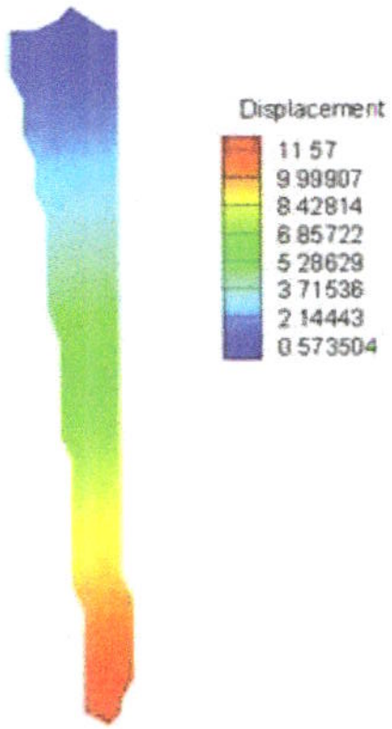	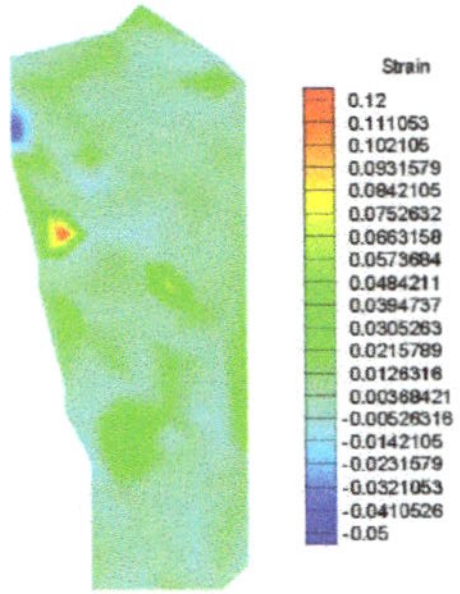	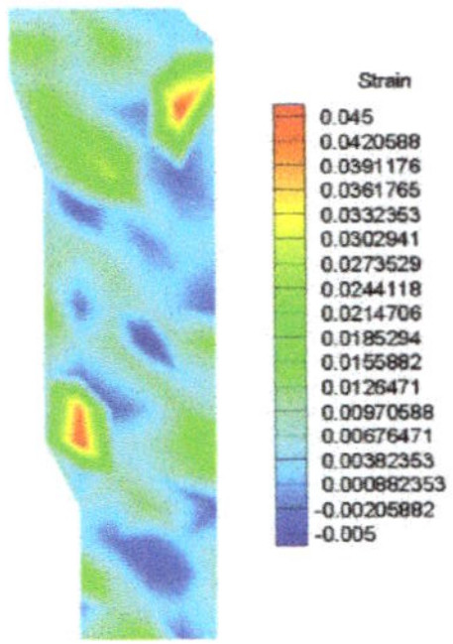	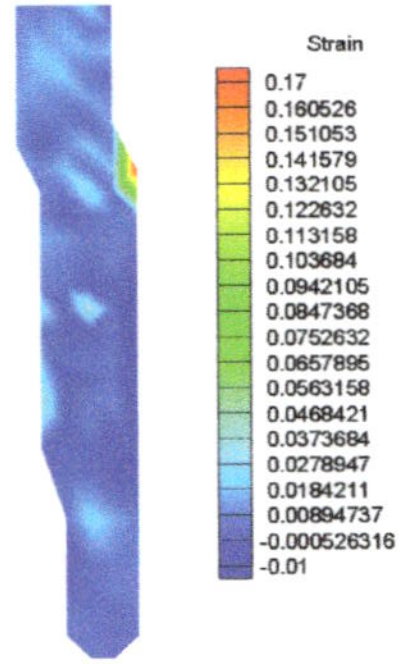

Table 1. *Cont.*

Blade Type	Application of DIC Tests	Displacements of the Blades Corresponding to the DIC Tests	Blade Surface Strain Values Corresponding to the DIC Tests		
			Blade Root Region	Blade Middle Region	Blade Tip Region
Commercial Blade	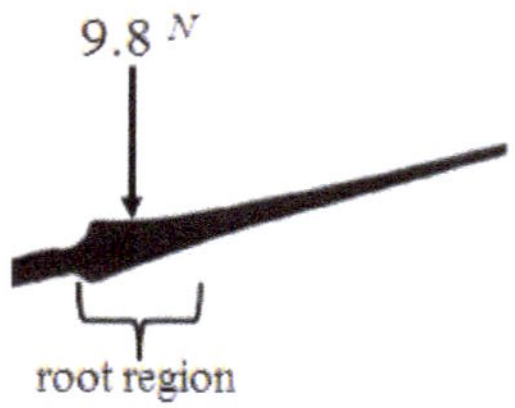	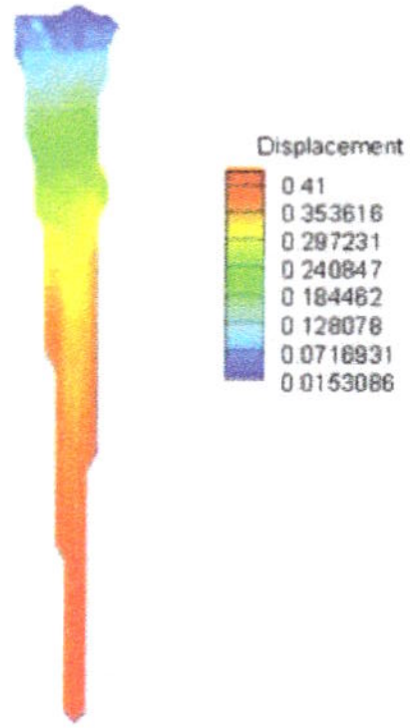	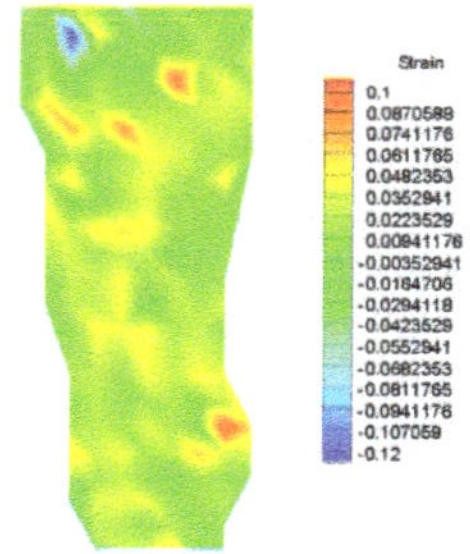	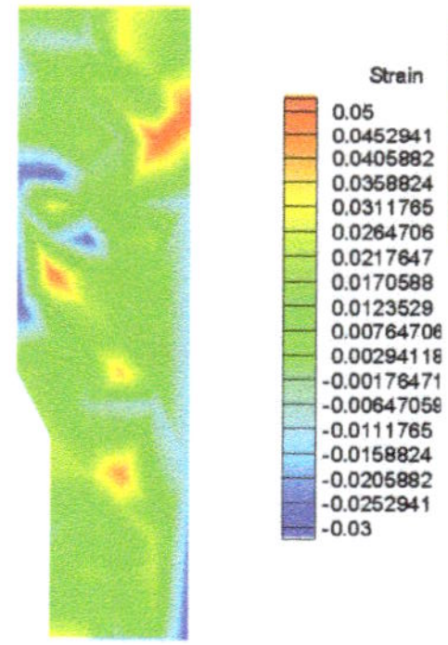	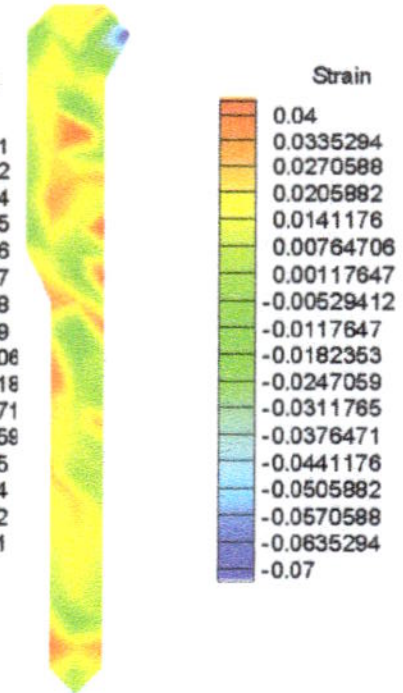
New Generation Blade	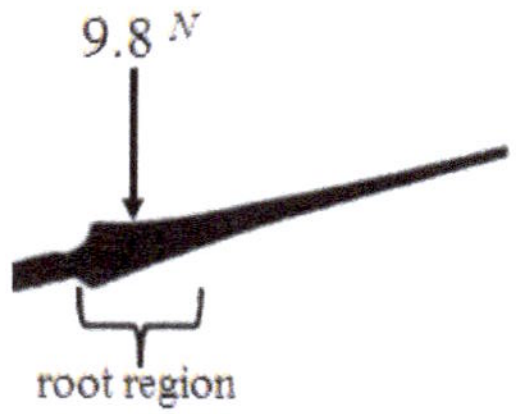	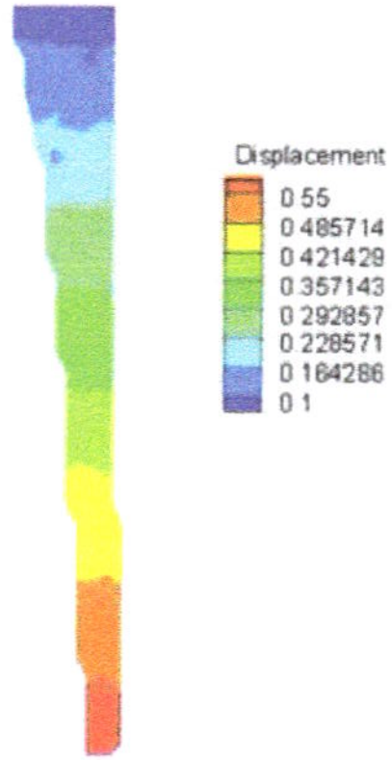	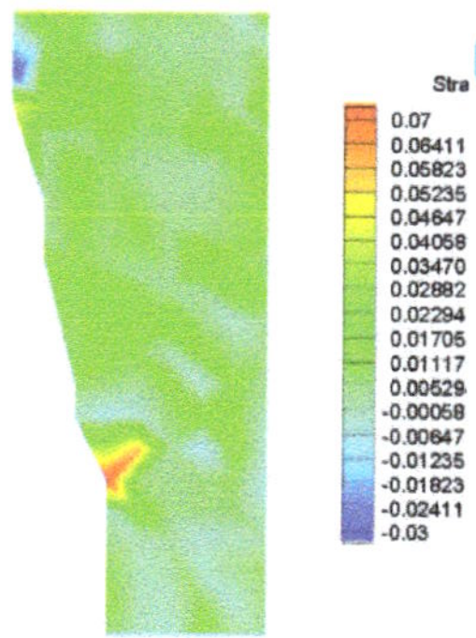	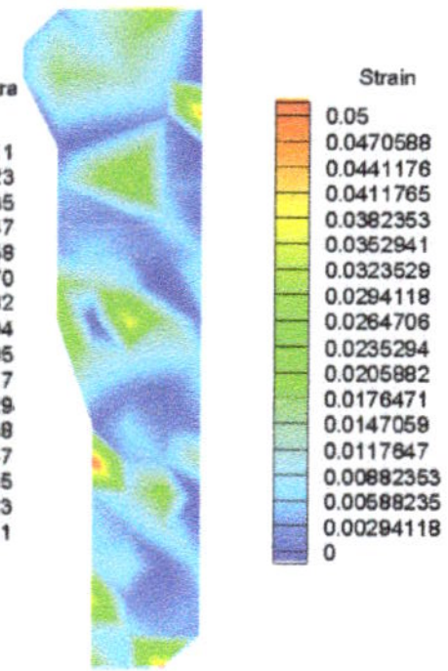	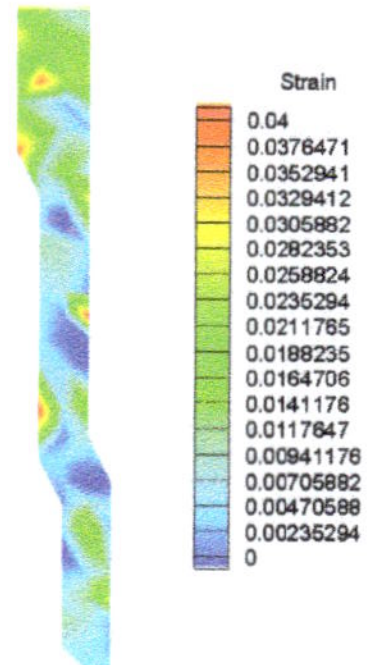

3.2. Results of Flow Analyses

Figure 7 shows the pressure distributions on the blade as the wind speed increases from 5 m/s to 20 m/s at different angles of attack (5° and 10°). In these distributions, it is seen that the flow separations increase with the increase of the angle of attack. In addition, with the wind speed increase, the total pressure values increased approximately 13.6 times for the angle of attack of 5°; this value is approximately 16.4 for the angle of attack of 10°. While the flow separation leaves the blade surface late in Figure 7a, it is seen in Figure 7e that the flow separation decreases with the effect of the inertia force. It is seen that the flow separation in Figure 7a is later than Figure 7b at the same wind speed. It can be concluded that, with the increase of the angle of attack, the flow is not able to cope with the reverse pressure gradient and viscous forces, and it cannot hold on to the blade surface. The separation of the flow from the blade surface reduces the aerodynamic performance of the blade.

According to the CRADLE analysis given in Table 2, Force-Y expresses the drag force (FD) on the blade, while Force-Z expresses the lift force (FL). When the wind speed value increases from 5 m/s to 20 m/s, it was observed that the lift force increased 15.23 and 15.09 times at the 50° and 10° angles of attack, respectively. It is also seen that the lift force is 0.91% more effective at the angle of attack of 5° than 10°.

Table 2. Lift and drag forces at varying speeds and angles of attack.

Wind Speed (m/s)	Force Type	Angle of Attack	
		5°	10°
5	Force-Y (N)	0.2809	0.3534740
	Force-Z (N)	−0.19985	−0.3127885
10	Force-Y (N)	1.118019	1.388722
	Force-Z (N)	0.86685	−1.2039
15	Force-Y (N)	2.459789	3.097886
	Force-Z (N)	−1.719546	−2.6772
20	Force-Y (N)	4.351231	5.480113
	Force-Z (N)	−3.042711	−4.72061

3.3. Blade Modeling Results

To recall, the curvatures of the blades were calculated according to Equation (1) by using the blade surface strain values obtained from the DIC tests. Then, these curvature data together with the moment diagram of the blades formed the moment–curvature inputs for the blades in modelling. These DIC tests could be verified by using these inputs in the material definition section in the blade modeling. For this verification, the blade displacement outputs from the models can be compared with the displacement data obtained from the experimental DIC tests (see Table 3). The errors in Table 3 show that the model results generally show a good agreement with the test results. In Table 3, it can be accepted that one of these errors is high. As mentioned before, it is thought that this high error may stem from a problem with the fixation of the new generation blade during one of the DIC tests.

After validating the models with experimental data, the pressure values obtained from the flow analyses at different wind speeds and angles of attack were applied to the root, middle and tip sections of the blades. With these new simulation models, the mechanical behavior of both commercial and new generation blades at different angles of attack and different wind speeds were investigated. The displacement values in the root, middle and tip regions of the wings, which are dependent on these behaviors, are shared in Table 4. In addition, to facilitate the comparison of these displacement values with each other, these values are also discussed in Figures 8–10. Figures 8–10 show the displacement values in the root, middle and tip regions of both blades, respectively. The "brown" and "blue colors" in these figures represent commercial and new generation blades, respectively.

Table 3. Displacements of both blades at the tip region corresponding to DIC tests and finite element modellings.

Blade Type	Displacements from DIC Tests			Displacements from Modelings			Error (%)		
	Load at Root Region (mm)	Load at Middle Region (mm)	Load at Tip Region (mm)	Load at Root Region (mm)	Load at Middle Region (mm)	Load at Tip Region (mm)	Load at Root Region (mm)	Load at Middle Region (mm)	Load at Tip Region (mm)
Commercial blade	0.41	14.0	60.0	0.3758	13.2656	57.7843	8	5	4
New generation blade	0.55	11.6	30.3	0.3191	10.8647	32.5191	42	6	7

Table 4. Modeling results for commercial and new generation blades for varying wind speeds and different angles of attack.

		Blade Root Region		Blade Middle Region		Blade Tip Region	
		α=5°	α=10°	α=5°	α=10°	α=5°	α=10°
Commercial Blade	$v = 5$ m/s	0.00271	0.00634	0.03329	0.07430	0.05536	0.11812
	$v = 10$ m/s	0.01992	0.02365	0.23492	0.27585	0.37492	0.43756
	$v = 15$ m/s	0.05482	0.05672	0.64525	0.67217	1.03456	1.07430
	$v = 20$ m/s	0.08716	0.06092	1.03066	0.71374	1.64834	1.14930
New Generation Blade	$v = 5$ m/s	0.00231	0.00540	0.02185	0.04940	0.03460	0.07550
	$v = 10$ m/s	0.01696	0.02014	0.15597	0.18361	0.23910	0.28010
	$v = 15$ m/s	0.04666	0.04829	0.42826	0.44586	0.65880	0.68433
	$v = 20$ m/s	0.07421	0.05186	0.68383	0.47379	1.05003	0.73111

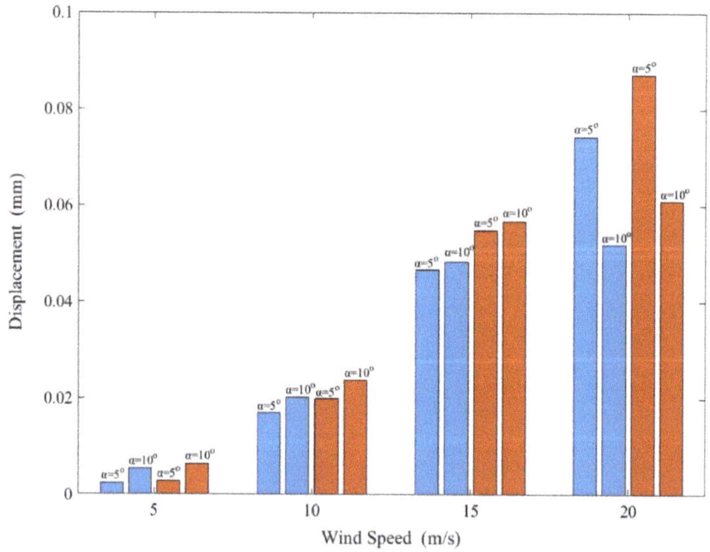

Figure 8. Modelling displacements of both blades at the root region. Red and blue colors are for commercial and new generation blades, respectively.

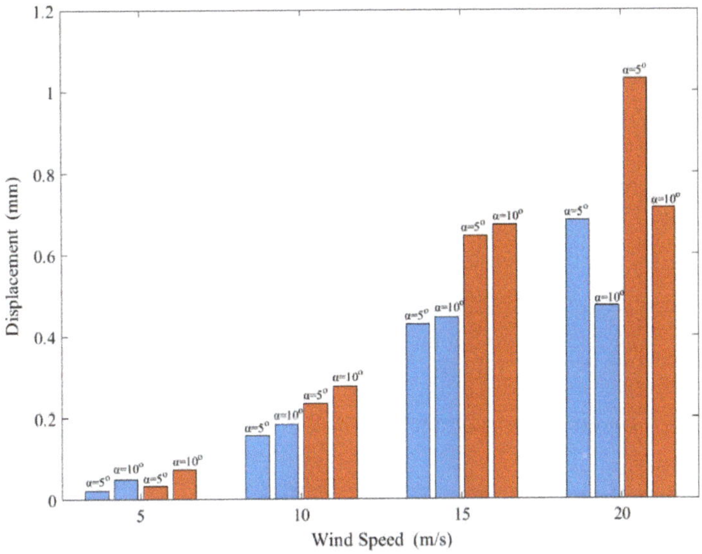

Figure 9. Modelling displacements of both blades at the middle region. Red and blue colors are for commercial and new generation blades, respectively.

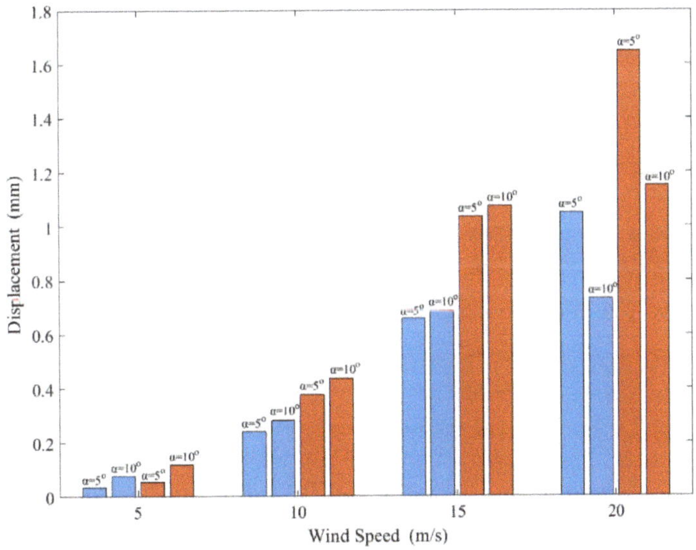

Figure 10. Modelling displacements of both blades at the tip region. Red and blue colors are for commercial and new generation blades, respectively.

According to the data in Figure 8, the displacement values of both wings increase at similar rates for both angles of attack (5° and 10°) with increasing wind speed from 5 m/s to 15 m/s. However, the effect of the angle of attack comes to the fore when the wind speed is 20 m/s. Therefore, when the wind speed is 15 m/s or 20 m/s at a 10° angle of attack, the variation of the displacement values of both blades remains small. In addition, when the blue and brown colors in Figure 8 are compared with each other, it is revealed that the displacement values of the root zone of the new generation blade are always calculated lower than the commercial one for each speed and each angle of attack. It is seen that the difference between the displacement values of the new generation and the commercial blades increases with the increase in wind speed.

The data in Figures 9 and 10 also show other displacement distributions compatible with the data in Figure 8. Therefore, according to the data in these figures, it is revealed that the displacement values of the middle and tip regions of the new generation blade are always lower than the commercial blade at different wind speeds and different angles of attack. This also indicates that the stiffness of the new generation blade is higher than the commercial one. Finally, in these figures, as in Figure 8, the difference between the displacement values of the new generation blade and the commercial blade increases with increasing wind speed.

4. Conclusions

The main purpose of this study was to provide the production of an innovative wind turbine blade with a new perspective. This new point of view is to compare the turbine blade structure to the reinforced concrete beam structure, which is frequently used in construction technology. Just as concrete is used as a matrix material in reinforced concrete beams and steel bars are used for reinforcing material, it is also possible to strengthen the recyclable light plastic matrix material in turbine blades with the help of carbon or glass bars. The use of recyclable materials in the production of this turbine blade also makes this new production technique more environmentally friendly. In addition, the fact that the amount of epoxy or resin used in traditional turbine blades is considerably higher than the trace amount of epoxy material used in the production of the new generation blade, emphasizes the importance of this new production.

In order to explain to the energy sector that the new generation turbine blade technology can be functional, it was compared with the commercial turbine blade of the same size and profile. Experimental DIC tests were applied for these comparisons and it can be seen that the new generation blade was stiffer. In addition, a simple modeling approach for both the commercial and the new generation blades is also presented in this study in order to attract the attention of the energy industry. In addition, the models of both blades were used with flow analysis and their behaviors at different speeds and different angles of attack were determined. All these findings demonstrate that the new generation blade has the possibility to exhibit sufficient strength in the field.

Author Contributions: Conceptualization, C.Ç. and M.S.G.; Methodology, C.Ç. and M.S.G.; Software, C.Ç., A.E. and M.S.G.; Validation, C.Ç. and M.S.G.; Formal analysis, C.Ç. and M.S.G.; Investigation, C.Ç., A.E. and M.S.G.; Writing—original draft, C.Ç., A.E. and M.S.G.; Writing—review & editing, C.Ç. and M.S.G.; Visualization, A.E. and M.S.G.; Supervision, C.Ç. and M.S.G.; Funding acquisition, C.Ç. and M.S.G. All authors have read and agreed to the published version of the manuscript.

Funding: This work was supported by R&D Innovation Program of KOSGEB (Small and Medium Enterprises Development Organization of Turkey), which is established in the Ministry of Industry and Technology of Republic of Turkey (grant number 1774102). Additionally, the authors would like to thank the Scientific and Technological Research Council of Turkey (TÜBİTAK) under the 2210-D and Erciyes University Scientific Research Projects (BAP) Unit-Turkey under (FKA-2020-10255).

Data Availability Statement: Not applicable.

Conflicts of Interest: The authors proclaim that they have no known conflicting financial interest or personal associations that could have appeared to affect the work stated in this paper.

References

1. Ergüner, A.; Özkan, R.; Koca, K.; Genç, M.S. Improvement of Mechanical Behaviour of Wind Turbine Blade Using Nanofluid-Graphene and/or Glass Fiber in Epoxy Resin. *J. Therm. Eng.* **2019**, *5*, 93–99.
2. Genç, M.S.; Gökçek, M. Evaluation of wind characteristics and energy potential in Kayseri, Turkey. *J. Energy Eng.* **2010**, *135*, 33–43. [CrossRef]

3. Genç, M.S. Economic analysis of large-scale wind energy conversion systems in central anatolian Turkey. In *Clean Energy Systems and Experiences*; IntechOpen: Rijeka, Croatia, 2010; pp. 131–154.
4. Genç, M.S. Economic viability of water pumping systems supplied by wind energy conversion and diesel generator systems in North Central Anatolia, Turkey. *J. Energy Eng.* **2011**, *137*, 21–35. [CrossRef]
5. Genç, M.S.; Karipoğlu, F.; Koca, K.; Azgın, Ş.T. Suitable site selection for offshore wind farms in Turkey's seas: GIS-MCDM based approach. *Earth Sci. Inform.* **2022**, *14*, 1213–1225. [CrossRef]
6. Karipoğlu, F.; Genç, M.S.; Koca, K. Determination of the most appropriate site selection of wind power plants based Geographic Information System and Multi-Criteria Decision-Making approach in Develi, Turkey. *Int. J. Sustain. Energy Plan. Manag.* **2021**, *30*. [CrossRef]
7. Akarsu, B.; Genç, M.S. Optimization of electricity and hydrogen production with hybrid renewable energy systems. *Fuel* **2022**, *324*, 124465. [CrossRef]
8. Kara, M.; Ercan, Y.; Yumuşak, R.; Cürebal, A.; Eren, T. Yenilenebilir Hibrit Enerji Santrali Uygulamasında Tesis Yer Seçimi. *Uluslararası Mühendislik Araştırma Ve Geliştirme Derg.* **2022**, *14*, 208–227. [CrossRef]
9. Türkiye Rüzgar Enerjisi Birliği, 2021 Faaliyet Raporu. Available online: https://tureb.com.tr//haber/2021-faaliyet-raporu/276 (accessed on 25 January 2023).
10. İzmir Kalkınma Ajansı, Rüzgâr Enerjisi Sektörü Ve İzmir Denizüstü Rüzgâr Enerjisi Yol Haritası, Temmuz 2021. Available online: https://izka.org.tr/wp-content/uploads/2021/08/ruzgar_enerjisi_sekto%CC%88ru%CC%88-izmir_denizustu_ruzgar_enerjisi_yol-haritasi-2.pdf (accessed on 4 March 2022).
11. Cooperman, A.; Eberle, A.; Lantz, E. Wind turbine blade material in the United States: Quantities, costs, and end-of-life options, Resources. *Conserv. Recycl.* **2021**, *168*, 105439. [CrossRef]
12. Liu, P.; Barlow, C.Y. Wind turbine blade waste in 2050. *Waste Manag.* **2017**, *62*, 229–240. [CrossRef]
13. Andersen, P.D.; Bonou, A.; Beauson, J.; Brøndsted, P. *Recycling of Wind Turbines, DTU International Energy Report 2014: Wind Energy—Drivers and Barriers for Higher Shares of Wind in the Global Power Generation Mix*; Technical University of Denmark: Lyngby, Denmark, 2014; pp. 91–97.
14. İzmir Kalkınma Ajansı, İzmir İli Rüzgâr Türbini Kanadı Geri Dönüşüm Tesisi, Ön Fizibilite Raporu, Haziran 2021. Available online: https://www.yatirimadestek.gov.tr/pdf/assets/upload/fizibiliteler/izmir-ili-ruzgar-turbini-kanadi-geri-donusum-tesisi-on-fizibilite-raporu2022.pdf (accessed on 27 January 2023).
15. Genç, M.S.; Açıkel, H.H.; Akpolat, M.T.; Özkan, G.; Karasu, İ. Acoustic control of flow over NACA 2415 aerofoil at low Reynolds numbers. *J. Aerosp. Eng.* **2016**, *29*, 04016045. [CrossRef]
16. Açıkel, H.H.; Genç, M.S. Flow control with perpendicular acoustic forcing on NACA 2415 aerofoil at low Reynolds numbers. *Proc. IMechE. Part G- J. Aerosp. Eng.* **2016**, *230*, 2447–2462. [CrossRef]
17. Genç, M.S.; Lock, G.; Kaynak, Ü. An experimental and computational study of low Re number transitional flows over an aerofoil with leading edge slat. In Proceedings of the 26th Congress of ICAS and 8th AIAA ATIO, AIAA-8877, Anchorage, AK, USA, 14–19 September 2008.
18. Açıkel, H.H.; Genç, M.S. Control of laminar separation bubble over wind turbine airfoil using partial flexibility on suction surface. *Energy* **2018**, *165*, 176–190. [CrossRef]
19. Genç, M.S.; Koca, K.; Demir, H.; Açıkel, H.H. Traditional and new types of passive flow control techniques to pave the way for high maneuverability and low structural weight for UAVs and MAVs. *Auton. Veh.* **2020**, 131–160. [CrossRef]
20. Genç, M.S.; Açıkel, H.H.; Koca, K. Effect of partial flexibility over both upper and lower surfaces to flow over wind turbine airfoil. *Energy Convers. Manag.* **2020**, *219*, 113042. [CrossRef]
21. Koca, K.; Genç, M.S.; Bayır, E.; Soğuksu, F.K. Experimental study of the wind turbine airfoil with the local flexibility at different locations for more energy output. *Energy* **2022**, *239*, 121887. [CrossRef]
22. Blasques, J.P.; Bitsche, R.D.; Fedorov, V.; Lazarov, B.S. Accuracy of an efficient framework for structural analysis of wind turbine blades. *Wind Energy* **2016**, *19*, 1603–1621. [CrossRef]
23. Tüfekci, M.; Genel, Ö.E.; Tatar, A.; Tüfekci, E. Dynamic Analysis of Composite Wind Turbine Blades as Beams: An Analytical and Numerical Study. *Vibration* **2021**, *4*, 1–15. [CrossRef]
24. Meng, H.; Jin, D.; Li, L.; Liu, Y. Analytical and numerical study on centrifugal stiffening effect for large rotating wind turbine blade based on NREL 5 MW and WindPACT 1.5 MW models. *Renew. Energy* **2022**, *183*, 321–329. [CrossRef]
25. Algolfat, A.; Wang, W.; Albarbar, A. Study of Centrifugal Stiffening on the Free Vibrations and Dynamic Response of Offshore Wind Turbine Blades. *Energies* **2022**, *15*, 6120. [CrossRef]
26. Thai, L.M.; Luat, D.T.; Ke, T.V. Finite element modeling for static bending analysis of rotating two-layer FGM beams with shear connectors resting on imperfect elastic foundations. *J. Aerosp. Eng.* **2023**, *in press*. [CrossRef]
27. Phung, V.M.; Ta, D.T.; Tran, V.K. Static Bending Analysis of Symmetrical Three-Layer FGM Beam With Shear Connectors Under Static Load. *J. Sci. Tech.* **2021**, *15*. [CrossRef]
28. Phung, M.V.; Nguyen, D.T.; Doan, L.T.; Van Duong, T. Numerical Investigation on Static Bending and Free Vibration Responses of Two-Layer Variable Thickness Plates with Shear Connectors. *Iran. J. Sci. Technol. Trans. Mech. Eng.* **2022**, *46*, 1047–1065. [CrossRef]
29. Ozsoy, N.; Ozsoy, M.; Mimaroglu, A. *Mechanical Properties of Chopped Carbon Fiber Reinforced Epoxy Composites, Special Issue of the 2nd International Conference on Computational and Experimental Science and Engineering*; Sakarya University: Serdivan, Turkey, 2016; Volume 130.

30. Siddharta; Kuldeep, G. Mechanical And Abrasive Wear Characterization Of Bidirectional And Chopped E-Glass Fiber Reinforced Composite Materials. *Mater. Des.* **2012**, *35*, 467–479. [CrossRef]
31. Barton, D.C.; Soden, P.D. Short-Term In-Plane Stiffness And Strength Properties Of CSM-Reinforced Polyester Laminate. *Composites* **1982**, *13*, 66–78. [CrossRef]
32. Hancox, N.L.; Mayer, R.M. *Design Data For Reinforced Plastics—A Guide For Engineers And Designers*; Chapman & Hall: London, UK, 1994.
33. Johnson, A.F. *Engineering Design Properties of GRP*; British Plastics Federation: London, UK, 1979.
34. Janeliukstis, R.; Chen, X. Review Of Digital Image Correlation Application To Large-Scale Composite Structure Testing. *Compos. Struct.* **2021**, *271*, 114143. [CrossRef]
35. Winstroth, J.; Schoen, L.; Ernst, B.; Seume, J.R. Wind Turbine Rotor Blade Monitoring Using Digital Image Correlation: A Comparison To Aeroelastic Simulations Of A Multi-Megawatt Wind Turbine. *J. Phys. Conf. Ser.* **2014**, *524*, 012064. [CrossRef]
36. Wu, R.; Zhang, D.; Yu, Q.; Jiang, Y.; Arola, D. Health Monitoring Of Wind Turbine Blades In Operation Using Three-Dimensional Digital Image Correlation. *Mech. Syst. Signal Process.* **2019**, *130*, 470–483. [CrossRef]
37. Mastrodicasa, D.; Lorenzo, E.D.; Manzato, S.; Peeters, B.; Guillaume, P. Full-Field Modal Analysis by Using Digital Image Correlation Technique, Rotating Machinery. *Opt. Methods Scanning LDV Methods* **2022**, *6*, 105–112.
38. Khadka, A.; Afshar, A.; Zadeh, M.; Baqersad, J. Strain Monitoring Of Wind Turbines Using a Semi-Autonomous Drone. *Wind. Eng.* **2022**, *46*, 296–307. [CrossRef]
39. Genç, M.S. Unsteady aerodynamics and flow-induced vibrations of a low aspect ratio rectangular membrane wing with excess length. *Exp. Therm. Fluid Sci.* **2013**, *44*, 749–759. [CrossRef]
40. Genç, M.S.; Demir, H.; Özden, M.; Bodur, T.M. Experimental analysis of fluid-structure interaction in flexible wings at low Reynolds number flows. *Aircr. Eng. Aerosp. Technol.* **2021**, *93*, 1060–1075. [CrossRef]
41. Beer, F.-P.; Johnston, E.-R.; Dewolf, J.-T.; Mazurek, D.-F. *Mechanics of Materials*, 7th ed.; McGraw-Hill Education: New York, NY, USA, 2015.
42. Karasu, İ.; Özden, M.; Genç, M.S. Performance assessment of transition models for three-dimensional flow over NACA4412 wings at low Reynolds numbers. *J. Fluids Eng.* **2018**, *140*, 121102. [CrossRef]
43. Demir, H.; Özden, M.; Genç, M.S.; Çağdaş, M. Numerical Investigation of Flow on NACA4412 Aerofoil with Different Aspect Ratios. In *EPJ Web of Conferences*; EDP Sciences: Ulysses, France, 2016; Volume 114.

energies

Article

Effect of Macroscopic Turbulent Gust on the Aerodynamic Performance of Vertical Axis Wind Turbine

Lakshmi Srinivasan [1], Nishanth Ram [1], Sudharshan Bharatwaj Rengarajan [1], Unnikrishnan Divakaran [1], Akram Mohammad [2] and Ratna Kishore Velamati [1,*]

1 Department of Mechanical Engineering, Amrita School of Engineering Coimbatore, Amrita Vishwa Vidyapeetham, Coimbatore 641112, India
2 Department of Aerospace Engineering, King Abdulaziz University, Jeddah 21589, Saudi Arabia
* Correspondence: v_ratnakishore@cb.amrita.edu; Tel.: +91-812-295-7821

Abstract: Vertical Axis Wind Turbines (VAWTs) have proven to be suitable for changing wind conditions, particularly in urban settings. In this paper, a 2D URANS (Unsteady Reynolds-Averaged Navier Stokes) numerical analysis is employed for an H-Darrieus VAWT. A turbulent domain is created through systemically randomising the inlet velocity to create macro-turbulence in front of the VAWT. The parameters for spatial and temporal randomisation of velocity and its effects on the turbine performance are studied for a mean free stream velocity, U_∞ = 10 m/s, and a tip speed ratio (TSR) of 4.1. The mean Coefficient of power (C_p) for randomised fluctuation of 2 m/s and half-cycle randomisation update frequency is 0.411 and for uniform inlet velocity is 0.400. The C_p vs. Tip Speed ratio plot suggests that the optimal tip speed ratio for operation is around 4.1 for this particular wind turbine of diameter 1 m, chord 0.06 m, and NACA 0018 airfoils. The effect of randomisation for tip speed ratio λ = 2.5, 3.3, 4.1, and 5.3 on the performance of the turbine is studied. Turbine wake recovers at a faster rate for macro-turbulent conditions and is symmetric when compared to wake generated by uniform velocity inlet. The maximum velocity deficit for a distance behind the turbine, x/d = 8 at TSR (λ) = 4.1 is 46% for randomised inlet and 64% for uniform inlet. The effect of randomisation for λ = 2.5 to 5.3 on the performance of the turbine is analysed. A time-varying gust based on International Electrotechnical Commission (IEC) Extreme Operating Gust is used to study the effect of fluctuating wind conditions in a turbulent environment. Since real-time conditions often exceed gust factors mentioned by IEC, winds with large gust factors such as 1.50, 1.64, and 1.80 are analysed. With an increase in gust amplitude, U_{gust} = 6 m/s to U_{gust} = 12 m/s on a free stream velocity of $U\infty$ = 10 m/s, the mean C_p decreases from 0.41 to 0.35 since the wind turbine operates under tip speed ratios outside optimal range due to large fluctuations in incoming velocity.

Keywords: vertical axis wind turbine; gust; macroturbulence; URANS

Citation: Srinivasan, L.; Ram, N.; Rengarajan, S.B.; Divakaran, U.; Mohammad, A.; Velamati, R.K. Effect of Macroscopic Turbulent Gust on the Aerodynamic Performance of Vertical Axis Wind Turbine. *Energies* **2023**, *16*, 2250. https://doi.org/10.3390/en16052250

Academic Editor: Paweł Ligęza

Received: 13 January 2023
Revised: 20 February 2023
Accepted: 22 February 2023
Published: 26 February 2023

1. Introduction

The recent trend in environmental degradation and climate change has led to a pressing need to shift towards renewable and sustainable sources of energy. With wind energy steadily gaining pace, the global wind energy market continues to be stable with about 591 GW in 2018, and approximately 51 GW installed worldwide every year. There has been an approximate 400% increase in power generated by wind turbines from 2008 to 2018 [1]. Though there are efficient methods to harness wind energy using large wind farms and offshore sites, there is still a void in the development of effective harnessing methods for urban conditions.

Unlike wind conditions in large wind sites, the urban winds are mainly characterised by unsteadiness in wind velocity and wind direction and have high turbulence intensity due to the interaction of winds with various physical obstacles such as buildings and trees, and localised temperature variations. Emejeamara et al. [2] measured wind velocities in urban environments and reported the presence of gustiness in urban winds and the need for their consideration in urban wind studies.

For energy extraction in large wind farms, conventional Horizontal Axis Wind Turbines (HAWT) are efficient choices. However, HAWTs do not prove to be as efficient as Vertical Axis Wind Turbines (VAWT) in urban areas. The power produced by HAWTs drops significantly when the direction of the wind changes, while that of the Vertical Axis Wind Turbines (VAWTs) remains constant irrespective of the directional change. VAWTs pose an advantage over HAWTs in urban conditions due to their omnidirectional characteristic [3–5]. As the angle of wind changed from 0° to 45° incident to HAWT, the voltage produced by the generator of wind turbines significantly dropped, according to Ishugah et al. [6]. Voltage further dropped to 0 V when the angle reached 90°. VAWTs, however, produced a steady voltage for all angles. When compared to a HAWT, a VAWT is suitable for urban environments because of less space requirement and more potential to produce power for the same swept area of the turbine [6]. VAWTs also have several other advantages over HAWTs due to their capability to operate at low tip speed ratios (TSR), low noise levels and ease of construction and maintenance, and thus are suitable choices for urban wind conditions [3,7,8].

In recent years, several kinds of research and analyses have been conducted on the performance and design of the VAWTs under various wind conditions. Many studies have focused on the effect of unsteady and fluctuating winds [9–19], impact of gust [20–26], and turbulence [27,28] on the performance of VAWTs.

Both numerical and experimental studies have been performed on the effects of unsteady winds on the performance of VAWTs. Danao et al. [18] experimentally studied these effects by employing 7% and 12% velocity amplitude fluctuations on a sinusoidal wind profile for a scaled wind tunnel model of a VAWT. Shahzad et al. [29] studied the effect of accelerating and decelerating airflow on VAWTs by steadily increasing the velocity from 4 m/s to 10 m/s and decreasing it back to 4 m/s. They established that even though the rate of acceleration is constant during acceleration and deceleration, their effects on VAWTs vary. Bhargav et al. [14] and Danao et al. [17] have numerically studied the performance dependency of the unsteadiness of the wind by establishing a relation between the wind velocity fluctuation amplitude and the fluctuation frequency. The fluctuating wind varies its velocity sinusoidal, and different analyses are performed by varying the fluctuation amplitude and frequency. These studies show that, as the fluctuation amplitude increases, the coefficient of power (C_P) values decreases significantly, and as the fluctuation frequency increases, the C_P value increases. Jafari et al. [30] used a wind model that generates quasi unsteady wind velocity by generating random velocities using the turbulence intensity of the wind, which was 20% in their study.

Studies comparing the computationally modelled sinusoidal wind and real-time unsteady winds have also been conducted. In a comparison study performed by Wekesa et al. [12], the performance results obtained computationally were compared to the empirically obtained results at specified locations, namely Marsabit and Garissa. Numerically obtained power densities of wind compared closely with the empirically obtained ones and had only marginal errors of 4% and 17%, respectively, at Marsabit and Garissa.

Gust winds are the winds with a sudden increase in wind speed, followed by a decrease in wind speed. The incoming winds, based on the direction, are classified into longitudinal, lateral, and upward [21]. Lee et al. [31] studied the influence of the vertical wind on the performance of a small vertical-axis wind turbine installed on the rooftop of a building and concluded that if the horizontal wind is greater than 8 m/s, the impact of the vertical wind and the impact of horizontal wind on the power output of the turbine is reduced.

The Extreme Operating Gust (EOG) standardised by the International Electrochemical Commission's (IEC) [32] is used in this paper. EOGs are constituted by a decrease, followed by a steep rise, a steep drop, and a rise back to the original wind speed value. There has been research on the effects of gust winds on VAWTs. Onol and Yesilyurt [23] studied the effect of the IEC's model of EOG on VAWTs through 2D URANS-based numerical simulations by plotting the power coefficient against the angle of attack for various Tip Speed Ratios. Wu et al. [20] studied the responses of VAWTs to lateral gustiness of wind using sinusoidal wind signals. They stated that gust influences are not only present in the immersion period but also in a much broader influencing period because of the mutual aerodynamic interaction between the gusts and rotor.

Even though the IEC has standardised gusts, several studies [22,33] have been performed comparing the original gust data and possible corrections that have to be made to the model to make it more accurate. Rakib et al. [22] compared the IEC 61400-2013 gust model to actual wind data for a 5 kW Aerogenesis turbine at Callaghan. It was observed that the average gust factor of experimental data was 26% higher and the rise-and-fall time was 21% shorter when compared to the IEC gust model. The frequency of occurrence of gust was also higher than the predictions by IEC. Thus, an increase in the mean amplitude of gust velocity, shorter rise, and fall time, and an increase in gust factor must be incorporated into the inlet gust model. The analyses on gust winds did not take into consideration the high turbulence levels in urban winds.

In order to reproduce the real-life turbulent urban wind conditions in numerical simulations, Balduzzi et al. [27] developed a numerical strategy for an unsteady RANS approach to generate macro-turbulence for wind energy applications. The study introduced macro-turbulence by spatially randomising the velocity values at the inlet. Further in the study, it was used to simulate the behaviour under turbulence of an H-Darrieus vertical axis wind turbine. The results produced by the simulations were also reported to match the experimental results.

The study of the wake is necessary for predicting aerodynamic behaviour downstream of the turbine. Extensive experimental and numerical studies on the effect of wind conditions on wake and wake characteristics of VAWTs are reported. Lam and Peng [34] studied wake characteristics of VAWT using 2D and 3D simulation models. They found that SST turbulence models are coherent with experimental results. Peng et al. [35] experimentally studied wake aerodynamics of a 5-straight-bladed VAWT through wind tunnel tests. They concluded that wake asymmetry was attributed to the larger number of vortices shed in the windward direction. In addition, a counter-rotating pair of vortices was found downstream that contributed to complete mixing and faster wake recoveries. Rezaeiha et al. [36] studied the effect of tip speed ratio on wake structures and its influence on wake parameters such as length of the turbine wake and velocity deficit.

The effect of macro-turbulence in winds and gustiness of winds on VAWTs has been separately studied in the past for a typical urban environment. However, VAWTs are likely to be installed in environments with wind profiles that have a super-imposition of both. Thus, this study analyses the effect that macro-turbulent gusty winds on the performance and wakes of VAWTs. In the current study, wind speeds are varied both spatially and temporally in an attempt to reproduce a typical urban setting under an Extreme Operating Gust (EOG) criteria. The temporal randomisation accounts for the unsteadiness in wind velocities, and spatial randomisation attempts to reproduce non-homogeneous eddies with larger length scale as present in urban winds. This work studies inlet randomisation parameters such as randomised fluctuation and randomisation update frequency on a steady inlet velocity and its effects on the upwind conditions and performance of the VAWT. It includes the effect of Tip Speed Ratio (TSR) on the performance of the VAWT and a study of wake turbine characteristics for different tip speed ratios for randomised inlet conditions. A comparison of uniform and randomised velocity inlet cases on turbine performance and wake is performed. The effect of gust parameters such as gust amplitude and gust time period are also studied for spatially randomised time-varying gust cases.

2. Problem Statement

The present study uses a 2-dimensional simulation of a 2-bladed H type VAWT with NACA0018 aerofoil. The height and diameter of the turbine are both equal to 1 m, and solidity (σ) is 0.12(σ = Nc/D). The shaft diameter (D) is 0.04 m, and the chord length of the aerofoil (c) is 0.06 m. N signifies the number of blades. The effect of struts is disregarded in this study. Table 1 shows the geometric features of the turbine considered for the study.

Table 1. Geometric features of the turbine considered for the study.

Axis of Rotation	Vertical
Blade profile	NACA 0018
Chord, c	0.06 m
Diameter of rotor	1 m
Frontal area of the rotor	1 m^2
Solidity, σ	0.12
Number of blades	2

Urban wind conditions are non-uniform spatially. Hence it is essential to understand the performance of the VAWT with spatially varying macro-turbulence.

The urban winds are mainly characterised by their gustiness, unsteadiness in wind velocity and direction, and high turbulence levels. Unlike the wind flow in an open wind farm, the wind in urban regions will have to make its way around significantly denser tall structures such as buildings and poles. This leads to a flow that is non-homogeneously filled with eddies of large length scales, which result in a 'macro-turbulent' wind flow. The velocity contour of such a macro-turbulent inflow is presented in Figure 1a. Along with the local fluctuations of wind speeds caused by urban structures, one can observe larger and characteristic fluctuations in the freestream velocity of the wind result. The International Electrotechnical Commission (IEC) has characterised different types of such fluctuations and one such wind fluctuation, characterised by sudden rise and fall of wind speeds, is an Extreme Operating Gust (EOG), represented by the velocity plot in Figure 1b. With gusts being a fairly common wind phenomenon, urban environments are also susceptible to them. Therefore, in an urban environment, one can observe local winds with smaller and high frequency velocity fluctuations compounded with larger and low frequency variations in wind due to gusts. Inflow wind speed, averaged over the frontal area of a VAWT rotor subjected to such gusty wind conditions, can be best represented by the black coloured plot of wind velocity in Figure 1b. These characteristics of urban winds are detrimental to both the performance and structural integrity of VAWTs. Though significant research has been performed on studying the effects of gusts on the performance of VAWTs, the local unsteadiness and turbulent nature of urban winds have been undermined. Thus, the focus of the present study is to analyse the effects of unsteady, macro-turbulent gusty winds on the performance of VAWTs. Randomisation has been performed spatially (Figure 1a) and temporally (Figure 1b) to account for unsteadiness in wind velocity and turbulence levels, respectively.

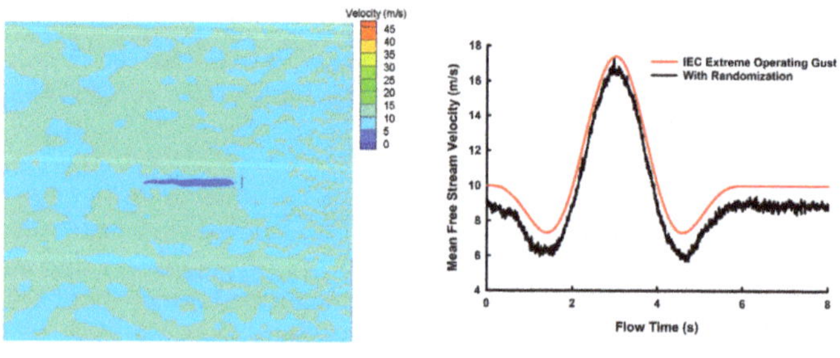

Figure 1. (**a**) Velocity contour of non-uniform flow upstream of the VAWT; (**b**) Temporal variation of average free stream wind velocity upstream of the VAWT.

3. Numerical Methodology

The simulations were carried out in commercial CFD package ANSYS Fluent 19.2. The flow is considered to be incompressible, as the velocity is negligible compared to the sonic speed. The flow is solved using incompressible 2D Unsteady Reynolds Averaged Navier Stokes (URANS)-based 4 equation transition SST model. The equations of conservation of mass and momentum solved for modelling the flow are given in Equations (1) and (2):

$$\nabla\cdot\vec{v} = 0 \tag{1}$$

$$\nabla\cdot\left(\vec{v}\,\vec{v}\right) = -\nabla p + \nabla\cdot\left(\bar{\bar{\tau}}\right) + \rho\vec{g} \tag{2}$$

$$\bar{\bar{\tau}} = (\mu + \mu_t)\left[\nabla\vec{v} + \left(\nabla\vec{v}\right)^T - \frac{2}{3}\nabla.\vec{v}I\right] \tag{3}$$

where ρ is the density [kg/m^3], $\vec{v}$ is the velocity [m/s], $\vec{g}$ is the acceleration due to gravity [m/s^2], p is the pressure [Pa], $\bar{\bar{\tau}}$ is the stress tensor [N/m^2], μ is the dynamic viscosity [N m^2/s], μ_t is the turbulent viscosity [m^2/s] closed by a suitable turbulence model, and I is an identity matrix.

The equations used for modelling turbulence, i.e., the equations for transition SST model are given in Equations (4) and (5):

$$\frac{\partial(\rho\gamma)}{\partial t} + \frac{\partial(\rho U_j\gamma)}{\partial x_j} = P_{\gamma 1} - E_{\gamma 1} + P_{\gamma 2} - E_{\gamma 2} + \frac{\partial}{\partial x_j}\left[\left(\mu + \frac{\mu_t}{\sigma_y}\right)\frac{\partial\gamma}{\partial x_j}\right] \tag{4}$$

$$\frac{\partial(\rho\overline{R}\theta_{\theta t})}{\partial t} + \frac{\partial(\rho U_j\overline{R}\theta_{\theta t})}{\partial x_j} = P_{\theta t} + \frac{\partial}{\partial x_j}\left[(\mu + \mu_t)\frac{\partial\overline{R}\theta_{\theta t}}{\partial x_j}\right] \tag{5}$$

In a comparative study of turbulence models by Darcozy et al. [37], the k-epsilon Realisable and k-omega SST models provide accurate results in locating the optimal tip speed ratio value for a VAWT for a 2D CFD analysis of an H-Darrieus-type VAWT. However, in a study by Rezaeiha et al. [38], various turbulence models were compared for accuracy in modelling turbulence. The study concluded that SST-based models are best suited for URANS-based VAWT simulations and transitional SST models for simulations in which the flow transitions from laminar to turbulent. Since the present study is concerned about transitional flows in which the range of the Reynolds number is from 1.2×10^6 to 2.0×10^6, the transition SST turbulent model is chosen for the simulations.

For convective term discretisation, the second-order upwind scheme is used. The Coupled algorithm is employed for handling the pressure and velocity coupling. The double-precision segregated solver with an implicit method, was utilised for solving the discretised algebraic equations.

3.1. Wind Inlet Conditions

The inlet velocity is defined using a custom User-defined Function (UDF) in ANSYS FLUENT to reproduce desired turbulent flow conditions inside the computational domain. UDF allows direct interaction with the solver through customisation of inlet boundary conditions and execution for a specified number of iterations. The velocity values are randomised, both temporally and spatially. Temporal randomisation is performed to account for the unsteadiness of wind velocities in urban conditions and spatial randomisation for the high turbulent wind conditions in urban areas. The velocity values fed to the solver will be randomised by maintaining the integrity of the original wind data, i.e., the mean and standard deviation of the wind data before and after randomisation will be the same.

The gust signal chosen for the study is an IEC standardised extreme operating gust and the variation of velocity of wind with time follows the Equation (6):

$$V(t) = (0.37 * A * (sin(3\pi t/T)) * (1 - cos(2\pi t/T)) \tag{6}$$

where A and T are gust amplitude and time period.

Accounting for the unsteadiness of wind velocities in urban wind conditions, the wind signal is randomised temporally. This is achieved by adding random values, within the range of random velocity fluctuation, to the IEC standard gust profile. The temporal randomisation scheme discretises the IEC gust signal based on the desired randomisation update frequency. It adds fluctuation magnitude to the original magnitude of wind velocity. The resultant velocity will thus be a sum of the original gust velocity and a random fluctuation added within the range of specified velocity fluctuation. This randomisation is better expressed through the figures in Figure 1b. Uniform wind case (where the mean velocity is kept constant at 10 m/s) is also randomised using the same method. The chosen velocity fluctuation for randomising the IEC gust profile temporally is 6 m/s and the randomisation update frequency is 90.

The spatial randomisation randomises the wind data spatially by assigning different velocities to each of the nodes at the velocity inlet of the computational domain. Both the X and Y components of the velocity vectors will be randomised. The X component gets assigned by different velocities to each of the nodes at the inlet. The randomisation in the Y component of velocity helps in changing the direction of the velocity vector, thus introducing macro-turbulence.

3.2. Computational Domain and Grid

The computational analyses performed in this study employ the use of 2D computational domains for computational simplicity. Rezaeiha et al. [39] compared the results of 2D and 2.5D simulations for low solidity wind turbines and established that 2D simulation results show good agreement with experimental results. The study on the effectiveness of 2D analyses for straight-bladed Darrieus VAWTs by Bianchini et al. [40] showed that the results from 2D CFD analyses are significantly accurate and reliable upon considering the placement of lateral boundaries at a distance far enough to simulate an open-air flow field.

The computational domain, as illustrated in Figure 2a, consists of a rotating inner domain where the turbine is located and a fixed rectangular domain surrounding the core. To provide an idea about the turbine geometry in 3D, a representative image is shown in Figure 2b. For the computational purposes, we have considered a 2D cross-sectional plane of the turbine perpendicular to the axis of rotation and have not considered the struts connecting the blades to the shaft. A sliding interface is employed between the fixed domain, and the rotating core enables rotation of the turbine. The meshing of the computational domain has been performed using commercial software GAMBIT v2.4. The inner domain is dynamic and made to rotate at an angular velocity of 82 rad/s to simulate the rotation of the turbine. The diameter of the inner mesh (dc) is 2.5 m, and the dimensions of the outer domain are 45 m and 40 m along the length (di + do) and breadth (W), respectively. With reference to Rezaeiha et al. [39], the turbine is placed at a distance of 15 m from the inlet (di). The side AB of the domain is the spatially varying velocity inlet, side CD is the pressure outlet, and BC and CD of the domain are walls as shown in the Figure 2a.

The side AB of the domain is a spatially varying velocity inlet that is given velocity values by a UDF hooked to the CFD solver, side CD is the pressure outlet with 1 atm pressure, and BC and AD of the domain are no-slip walls as shown in Figure 2a. The blades have a no-slip wall boundary condition with 628 nodes and a grid length of 5×10^{-5} m with a growth factor of 1.22. The computational domain totally consists of around 0.8 million cells. The grid independence study is performed by comparing the turbulence intensity and the turbulent length scales acquired by running simulations in the present grid and a finer grid for both randomised velocity inlet cases. Figure 3a shows the refinement of the mesh downstream of the turbine for performing accurate wake studies. The blade angle is measured in a counter-clockwise direction, which forms the basis for choosing the time step interval. Figure 3b represents the rotating domain of the mesh, and Figure 3c shows the region around the blade that is finely meshed to analyse blade-flow interactions with better accuracy.

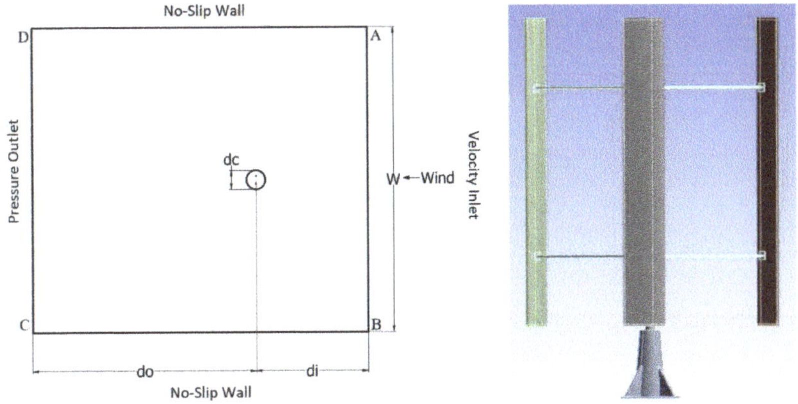

(a) (b)

Figure 2. (**a**) Computational domain ABCD; (**b**) A representation of straight-bladed VAWT with 3 blades.

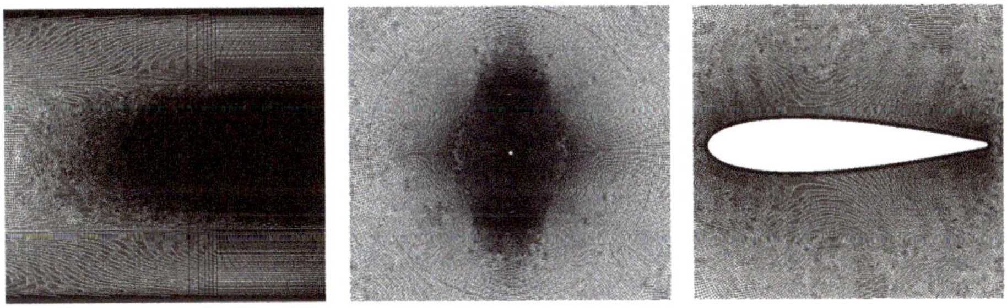

(a) (b) (c)

Figure 3. Details of computational grid used in the present work. (**a**) Overall grid of the entire domain; (**b**) Mesh in the rotation part of domain (**c**) grid around the wind turbine blade.

3.3. Computational Methodology

The explicit relaxation factors for momentum and pressure are both 0.75. The under-relaxation factors for turbulent kinetic energy, dissipation rate, and intermittency are 0.4. The residuals for convergence are 10^{-5} for the continuity equation and 10^{-3} for velocity, turbulent kinetic energy, dissipation rate, and other variables. Time step size is taken to be time taken by the rotor to complete 0.5° of rotation. The results are read during the cycles 60–120 turbine cycles for the steady mean velocity case at tip speed ratio (λ) of 4.1. For the IEC gust cases, the results are read 120–160 turbine cycles for gust time period 3 s, 120–200 turbine cycles for time periods 6 s, and 120–200 turbine cycles 10.5 s. This allows time for the gust signal to reach the turbine and enough time for proper wake study.

4. Validation and Grid Independence

For any results obtained by numerical simulation, validation is essential and mandatory for the results of the same to be considered for further analysis. The study involves a numerical analysis of vertical axis wind turbine. Hence, a validation against an experimental result is carried out. The CFD results of Rezaeiha et al. [36] and experimental results of

Castelli et al. [41] are plotted against the CFD results obtained for a 3-bladed vertical axis wind turbine in Figure 4. It is found that at lower tip speed ratio, the numerical results are in line with the experimental results. This percentage of error is minimal. Grid independence is conducted by running three different grid cases with 0.4 million, 0.8 million, and 1.4 million grid points represented by grid 1, grid 2, and grid 3, respectively, in Figure 5. As the results given by grid 2 were at par with the results predicted by grid 3, which is almost double the grid points, it was decided to do the remaining set of simulations using grid 2.

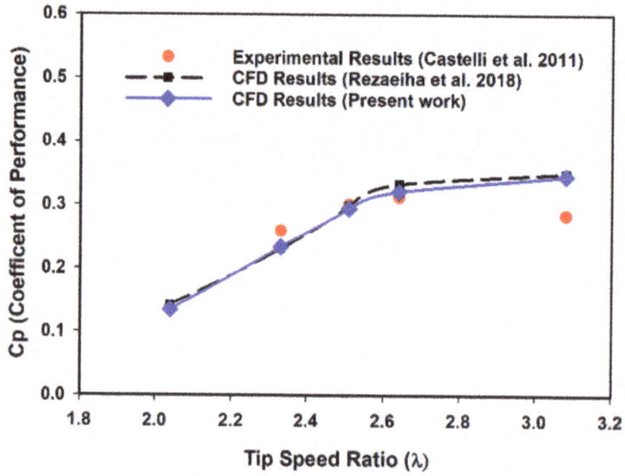

Figure 4. Validation of numerical model with experimental results [36,41] and other standard numerical results.

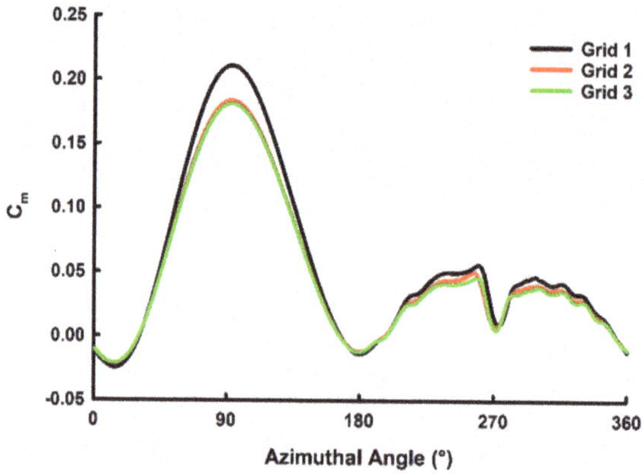

Figure 5. Grid independence study conducted using three different grid numbers.

5. Results and Discussion

The present study analyses the effect of inlet velocity randomisation parameters such as randomisation fluctuation and randomisation update frequency on the performance of a VAWT at a constant tip speed ratio of 4.1. The effect of Tip Speed Ratio on the coefficient of power and wake structure was studied. Further, the effect of gust parameters such as gust amplitude and gust time period for an IEC Extreme Operating Gust was investigated.

5.1. Effect of Randomisation Parameters on the VAWT Performance

5.1.1. Effect of Randomised Fluctuation

In Figure 6a, the variation of coefficient of power of the wind turbine due to the effect of randomised fluctuation of 2 m/s, 4 m/s, and 6 m/s at a constant tip speed ratio of 4.1 is shown.

The randomised fluctuation magnitude is applied on an inlet mean velocity magnitude of 10 m/s. The value of C_p keeps fluctuating due to frequent updates of velocity at the inlet boundary, for every half rotation of the turbine. It is observed that as the randomisation fluctuation of velocity decreases from 6 m/s to 2 m/s, the mean value of C_p, represented by the dashed line, increases by a small margin of 0.01. This is because, for lower randomised fluctuations, the range of velocity values at all the inlet nodes is less compared to higher randomised fluctuations. For randomised fluctuation of 2 m/s, the velocity values vary from 12 m/s to 8 m/s, whereas for randomised fluctuation of 6 m/s, the velocity values vary from 16 m/s to 4 m/s. Since the velocities before the turbine could be of small magnitude in higher randomised fluctuation cases, the value of C_p produced is also low. The maximum and minimum bounds of C_p are represented by dotted lines in Figure 6a. The decrease in range values of C_p by 0.06 and a standard deviation of C_p by 0.01 from the randomised fluctuation of 6 m/s to 2 m/s is due to the same reason. The mean C_p values of cases with randomised inlet velocity are always higher than uniform velocity cases due to the effect of upstream macro-turbulence. The mean C_p value for randomisation fluctuation of 2 m/s is 0.411 and mean C_p for uniform velocity is 0.400 for $\lambda = 4.1$.

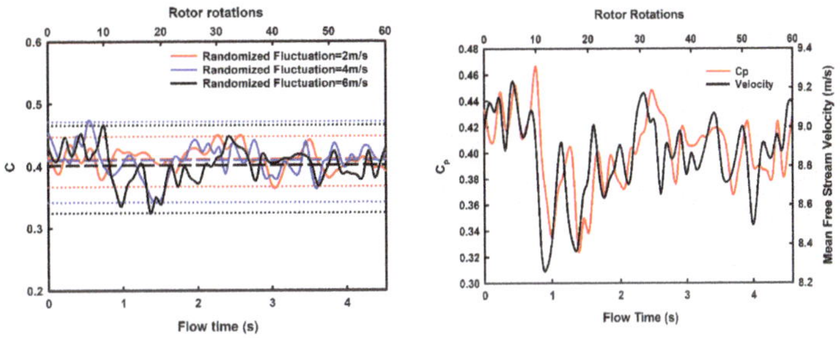

(a) (b)

Figure 6. (**a**) Variation of Cp for various randomised fluctuation; the mean and minimum maximum is represented in the graph with the help of dashed lines and dotted lines respectively; (**b**) Effect of free stream velocity on Cp.

The effect of randomisation inlet velocity on the coefficient of power is shown in Figure 6b. It is noticed that as the mean free stream velocity varies, C_p follows a similar trend. There is a delay in C_p with respect to the free stream velocity through most peaks and troughs when plotted against flow time. The value of C_p increases as the magnitude of velocity in front of the turbine increases, and C_p drops as velocity magnitude decreases. As the magnitude of fluctuation of velocity increases, the time lag for C_p increases. It is also noticed that in regions with a higher frequency of fluctuation in velocity, the variation in C_p does not reflect the frequency of velocity fluctuation.

Figure 7 describes the variation of the Coefficient of Moment (C_m) for a single blade over an entire turbine rotation. In general, the C_m values peak at 90° where a dynamic stall occurs. During downwind conditions, a small variation in C_m is noticed and can be attributed to the interaction of blades with the wake created by the central shaft. There is a substantial variation in peak C_m values for different turbine cycles due to variation of wind velocity due to spatial randomisation and different randomised fluctuations of the inlet velocity. The range of peak C_m values is of higher magnitude for randomised fluctuation of 6 m/s (Figure 7a) than for randomised fluctuation of 2 m/s (Figure 7b).

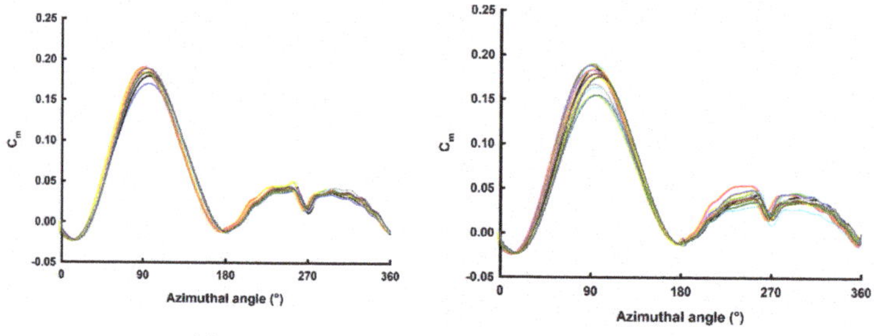

(**a**) (**b**)

Figure 7. Moment coefficient for various rotational cycles. (**a**) Randomised Fluctuation = 2 m/s; (**b**) Randomised Fluctuation = 6 m/s. Coefficient of Moment at various time instances (rotational cycles) are represented through different coloured lines. Due to randomized wind conditions, values vary every rotational cycle.

5.1.2. Effect of Randomisation Update Frequency

The effect of update frequency of inlet profile for updates at every 45, 90,180, and 360 time-steps on the variation of C_p is studied. The tip speed ratio is maintained at 4.1, and the randomised fluctuation magnitude is 6 m/s on an inlet mean velocity magnitude of 10 m/s for all simulations. Figure 8a shows the variation of mean free stream wind velocity (U_∞) for all randomisation update frequency cases. The velocity values are taken at 0.75 d upstream of wind turbine and averaged over a width of 1.5 d. There is a decrease in the magnitude of fluctuation of wind velocity from an update frequency of 360 time-steps to 45 time-steps. The standard deviation of fluctuating velocity is 0.21 m/s for update frequency of 360 time-steps and 0.11 m/s for update frequency of 45 time-steps. The range of velocity fluctuation is 0.4 m/s higher in the case of update frequency of 360 time-steps compared to 45 time-steps. The flow field upstream of the turbine is more uniform for update frequency of 45 time-steps (Figure 8c) than for update frequency of 360 time-steps (Figure 8b), which is characterised by a higher magnitude of macro-turbulence. As the update interval decreases, the flow field in front of the turbine becomes uniform.

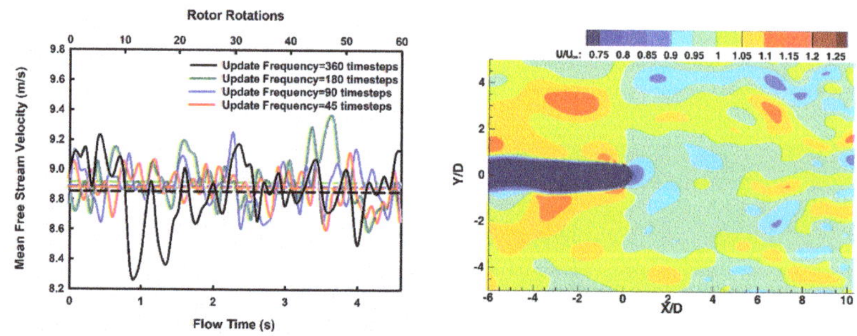

(**a**) (**b**)

Figure 8. *Cont.*

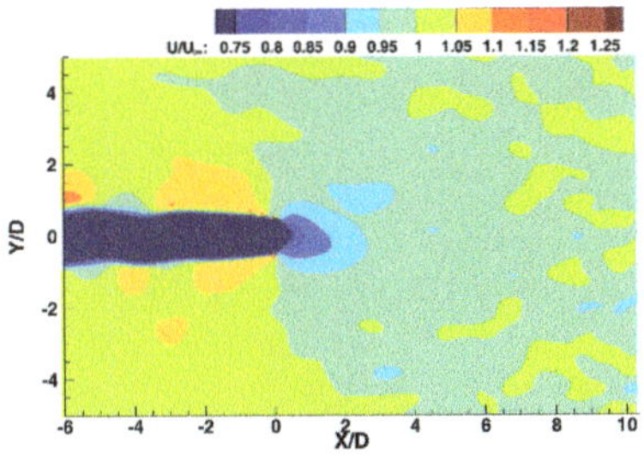

(c)

Figure 8. (**a**) Effect of update frequency on the average wind velocity upstream of VAWT; (**b**) Snapshots of flow field upstream of VAWT for Randomisation Update Frequency = 360 time-steps; (**c**) Snapshots of flow field upstream of VAWT for Randomisation Update Frequency = 45 time-steps.

In Figure 9, it is observed that there are frequent C_p peaks in analysed data for an update frequency of 45 time-steps when compared to an update frequency of 360 time-steps. For cases of slower randomisation updates (update frequencies of 180 time-steps and 360 time-steps), the velocity at the inlet diffuses and becomes less spatially random by the time wind reaches the turbine. The effect of these velocities is prominent on the blade because of larger update intervals, and thus higher peaks and troughs of C_p are achieved. Thus, lower update frequencies such as 180 time-steps and 360 time-steps have a higher magnitude of fluctuations of C_p with respect to the mean C_p, hence larger values of the standard deviation of C_p, 0.24 and 0.28, respectively.

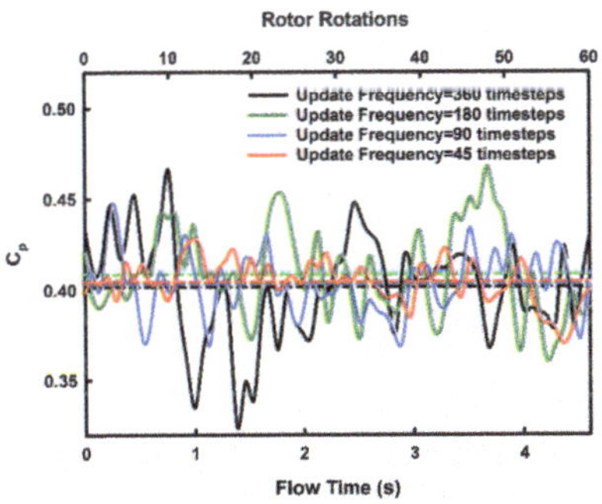

Figure 9. Variation of C_p for various randomisation update frequencies. Corresponding mean C_p is represented through dashed lines.

The range of variation of C_p is 0.11 for update frequency of 180 time-steps and 0.14 for the update frequency of 360 time-steps. On the contrary, the spatial randomness of velocity is maintained before the turbine for faster updates of 45 and 90 time-steps. The number of time-steps for the inlet wind is not sufficient to develop large fluctuations on C_p. Consequently, update frequencies of 45 and 90 time-steps produce C_p fluctuations of smaller magnitude closer to the mean C_p generated by the wind, i.e., the standard deviations of C_p are 0.12 and 0.16, respectively. The range of variation of C_p is 0.06 for update frequency of 45 time-steps and 0.07 for the update frequency of 90 timesteps. The fluctuations of the C_p value from the mean are highest in case of an update frequency of 360 time-steps and gradually decrease for quicker updates of 180, 90, and 45 time-steps.

Figure 10 shows the variation of the Coefficient of Moment (C_m) with respect to the azimuthal angle for a single blade. It is observed that the C_m plot follows a similar pattern in both update frequencies of 45 time-steps (Figure 10a) and 360 time-steps (Figure 10b) except at peak C_m values at 90° and downstream wake affected region after 180°.

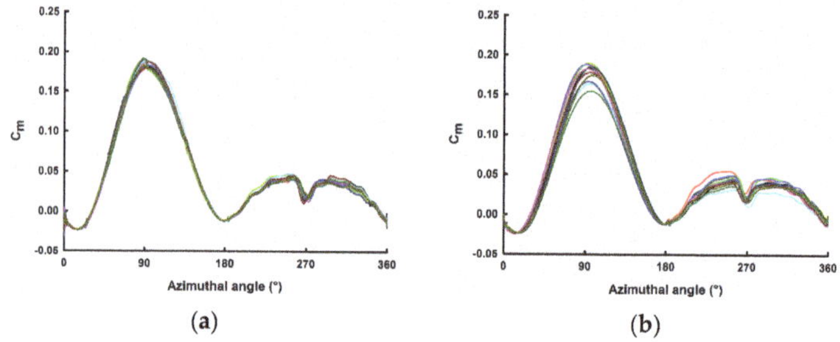

Figure 10. Coefficient of Moment (C_m) for various rotational cycles. (**a**) Randomisation Update Frequency = 45 time-steps; (**b**) Randomisation Update Frequency = 360 time-steps. Coefficient of Moment at various time instances (rotational cycles) are represented through different coloured lines. Due to randomized wind conditions, values vary every rotational cycle.

5.2. Effect of Tip Speed Ratio on Coefficient of Power

The average wind velocity upstream of the turbine for tip speed ratios 2.5, 3.3, 4.1, and 5.3 is shown in Figure 11. Tip Speed ratio (λ) is the ratio between the rotational speed of the wind turbine and the free stream velocity. The mean wind velocity at the inlet for all cases is maintained at 10 m/s with a randomised fluctuation of 6 m/s; the rotational speed of the turbine varies with the tip speed ratio. As the wind approaches the turbine, it is noticed that with a decrease in tip speed ratio, the mean velocity of the wind increases, taken at 0.75 d upwind, and averaged over a width of 1.5 d (Figure 11a). Since the relative velocity of the turbine increases with the tip speed ratio, resistance due to turbine will increase, blocking the incoming wind. Hence, the mean velocity values are 9.48 m/s for λ = 2.5 and 8.56 m/s for λ = 5.3. The magnitude of fluctuation of velocity also decreases with an increase in the tip speed ratio. The standard deviation of fluctuating velocity is 0.30 m/s for λ = 2.5 and 0.20 m/s for λ = 5.3.

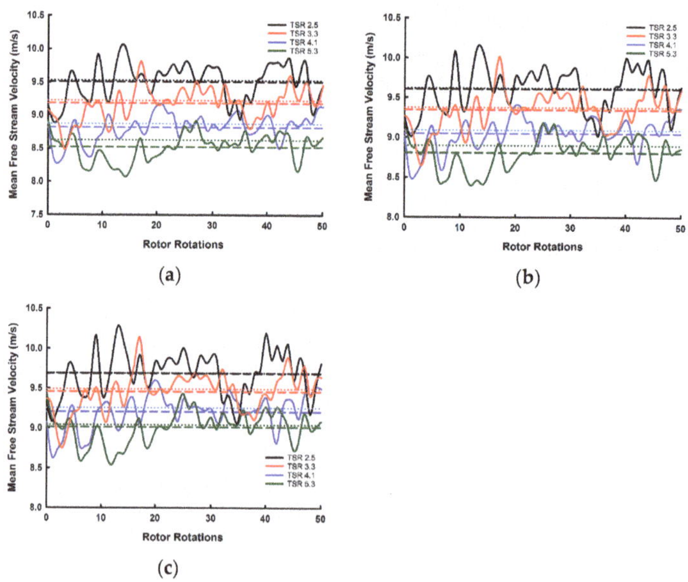

Figure 11. Effect of tip speed ratio on the average wind velocity upstream of VAWT. (**a**) 0.75 d; (**b**) 1 d; (**c**) 1.25 d. The mean and minimum-maximum is represented in the graph with the help of dashed lines and dotted lines respectively.

The dotted line in Figure 11a represents the uniform velocity condition, and the dashed line represents the mean of randomised velocity condition. The mean of randomised velocity is of similar value to the magnitude of uniform velocity. The mean of randomised velocity for $\lambda = 2.5$ is 9.48 m/s and 9.53 m/s for the uniform velocity case. The fluctuation of inlet velocity is also taken at 1 d upwind of the turbine and 1.25 d upwind of the turbine averaged over a width 1.5 d (Figure 11b). The velocity fluctuations follow a similar variation for all tip speed ratios at the three upwind locations. As the distance from the turbine decreases, the mean of randomised velocity also decreases. The mean of randomised velocity for $\lambda = 2.5$ is 9.69 m/s at 1.25 d, 9.62 m/s at 1 d, and 9.53 m/s at 0.75 d.

Figure 12 shows the effect of tip speed ratio on the coefficient of power of the VAWT. In general, the C_p value gradually increases until a particular tip speed ratio and then steadily decreases. The C_p of uniform flow steadily increases to a maximum of 0.399 at $\lambda = 4.1$, then decreases for higher tip speed ratios. The randomised wind C_p follows the same trend and increases up to a maximum value of 0.401 at $\lambda = 4.1$. The C_p values at $\lambda = 3.3$ and 4.1 are comparable in both wind cases. At $\lambda = 2.5$, $C_p = 0.160$ for randomised wind and $C_p = 0.113$ for uniform wind. There is also a considerable difference at $\lambda = 5.3$: $C_p = 0.283$ for randomised wind, and $C_p = 0.324$ for uniform wind. This difference in C_p value between randomised and uniform wind, at higher C_p and lower C_p tip speed ratio, is due to the steep C_p-λ slope.

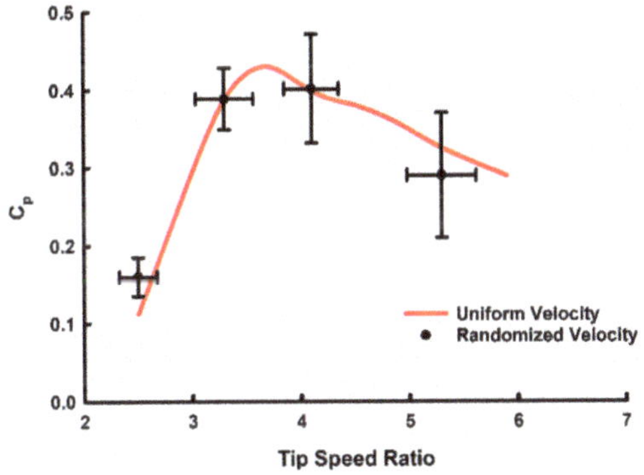

Figure 12. Effect of tip speed ratio on Coefficient of Performance (C_p) with random wind conditions.

Figure 12 also shows the error values in C_p and tip speed ratio for randomised wind. It is noticed that the error bar in C_p slightly increases from $\lambda = 2.5$ to $\lambda = 5.3$, i.e., the performance of the VAWT is highly sensitive at higher tip speed ratios. The error bar for λ also increases from $\lambda = 2.5$ to $\lambda = 5.3$. The higher rotor velocity causes more error in tip speed ratio though the magnitude of wind velocity before the turbine is less at higher tip speed ratios (Figure 11), which is similar to results observed by Balduzzi et al. [27].

It is essential to find an operational range of tip speed ratio to extract maximum power from available wind. It can be concluded that for a 2-bladed turbine of 1 m diameter and solidity of 0.12, for a steady wind case, it is beneficial to operate the turbine at a tip speed ratio around 4.1.

In Figure 13, the variation of coefficient of moment (C_m) is given for a single blade for a rotor rotation for various tip speed ratios. It is noticed that at $\lambda = 2.5$ (Figure 13a), the C_m vs. azimuthal angle plot deviates from Figure 7. The peak C_m occurs before 90°, and there is a sudden drop in moment coefficient to negative values after dynamic stall due to induced vortices of the previous blade. Since $\lambda = 3.3$ (Figure 13b) and $\lambda = 4.1$ (Figure 13c) lie in the beneficial operation tip speed ratio range, the peak C_m values at 90° are higher than that of $\lambda = 2.5$ and $\lambda = 5.3$ (Figure 13d).

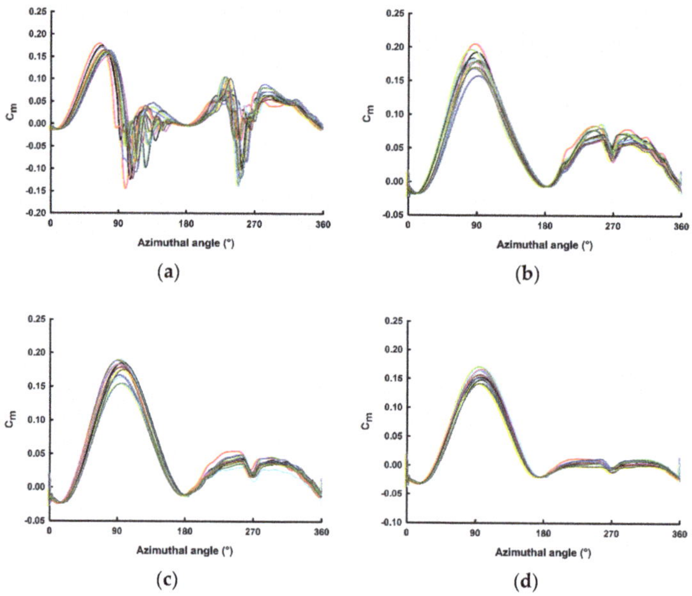

Figure 13. Coefficient of Moment (C_m) for various rotational cycles. (**a**) $\lambda = 2.5$; (**b**) $\lambda = 3.3$; (**c**) $\lambda = 4.1$; (**d**) $\lambda = 5.3$. Coefficient of Moment at various time instances (rotational cycles) are represented through different coloured lines. Due to randomized wind conditions, values vary every rotational cycle.

During downstream conditions, vortex shedding from the central shaft contributes to higher moment coefficient values and more considerable variations for $\lambda = 2.5$ compared to other tip speed ratios. Since the relative velocity is low for $\lambda = 2.5$, the wake produced by the central shaft of the turbine contributes heavily to the moment production when the blade is downstream. The moment production downstream gradually decreases for $\lambda = 2.5$ to $\lambda = 4.1$ and is almost nil for $\lambda = 5.3$ due to the low interaction of the blade with wake structures from the central shaft.

5.3. Turbine Wake Study

Wake is generated from the interaction of wind with the blades and central shaft downstream of the turbine. The study of wake characteristics of a VAWT is essential to understand how wake structures evolve from near the turbine to far wake regions. The wake is characterised by velocity deficit and turbulence intensity caused by the power extraction by a VAWT. Such studies are critical in determining the wind farm layout, as the drastic reduction in wind velocity and the increased turbulence affects power production and causes fatigue loading on other turbines downstream. The downstream is distinguished into two regions—near wake and far wake.

Figures 14 and 15 show time-averaged velocities over 50 turbine cycles normalised with mean free stream velocity along $-0.75\ d \leq y/d \leq +0.75\ d$ for near wake and far wake cases, respectively. Wake Structures were plotted for four tip speed ratios 2.5, 3.3, 4.1, 5.3. Wake self-induction, i.e., reduction in the stream-wise velocity as the wake travels downstream, is significant in all tip speed ratios. Velocity deficit refers to the decreased wake velocity compared to free stream velocity. It signifies the percentage loss in velocity at downstream distances with respect to free stream velocity. It is observed that the velocity deficit in wake increases with an increase in λ as represented in Figure 16. At higher tip speed ratios, since the relative velocity between the turbine and wind is high, the turbine causes blockage. Hence, streamwise velocity behind the turbine is highly reduced owing to a larger velocity deficit. This is significant in near-wake cases, $x/d = 2.5$ and $x/d = 4$. The magnitude of the velocity deficit is comparable for near-wake cases. The maximum velocity deficit at x/d = 2.5 for λ = 2.5 is 34% and for λ = 5.3 is 78% (Figure 14a). At x/d = 4, the maximum velocity deficit for λ = 2.5 is 35% and for λ = 5.3 is 78% (Figure 14b). For far-wake conditions, the maximum velocity deficit decreases for all tip speed ratios, with an increase in the distance behind the turbine, from x/d = 6 to x/d = 10. The maximum velocity deficit at x/d = 6 for λ = 2.5 is 32% and for λ = 5.3 is 56% (Figure 15a). At x/d = 10, maximum velocity deficit for λ = 2.5 is 25% and for λ = 5.3 is 31% (Figure 15c). It is observed that in all cases of far wake, λ = 4.1 has a larger magnitude of velocity deficit compared to λ = 5.3. At x/d = 8, the maximum velocity deficit for λ = 4.1 is 46% and for λ = 5.3 is 38% (Figure 15c).

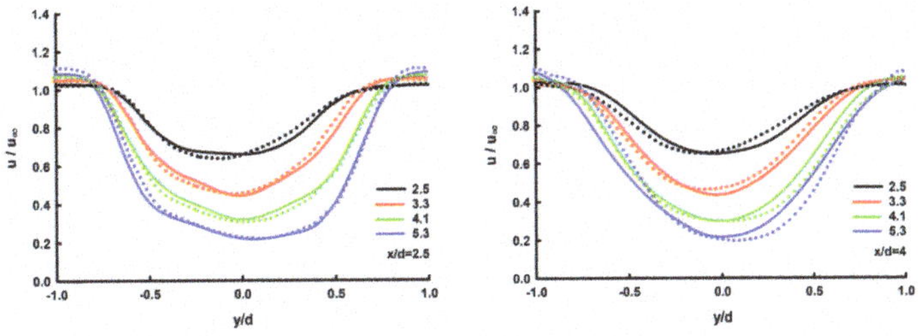

(a) (b)

Figure 14. Time-averaged normalised stream-wise wake velocity along −0.75 d ≤ y/d ≤ +0.75 d for different tip speed ratios at near-wake distances. (**a**) x/d = 2.5; (**b**) x/d = 4. Instantaneous normalised stream-wise wake velocity profiles are represented using dotted lines.

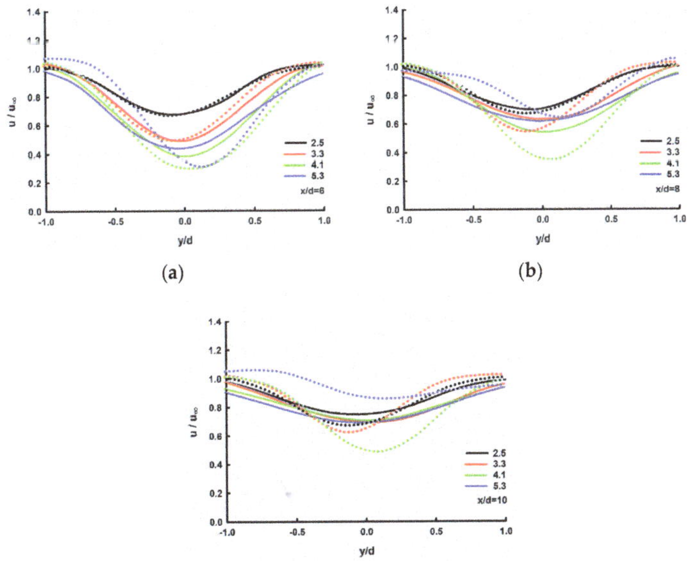

(a) (b) (c)

Figure 15. Time averaged normalised stream-wise wake velocity along −0.75 d ≤ y/d ≤ +0.75 d for different tip speed ratios at near-wake distances. (**a**) x/d = 6; (**b**) x/d = 8; (**c**) x/d = 10. Instantaneous normalised stream-wise wake velocity profiles are represented using dotted lines.

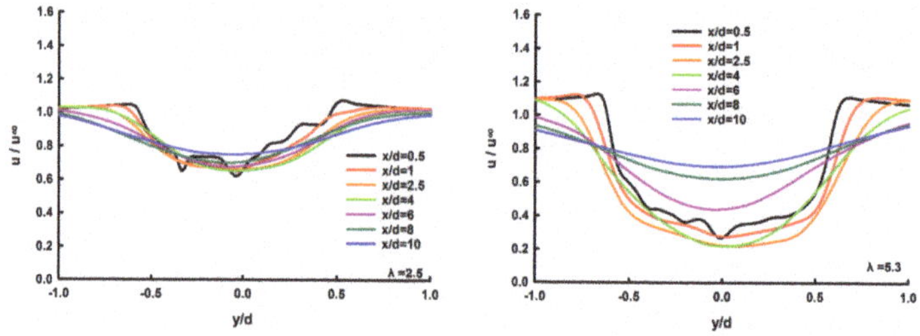

(a) (b)

Figure 16. Time averaged normalised stream-wise wake velocity along -0.75 d $\leq$ y/d $\leq$ +0.75 d for all wake distances for (**a**) λ = 2.5; (**b**) λ = 5.3.

For near-wake cases (Figure 14a,b), it is observed that there is a small magnitude of wake asymmetry towards the windward side (y < 0) for all tip speed ratios. The maximum velocity deficit position shifts towards the windward direction with an increased distance from the turbine downwind, i.e., from x/d = 2.5 to x/d = 10. Wake asymmetry is caused in the windward side as the blade travels against an adverse pressure gradient, contributing to stronger vortex shedding than the leeward side (y > 0). Since the blade moves against the wind, causing a blockage, a low pressure in the windward is created, inducing faster blade downstream movement. Tip speed ratios λ = 3.3 and λ = 4.1 do not show much wake asymmetry, whereas λ = 2.5 and λ = 5.3 show comparatively larger shifts in maximum velocity deficit positions in both near-wake and far-wake cases.

In the case of wake structures formed by uniform velocity inlet, represented by the dotted line, the trend is similar for near-wake cases. The maximum velocity deficit for λ = 4.1 at x/d = 2.5 is 69% for uniform inlet and 68% for randomised inlet (Figure 14a). Since wake data are extracted very close to the turbine, turbulence mixing occurs for both inlet cases, irrespective of macro-turbulence in randomised cases. As the distance behind the turbine increases, there is considerable variation between wake plots of the randomised and uniform inlet, leading to faster wake recovery, whereas for uniform inlet case, the mixing is slower, wake recovery is slower, and wake asymmetry is high. The maximum velocity deficit for uniform velocity is hence higher than randomised velocity for λ = 2.5, 3.3, 4.1 for far wake distances. The maximum velocity deficit at x/d = 8 at λ = 4.1 is 46% for randomised inlet and 64% for uniform inlet (Figure 15c). For λ = 5.3, the velocity deficit is higher for randomised inlet at x/d = 8 and x/d = 10.

5.4. Effect of Gust Amplitude on the VAWT Performance

Figure 17a shows the variation cycle-averaged randomised free stream velocity, U_∞ for 4 different randomised cases of Gust Amplitude, U_{gust} = 6 m/s, 8 m/s, 10 m/s, and 12 m/s, where one gust cycle corresponds to 80 rotor rotations. The respective gust factors corresponding to these U_{gust} values are 1.34, 1.50, 1.64, and 1.80. According to IEC 61400-2-Small wind turbines [32], the average gust factor in an extreme wind case is 1.44 and can be much higher in an extreme operating gust case. The fluctuation of U_∞ is minimum for U_{gust} = 6 m/s and maximum for U_{gust} = 12 m/s, where their maximum peak velocities are 14.4 m/s and 18.9 m/s. The difference in their maximum peak velocity is 4.3 m/s. The minimum peak velocity is 6.78 m/s corresponding to U_{gust} = 12 m/s and 8.4 m/s corresponding to U_{gust} = 6 m/s. Figure 17b represents the instantaneous randomised U_∞ values for the four gust cases taken at every 20 timesteps.

Figure 17c represents the variation of instantaneous U_∞ for randomised gust cases, U_{gust} = 6 m/s, 10 m/s, and 12 m/s in comparison with instantaneous U_∞ for uniform gust cases. Both randomised and uniform gust follow the same profile. However, randomised gust shows a large fluctuation in velocity due to the randomisation procedure at the inlet of the domain.

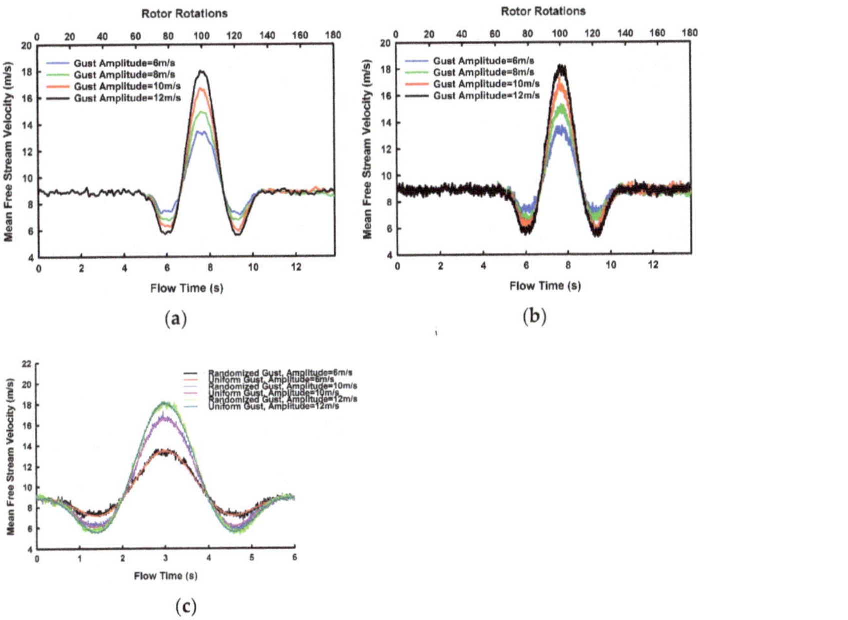

Figure 17. (**a**) The variation in cycle-averaged U∞ for Ugust = 6 m/s, 8 m/s, 10 m/s and 12 m/s; (**b**) Instantaneous variation of U∞ for Ugust = 6 m/s, 8 m/s, 10 m/s and 12 m/s; (**c**) Comparison of Randomised Gust velocity and Uniform gust velocity.

The magnitude of randomised fluctuation is 6 m/s. The continuous spikes in the velocity profile subsequently affect the C_p values of the wind turbine. Figure 18a shows the variation of cycle-averaged λ for various randomised U_{gust} cases. The minimum fluctuation of λ occurs for U_{gust} = 6 m/s case with maximum and minimum values of λ being 4.89 and 2.84, respectively. The maximum fluctuation occurs in U_{gust} = 12 m/s with maximum and minimum values of λ values at 6.04 and 2.18, respectively. Figure 18b represents the instantaneous randomised λ values for the four gust cases taken at every 20 timesteps.

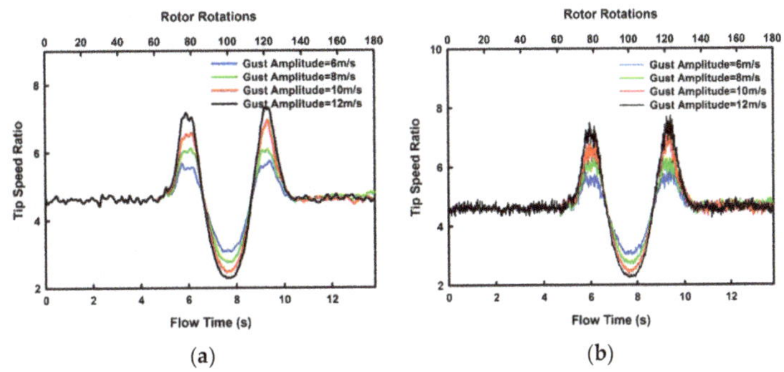

Figure 18. (**a**) The variation in cycle-averaged tip speed ratio for Ugust = 6 m/s, 8 m/s, 10 m/s, and 12 m/s; (**b**) Instantaneous variation of tip speed ratio for Ugust = 6 m/s, 8 m/s, 10 m/s and 12 m/s.

In Figure 19, the variation of C_m vs. azimuthal angle for a single blade for U_{gust} cases of 6 m/s, 8 m/s, 10 m/s, and 12 m/s is shown. It is observed that power produced by upstream wind is greater than the power produced by downstream wind. Unlike the case with a steady inlet velocity of 10 m/s shown in Figure 8a, the C_m vs. azimuthal angle for U_{gust} cases shown in Figure 17 deviates from the general profile.

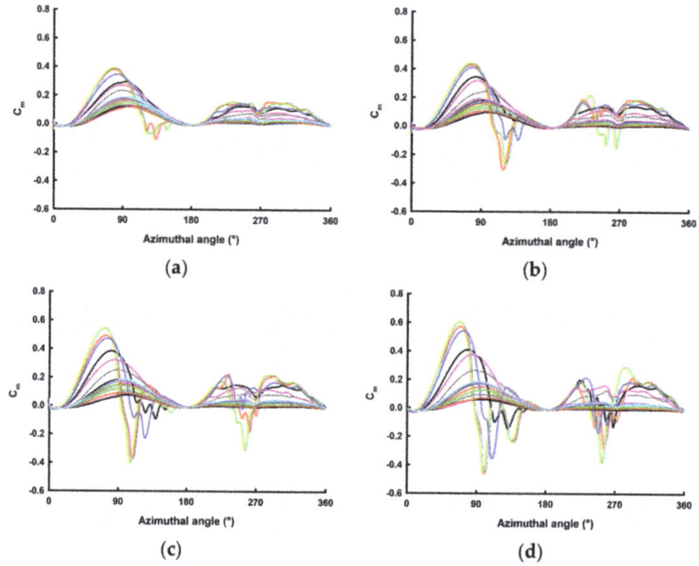

Figure 19. Variation of Coefficient of Moment C_m for azimuthal Angle for (**a**) Ugust = 6 m/s; (**b**) Ugust = 8 m/s; (**c**) Ugust = 10 m/s; (**d**) Ugust = 12 m/s. Coefficient of Moment at various time instances (rotational cycles) are represented through different coloured lines. Due to randomized wind conditions, values vary every rotational cycle.

The deviation occurs for turbine cycles where free stream velocity is greater than 14 m/s, i.e., the tip speed ratio is less than 2.92. This is attributed to induced vortexes of the leading and trailing edge that occur at high velocities (low λ values) that contribute to positive aerodynamic performance creating a larger lift. In such cases, the instantaneous moment coefficients begin to increase at a smaller azimuthal angle, and the C_m peaks sooner and higher. The maximum C_m value for U_{gust} = 6 m/s is 0.39 corresponding to peak velocity of 14.4 m/s (Figure 19a) and the maximum C_m value for U_{gust} = 12 m/s is 0.61 corresponding to peak velocity of 18.8 m/s (Figure 19d); both cases at cycle 160. The moment coefficients after dynamic stall drop steeply to a negative value. During downstream conditions, vortex shedding and subsequent flow separation contribute positively to higher moment coefficient values compared to the rest of the cycles. Cycles with higher tip speed ratios have a lower value of peak C_m during the upwind condition and have a broad operational range, i.e., C_m values are not negative.

Figure 20 explains the variation of C_p for flow time for one complete gust cycle. It is observed that, from a mean value of 0.433, there is a sudden drop of C_p to a minimum value. This is attributed to a decreasing U_∞ value and increasing λ value. From Figure 10, it can be concluded that for a particular rotational speed of the wind turbine, the magnitude of C_p increases until at a certain λ value and then drops after reaching a maximum. According to the gust profile, since velocity initially drops, there is a drop in C_p. After reaching the minimum C_p, the velocity of gust increases, decreasing the λ value and subsequently contributing to a steep increase in the C_p value. In the region where U_∞ is comparatively high, the λ and C_p values both decrease. Again, when the U_∞ values drop, the magnitude of C_p is higher corresponding to the same U_∞ previously (i.e., before C_p drop). This is due to the contribution of induced vortexes formed by previous cycles' high-velocity U∞. The mean value of C_p over the complete gust cycles is 0.35 for U_{gust} = 12 m/s and 0.41 for U_{gust} = 6 m/s. This can be attributed to the high U_∞ fluctuations of cases with high gust amplitude.

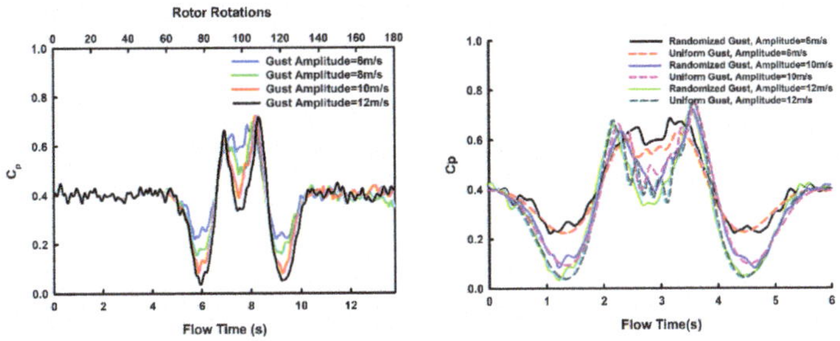

(a) (b)

Figure 20. (**a**) Variation of C_p for flow time for Ugust = 6 m/s, 8 m/s, 10 m/s, and 12 m/s; (**b**) Comparison of C_p of randomised gust velocity case with uniform gust velocity case against flow time.

5.5. Effect of Gust Time Period on the VAWT Performance

The effect of time period is investigated for U_{gust} = 10 m/s, and gust factor 1.64 is studied for gust time periods T = 3 s, 6 s, 10.5 s. According to Rakib et al. [22], the rise and fall time of a gust event should lie between 3 s to 5.6 s. Figure 21 shows the fluctuation of U_∞ for different gust time periods. The maximum and minimum velocities are 17.4 m/s and 7.3 m/s for all three cases. Similarly, Figure 22 shows the fluctuations of λ for different gust periods. The minimum and maximum are 2.35 and 5.60 for all three cases. It takes 40 rotor rotations at 82 rad/s to finish one complete gust cycle for the gust time period 3 s and 140 rotor rotations at 82 rad/s to finish one complete gust cycle for the gust time period 10.5 s.

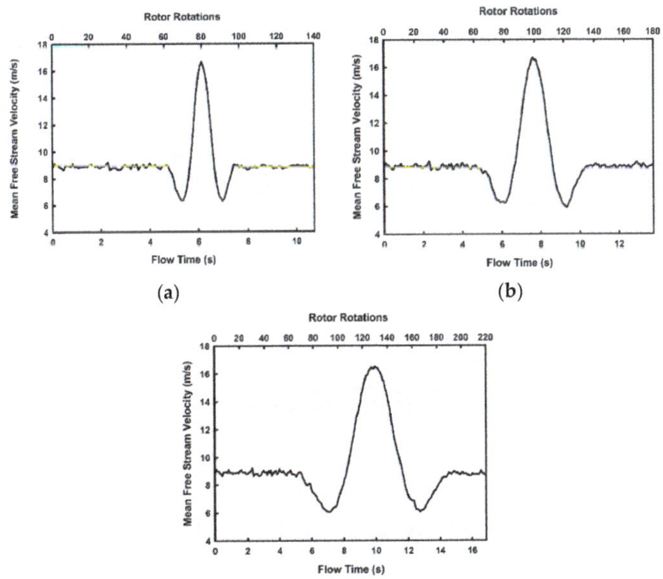

(a) (b) (c)

Figure 21. The variation of cycle-averaged mean free stream velocity (U∞) for gust time periods (**a**) 3 s; (**b**) 6 s; (**c**) 10.5 s.

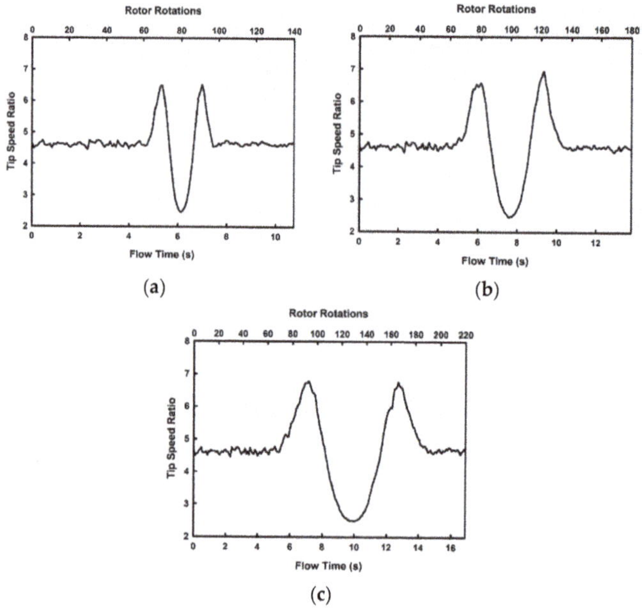

Figure 22. The variation of tip speed ratio (λ) for gust time periods (**a**) 3 s; (**b**) 6 s; (**c**) 10.5 s.

Figure 23 shows the variation of C_p for flow time for one complete gust cycle for the same U_{gust} = 10 m/s and gust time periods 3 s, 6 s, and 10.5 s, respectively. The trend is similar in all three cases, but the max C_p values are 0.64 for a time period of 3 s and 0.71 for the time periods of 6 s and 10.5 s. This difference is due to the rapid rise and fall time of the gust signal for the time period 3 s case. In Figure 22c, it is observed that there are many fluctuations of C_p from the flow time 9 s to 10.5 s. The variation of C_m with the azimuthal angle for different turbine cycles is similar in all gust time period cases as represented in Figure 24, attributing to the same gust amplitude of 10 m/s.

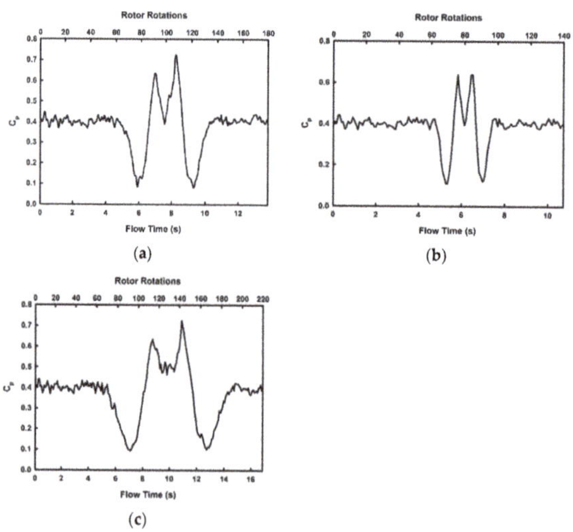

Figure 23. The variation of C_p for flow time for one complete gust cycle gust time periods (**a**) 3 s; (**b**) 6 s; (**c**) 10.5 s.

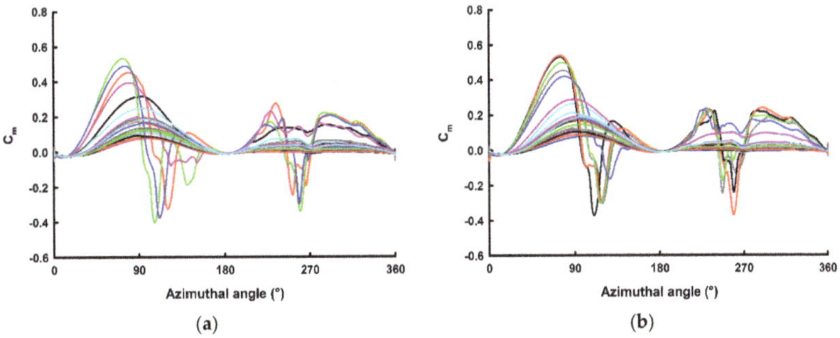

Figure 24. Variation of Coefficient of Moment (C_m) with azimuthal angle for gust time periods (**a**) 3 s; (**b**) 10.5 s. Coefficient of Moment at various time instances (rotational cycles) are represented through different coloured lines. Due to randomized wind conditions, values vary every rotational cycle.

6. Conclusions

In this study, a 2D Unsteady RANS Transition SST model was used to study the performance of a small-scale H-type Vertical Axis Wind Turbine with an NACA 0018 airfoil. A spatially randomised inlet velocity was generated to produce a macro-turbulent domain upstream of the turbine. The study focused on understanding the effect of the magnitude of randomness and the frequency of change in randomness in upstream wind on the coefficient of power (C_p) and the coefficient of moment (C_m).

1. The effect of inlet velocities randomisation parameters such as input randomised fluctuation of 6 m/s, 4 m/s, and 2 m/s were investigated. The average C_p value and magnitude of fluctuation of velocity was higher, with an increase in the magnitude of randomised fluctuation.
2. The effect of randomisation update frequencies of 45, 90,180, and 360 time steps was also studied. There was a decrease in the magnitude of velocity fluctuations from the update frequency 360 time-steps to the update frequency of 45 time-steps. The flow field upstream of the turbine was more uniform for the update frequency of 45 time-steps than for the update frequency of 360 time-steps, which is characterised by a higher magnitude of macro-turbulence. As the update interval decreases, the flow field in front of the turbine becomes uniform. The magnitude of C_p fluctuations also decreases with the size of the update interval.
3. The effect of the tip speed ratio on a wind turbine for randomised inlet conditions was also studied. The magnitude of fluctuation of velocity also decreased with an increase in tip speed ratio. Though randomised and uniform flow follows the same C_p-λ trend, the slope was steeper for randomised, and there was a visible difference in the C_p values at $\lambda = 2.5$ and $\lambda = 5.3$. The performance of the VAWT was highly sensitive at higher tip speed ratios. Thus, the operational range of tip speed ratio to extract maximum power from available wind should be around 4.1 for a 2 bladed turbine of 1 m diameter and solidity of 0.12.
4. The variation of C_m given for a single blade for a rotor rotation for various tip speed ratios was also studied. It was noticed that at lower tip speed ratios, the peak C_m occurred before 90° and a drop in moment coefficient to negative values after dynamic stall was observed. During downstream conditions, vortex shedding from the central shaft contributed to higher moment coefficient values and larger variations for $\lambda = 2.5$ compared to other tip speed ratios.
5. The effect of tip speed ratio on the turbine wake recovery was also studied. As the tip speed ratio increased, the velocity deficit in the wake structure also increased. Streamwise velocity behind the turbine was highly reduced. For near wake cases, x/d = 2.5 and x/d = 4, the velocity deficit values were comparable for all tip speed ratios. As the downstream distance increased, the velocity deficit decreased for all

tip speed ratios. Wake recovery was faster in the case of lower tip speed ratios. Wake generated by uniform velocity inlet had slower wake recovery compared to randomised velocity inlet. Macro-turbulence generated by randomisation enhanced the mixing of flow downwind of the turbine, allowing faster recovery.

6. Further, a study was performed on the influence of gust parameters such as gust amplitude and gust time period on the coefficient of power of the VAWT, for an IEC extreme operating gust. Since the average gust factor is 1.44 in extreme cases, 4 different randomised cases of U_{gust} = 6 m/s, 8 m/s, 10 m/s, and 12 m/s with gust factors of 1.34, 1.50, 1.64, and 1.80, respectively, were considered. A comparison of U_{∞} for uniform gust cases and randomised gust cases showed that the spatial fluctuation in the velocity profile of randomised gust subsequently affected the C_p values of the wind turbine. The tip speed ratio vs. flow time plots also showed fluctuations for all gust cases. Unlike the case with a steady inlet velocity of 10 m/s, the C_m vs. azimuthal angle for U_{gust} deviated from the general profile due to continuously varying tip speed ratios. The deviation occurred for turbine cycles where free stream velocity was greater than 14 m/s, i.e., the tip speed ratio was less than 2.92. This was clearly observed in gust cases with higher gust magnitude, U_{gust} = 10 m/s and U_{gust} = 12 m/s. As tip speed ratios increased, cycles had a lower value of peak C_m during the upwind condition and C_m values were not negative.
7. The effect of time period was investigated for U_{gust} = 10 m/s, and gust factor 1.64 was studied for gust time periods T = 3 s, 6 s, 10.5 s. All cases showed similar values of maximum and minimum velocities and tip speed ratios, spread over the respective time periods. The C_p plot for 10.5 s gust showed many fluctuations in the region where velocity rises to the maximum.

Author Contributions: Conceptualization, R.K.V. and U.D.; methodology, L.S. and N.R.; software, R.K.V. and A.M.; formal analysis, L.S., N.R., U.D. and S.B.R.; resources, R.K.V. and A.M.; writing—original draft preparation, L.S., N.R. and S.B.R.; writing—review and editing, R.K.V., A.M. and U.D.; supervision, R.K.V. and U.D. All authors have read and agreed to the published version of the manuscript.

Funding: This research received no external funding.

Data Availability Statement: Not applicable.

Acknowledgments: A part of the computations were performed on the Aziz Supercomputer at King Abdulaziz University's High-Performance Computing Center (http://hpc.kau.edu.sa/). The authors acknowledge the computer time and technical support provided by the center.

Conflicts of Interest: The Authors declare no conflict of interest.

References

1. Kusch-Brandt, S. Renewables 2019 Global Status Report. 2019, Volume 8. 9783981891140. Available online: https://www.ren21.net/wp-content/uploads/2019/05/2018-2019-Annual-Report_FINAL_low-res.pdf (accessed on 12 January 2023).
2. Emejeamara, F.C.; Tomlin, A.S.; Millward-Hopkins, J.T. Urban wind: Characterisation of useful gust and energy capture. *Renew. Energy* **2015**, *81*, 162–172. [CrossRef]
3. Johari, M.K.; Jalil, M.A.A.; Shariff, M.F.M. Comparison of horizontal axis wind turbine (HAWT) and vertical axis wind turbine (VAWT). *Int. J. Eng. Technol.* **2018**, *7*, 74–80. [CrossRef]
4. Pagnini, L.C.; Burlando, M.; Repetto, M.P. Experimental power curve of small-size wind turbines in turbulent urban environment. *Appl. Energy* **2015**, *154*, 112–121. [CrossRef]
5. Toja-Silva, F.; Colmenar-Santos, A.; Castro-Gil, M. Urban wind energy exploitation systems: Behaviour under multidirectional flow conditions—Opportunities and challenges. *Renew. Sustain. Energy Rev.* **2013**, *24*, 364–378. [CrossRef]
6. Ishugah, T.F.; Li, Y.; Wang, R.Z.; Kiplagat, J.K. Advances in wind energy resource exploitation in urban environment: A review. *Renew. Sustain. Energy Rev.* **2014**, *37*, 613–626. [CrossRef]
7. Tummala, A.; Velamati, R.K.R.K.; Sinha, D.K.; Indraja, V.; Krishna, V.H.; Subramanian, A.; Yogesh, S.A.; Sivanandan, H.; Giri, A.; Vasudevan, M.; et al. Progress and recent trends of wind energy technology. *Renew. Sustain. Energy Rev.* **2013**, *21*, 456–468. [CrossRef]
8. Britter, R.E.; Hanna, S.R. Flow and dispersion in urban areas. *Annu. Rev. Fluid Mech.* **2003**, *35*, 469–496. [CrossRef]

9. Mantravadi, B.; Unnikrishnan, D.; Sriram, K.; Mohammad, A.; Vaitla, L.; Velamati, R.K. Effect of solidity and airfoil on the performance of vertical axis wind turbine under fluctuating wind conditions. *Int. J. Green Energy* **2019**, *16*, 1329–1342. [CrossRef]
10. Tavernier, D.D.; Ferreira, C. The need for dynamic inflow models for vertical axis wind turbines. *J. Phys. Conf. Ser.* **2019**, *1356*. [CrossRef]
11. McIntosh, S.C.; Babinsky, H.; Bertényi, T. Optimizing the energy output of vertical axis wind turbines for fluctuating wind conditions. In Proceedings of the Collect Tech Pap—45th AIAA Aerospace Sciences Meeting, Reno, NV, USA, 8–11 January 2007; Volume 23, pp. 16202–16214. [CrossRef]
12. Wekesa, D.W.; Wang, C.; Wei, Y.; Danao, L.A.M. Analytical and numerical investigation of unsteady wind for enhanced energy capture in a fluctuating free-stream. *Energy* **2017**, *121*, 854–864. [CrossRef]
13. Battisti, L.; Benini, E.; Brighenti, A.; Soraperra, G.; Castelli, M.R. Simulating the dynamic behavior of a vertical axis wind turbine operating in unsteady conditions. *J. Phys. Conf. Ser.* **2016**, *753*. [CrossRef]
14. Bhargav, M.; Kishore, V.R.; Laxman, V. Influence of fluctuating wind conditions on vertical axis wind turbine using a three dimensional CFD model. *J. Wind Eng. Ind. Aerodyn.* **2016**, *158*, 98–108. [CrossRef]
15. Wekesa, D.W.; Wang, C.; Wei, Y.; Kamau, J.N.; Danao, L.A.M. A numerical analysis of unsteady inflow wind for site specific vertical axis wind turbine: A case study for Marsabit and Garissa in Kenya. *Renew. Energy* **2015**, *76*, 648–661. [CrossRef]
16. Bausas, M.D.; Danao, L.A.M. The aerodynamics of a camber-bladed vertical axis wind turbine in unsteady wind. *Energy* **2015**, *93*, 1155–1164. [CrossRef]
17. Danao, L.A.; Edwards, J.; Eboibi, O.; Howell, R. A numerical investigation into the influence of unsteady wind on the performance and aerodynamics of a vertical axis wind turbine. *Appl. Energy* **2014**, *116*, 111–124. [CrossRef]
18. Danao, L.A.; Eboibi, O.; Howell, R. An experimental investigation into the influence of unsteady wind on the performance of a vertical axis wind turbine. *Appl. Energy* **2013**, *107*, 403–411. [CrossRef]
19. Scheurich, F.; Brown, R.E. Modelling the aerodynamics of vertical-axis wind turbines in unsteady wind conditions. *Wind Energy* **2013**, *16*, 91–107. [CrossRef]
20. Wu, Z.; Wang, Q.; Bangga, G.; Huang, H. Responses of vertical axis wind turbines to gusty winds. *Proc. Inst. Mech. Eng. Part A J. Power Energy* **2020**, *235*, 81–93. [CrossRef]
21. Wu, Z.; Bangga, G.; Cao, Y. Effects of lateral wind gusts on vertical axis wind turbines. *Energy* **2019**, *167*, 1212–1223. [CrossRef]
22. Rakib, M.I.; Evans, S.P.; Clausen, P.D. Measured gust events in the urban environment, a comparison with the IEC standard. *Renew. Energy* **2020**, *146*, 1134–1142. [CrossRef]
23. Onol, A.O.; Yesilyurt, S. Effects of wind gusts on a vertical axis wind turbine with high solidity. *J. Wind Eng. Ind. Aerodyn.* **2017**, *162*, 1–11. [CrossRef]
24. Kwon, D.K.; Kareem, A.; Butler, K. Gust-front loading effects on wind turbine tower systems. *J. Wind Eng. Ind. Aerodyn.* **2012**, *104–106*, 109–115. [CrossRef]
25. Cheng, P.W.; Bierbooms, W.A.A.M. Distribution of extreme gust loads of wind turbines. *J. Wind Eng. Ind. Aerodyn.* **2001**, *89*, 309–324. [CrossRef]
26. Bierbooms, W.; Cheng, P.W. Modelling of Extreme Gusts for Design Calculations. In Proceedings of the EUWEC 96, Göteborg, Sweden, 20–24 May 2000; pp. 842–845.
27. Balduzzi, F.; Zini, M.; Ferrara, G.; Bianchini, A. Development of a computational fluid dynamics methodology to reproduce the effects of macroturbulence on wind turbines and its application to the particular case of a VAWT. *J. Eng. Gas Turbines Power* **2019**, *141*, 1–12. [CrossRef]
28. KC, A.; Whale, J.; Urmee, T. Urban wind conditions and small wind turbines in the built environment: A review. *Renew. Energy* **2019**, *131*, 268–283. [CrossRef]
29. Shahzad, A.; Asim, T.; Mishra, R.; Paris, A. Performance of a vertical axis wind turbine under accelerating and decelerating flows. *Procedia CIRP* **2013**, *11*, 311–316. [CrossRef]
30. Jafari, M.; Razavi, A.; Mirhosseini, M. Effect of Steady and Quasi-Unsteady Wind on Aerodynamic Performance of H-Rotor Vertical Axis Wind Turbines. *J. Energy Eng.* **2018**, *144*, 04018065. [CrossRef]
31. Lee, K.Y.; Tsao, S.H.; Tzeng, C.W.; Lin, H.J. Influence of the vertical wind and wind direction on the power output of a small vertical-axis wind turbine installed on the rooftop of a building. *Appl. Energy* **2018**, *209*, 383–391. [CrossRef]
32. *IEC Standard 61400-2*; International Standard—Part 2: Small Wind Turbines. International Electrotechnical Commission: London, UK, 2006; Volume 2006, ISBN 2831886376.
33. Stork, C.H.J.; Butterfield, C.P.; Holley, W.; Madsen, P.H.; Jensen, P.H. Wind conditions for wind turbine design proposals for revision of the IEC 1400-1 standard. *J. Wind Eng. Ind. Aerodyn.* **1998**, *74–76*, 443–454. [CrossRef]
34. Lam, H.F.; Peng, H.Y. Study of wake characteristics of a vertical axis wind turbine by two- and three-dimensional computational fluid dynamics simulations. *Renew. Energy* **2016**, *90*, 386–398. [CrossRef]
35. Peng, H.Y.; Lam, H.F.; Lee, C.F. Investigation into the wake aerodynamics of a five-straight-bladed vertical axis wind turbine by wind tunnel tests. *J. Wind Eng. Ind. Aerodyn.* **2016**, *155*, 23–35. [CrossRef]
36. Rezaeiha, A.; Montazeri, H.; Blocken, B. Characterization of aerodynamic performance of vertical axis wind turbines: Impact of operational parameters. *Energy Convers. Manag.* **2018**, *169*, 45–77. [CrossRef]
37. Daróczy, L.; Janiga, G.; Petrasch, K.; Webner, M.; Thévenin, D. Comparative analysis of turbulence models for the aerodynamic simulation of H-Darrieus rotors. *Energy* **2015**, 1–11. [CrossRef]

38. Rezaeiha, A.; Montazeri, H.; Blocken, B. On the accuracy of turbulence models for CFD simulations of vertical axis wind turbines. *Energy* **2019**, *180*, 838–857. [CrossRef]
39. Rezaeiha, A.; Kalkman, I.; Blocken, B. CFD simulation of a vertical axis wind turbine operating at a moderate tip speed ratio: Guidelines for minimum domain size and azimuthal increment. *Renew. Energy* **2017**, *107*, 373–385. [CrossRef]
40. Bianchini, A.; Balduzzi, F.; Bachant, P.; Ferrara, G.; Ferrari, L. Effectiveness of two-dimensional CFD simulations for Darrieus VAWTs: A combined numerical and experimental assessment. *Energy Convers. Manag.* **2017**, *136*, 318–328. [CrossRef]
41. Castelli, M.R.; Englaro, A.; Benini, E. The Darrieus wind turbine: Proposal for a new performance prediction model based on CFD. *Energy* **2011**, *36*, 4919–4934. [CrossRef]

MDPI
St. Alban-Anlage 66
4052 Basel
Switzerland
Tel. +41 61 683 77 34
Fax +41 61 302 89 18
www.mdpi.com

Energies Editorial Office
E-mail: energies@mdpi.com
www.mdpi.com/journal/energies

www.ingramcontent.com/pod-product-compliance
Lightning Source LLC
Chambersburg PA
CBHW061516210926
51850CB00008B/1943
* 9 7 8 3 0 3 6 5 7 3 0 6 9 *